Microheterogeneity of Glycoprotein Hormones

Editors

Brooks A. Keel
Women's Research Institute
Department of Obstetrics and Gynecology
University of Kansas School of Medicine
Wichita, Kansas

H. Edward Grotjan, Jr.
Animal Science Department
University of Nebraska
Lincoln, Nebraska

CRC Press
Taylor & Francis Group
Boca Raton London New York

CRC Press is an imprint of the
Taylor & Francis Group, an **informa** business

First published 1989 by CRC Press
Taylor & Francis Group
6000 Broken Sound Parkway NW, Suite 300
Boca Raton, FL 33487-2742

Reissued 2018 by CRC Press

Library of Congress Cataloging-in-Publication Data

Microheterogeneity of glycoprotein hormones.

   Includes bibliographies and index.
   1. Glycoprotein hormones.   I. Keel, Brooks A.
II. Grotjan, H. Edward.   [DNLM: 1. Glycoproteins--
analysis.   2. Gonadotropins--analysis.   WK 900 M628]
QP572.G58M53   1989   599'.01927   87-30015
ISBN 0-8493-4959-1

A Library of Congress record exists under LC control number: 87030015

ISBN 13: 978-1-315-89550-5 (hbk)
ISBN 13: 978-1-351-07460-5 (ebk)

Visit the Taylor & Francis Web site at http://www.taylorandfrancis.com and the
CRC Press Web site at http://www.crcpress.com

PREFACE

Early attempts to isolate, purify, and characterize the glycoprotein hormones led to the observation that the hormones, like other glycoproteins, existed as a heterologous population of molecules. The pioneering work of Drs. Bogdanove, Peckham, and others during the early to mid 1970s examined the effects of the gonadal steroid milieu on glycoprotein hormone microheterogeneity and introduced the concept of "andro" and "gyno" heterogeneic forms of pituitary hormones. Evidence soon began to accumulate suggesting that the overall biological activity of glycoprotein hormones was somehow related to their pleomorphic nature. These studies resulted in the concept that the pituitary, depending upon the hormonal environment, cannot only vary the quantity (amount) synthesized but can also modulate the quality (biological activity) by altering the type of hormone produced. This presumably represents a mechanism to fine tune target-organ responses.

Although a wealth of information exists concerning the microheterogeneity of pituitary and placental glycoprotein hormones, an up-to-date, single volume which addresses the heterogenous nature of all of these hormones was not available. The objective of this project was to comprehensively review the current knowledge concerning the characterization of glycoprotein hormone microheterogeneity, the relationships between biological activity and microheterogeneity, the endocrinological control mechanisms involved in the production of these forms and the underlying biochemical basis for glycoprotein hormone microheterogeneity. To this end we have invited a group of internationally recognized scientists to review the current knowledge in their respective areas of expertise and to contribute representative data. We have allowed these experts freedom in terms of the organization and presentation of their information which has resulted in a most interesting blend of presentation style and scientific interpretation of the available information. The book has been generally organized such that the heterogeneity of each hormone from a variety of species is covered in detail. In addition, the peptide components and oligosaccharide structures of glycoprotein hormones are reviewed and the heterogeneity of uncombined alpha and beta subunits is discussed. Finally, because of the unique biochemical and biological characteristics of equine luteinizing hormone and due to very recent data demonstrating glycosylated forms of prolactin, we have devoted chapters to these new and exciting topics. The reader, therefore, has the option of choosing the chapter or chapters of interest which also provides the necessary background information.

We believe that this single text represents a comprehensive up-to-date review of the current state of knowledge concerning an area of interest to a variety of investigators. Clearly, no single work can cover every aspect in detail. It is our hope, however, that this text may serve as a review and as a foundation for future study.

Brooks A. Keel
H. Edward Grotjan, Jr.

THE EDITORS

Brooks A. Keel, Ph.D. is Assistant Professor of Obstetrics and Gynecology and Clinical Assistant Professor of Pathology at the University of Kansas School of Medicine — Wichita. Dr. Keel serves as Scientific Director of The Women's Research Institute in Wichita, Kansas.

Dr. Keel received his B.S. degree in biology from Augusta College in 1978 and his Ph.D. in endocrinology from the Medical College of Georgia in 1982. Dr. Keel completed three years of postdoctoral training in reproductive endocrinology at the University of Texas Health Science Center in Houston and the University of South Dakota School of Medicine in Vermillion. He accepted his current position in 1985.

Dr. Keel is a member of the Endocrine Society, the Society for the Study of Reproduction, the American Society of Andrology and the American Fertility Society. He serves on the editorial board for Archives of Andrology and Advances in Contraceptive Delivery Systems. Dr. Keel has published more than 40 publications in national and international journals and has served as editor for several books in the area of reproductive endocrinology.

Dr. Keel's research interests include the study of the physiological role and biochemical basis for glycoprotein heterogeneity, and the mechanisms controlling pituitary-testicular function. His grant support is from the Wesley Medical Research Institutes, the Women's Research Institute, and the National Institutes of Health.

H. Edward Grotjan, Jr., Ph.D., is currently an Associate Professor in the Animal Science Department at the University of Nebraska in Lincoln. Dr. Grotjan received his B.S. degree in Agricultural Biochemistry in 1969 and his M.S. degree in Reproductive Physiology in 1971 from the University of Missouri in Columbia. He completed a Ph.D. in Physiology at the University of Kansas Medical Center in 1975 and then took a postdoctoral fellowship in the Department of Reproductive Medicine and Biology at the University of Texas Medical School in Houston. In 1977 he joined that Department as an Assistant Professor. In 1983 he moved to the University of South Dakota School of Medicine and in 1988 he moved to the Animal Science Department of the University of Nebraska in Lincoln. The current book was edited while he was on the faculty of the University of South Dakota School of Medicine. He currently holds research support from the National Institutes of Health and while in Texas he held support from the Robert A. Welch Foundation.

Dr. Grotjan's primary research interests are in the biosynthesis and secretion of the pituitary gonadotropins, in the function of testicular Leydig cells, and in pituitary-gonadal feedback relationships.

CONTRIBUTORS

Werner F. P. Blum
Children's Hospital
University of Tübingen
Tübingen, West Germany

George R. Bousfield
Department of Biochemistry and
  Molecular Biology
M.D. Anderson Hospital and Tumor
  Institute
University of Texas System Cancer
  Center
Houston, Texas

Laurence A. Cole
Assistant Professor
Department of Obstetrics and Gynecology
Yale University
New Haven, Connecticut

Wayne Lecky Gordon
Research Associate
Department of Biochemistry and
  Molecular Biology
M.D. Anderson Hospital and Tumor
  Institute
University of Texas System Cancer
  Center
Houston, Texas

H. Edward Grotjan, Jr.
Associate Professor
Animal Science Department
University of Nebraska
Lincoln, Nebraska

Derek Gupta
Professor and Director
Department of Diagnostic Endocrinology
University Children's Hospital
Tübingen, West Germany

Brooks A. Keel
Scientific Director
The Women's Research Institute
Department of Obstetrics and
Gynecology
University of Kansas School of Medicine
Wichita, Kansas

Urban J. Lewis
Chairman
Lutcher Brown Department of
  Biochemistry
Whittier Institute for Diabetes and
  Endocrinology
La Jolla, California

J. G. Loeber
Laboratory for Clinical Chemistry
National Institute of Public Health and
  Environmental Protection
Bilthoven, The Netherlands

Edith Markoff
Assistant Member
Lutcher Brown Department of
  Biochemistry
Whittier Institute for Diabetes and
  Endocrinology
La Jolla, California

Robert L. Matteri
Assistant Scientist
Wisconsin Regional Primate Research
  Center
University of Wisconsin
Madison, Wisconsin

Harold Papkoff
Professor
Hormone Research Laboratory
University of California
San Francisco, California

David M. Robertson
Department of Anatomy
Monash University
Melbourne, Victoria, Australia

Yagya Nand Sinha
Senior Member
Lutcher Brown Department of
  Biochemistry
Whittier Institute for Diabetes and
  Endocrinology
La Jolla, California

Hiromu Sugino
Research Associate
Department of Biochemistry and
 Molecular Biology
M.D. Anderson Hospital and Tumor
 Institute
University of Texas System Cancer Center
Houston, Texas

L. A. van Ginkel
Laboratory for Clinical Chemistry
National Institute of Public Health
 and Environmental Protection
Bilthoven, The Netherlands

Darrell N. Ward
Anise Sorrell Professor of Biochemistry
Department of Biochemistry and
 Molecular Biology
M.D. Anderson Hospital and Tumor
 Institute
University of Texas System Cancer
 Center
Houston, Texas

TABLE OF CONTENTS

Chapter 1

CHEMISTRY OF THE PEPTIDE COMPONENTS OF GLYCOPROTEIN HORMONES

Darrell N. Ward, George R. Bousfield, Wayne L. Gordon, and Hiromu Sugino

TABLE OF CONTENTS

I. INTRODUCTION

In this chapter we will consider our present knowledge of the protein amino acid sequences of the glycoprotein hormones. We will address this subject from a global overview, with particular emphasis on sources of heterogeneity related to the polypeptide portion of the molecule. We will leave to the authors of subsequent chapters all consideration of carbohydrate-derived heterogeneity.

There have been several reviews of the glycoprotein hormone structures. The review by Ward[1] was directed at functional group substitutions in order to pinpoint essential structural elements. The review by Pierce and Parsons[2] dealt with structural relationships and considered the then-available recombinant DNA studies on this group of hormones. Gordon and Ward[3] provided an update on functional group studies, but limited it to the hormones interacting with the luteinizing hormone (lutropin, LH) receptor. In the same volume, Strickland et al.[4] updated structural information on LH and human (h) chorionic gonadotropin (choriogonadotropin, CG). Bahl and Wagh[5] reviewed the carbohydrate structures of the glycoprotein hormones. Fiddes and Talmadge[6] reviewed the gene structure and evolution of the human glycoprotein hormones.

The number of glycoprotein hormones for which we have partial or complete amino acid sequence information has grown steadily since the historic first report of the bovine (b) thyroid stimulating hormone (thyrotropin, TSH) sequence by Pierce and Liao.[7] We have summarized those glycoprotein hormones for which complete amino acid sequence proposals are available in tabular form. Table 1 presents the summary of the alpha subunit sequences with the total sequence identities in terms of individual residues compared position by position in the table. The table provides the comparative identities for each species of hormone known. The alpha subunit is also known as the common subunit since much of the information, until very recently, has indicated that in a given species the alpha subunit is controlled by a single gene[8-13] and the same subunit is part of the dimeric structure of TSH, LH, follicle stimulating hormone (follitropin, FSH), and in those species that it pertains, CG. The earliest notation of a common subunit came from the studies of Pierce and colleagues.[14,15] However, recent studies suggest that the common subunit may, in fact, have some different properties (probably in the carbohydrate moieties) depending on whether it was derived from LH or FSH, at least in the equine (e) species,[16,17] but the structural basis for the reported differences has not been described. In spite of this, the generalization still appears valid that the alpha subunit amino acid sequence within a given species is identical in the various glycoprotein hormone heterodimeric pairs that comprise these hormones, i.e., the alpha-beta subunit combinations for the respective hormone.

In Table 1 we have presented the identical residues in comparison with the various species of alpha subunits now known. However, not all the hormones have identical processing, or gene sequence length, apparently, and some alpha subunits are longer than others. One structural feature that seems invariant in the glycoprotein hormone subunits is the placement of the 1/2-cystine residues that ultimately define the disulfide cross-linkage within each subunit. (We will mention a two-residue frameshift in the TSH beta subunits that is an exception to this rule, vide infra.) The identities tabulated in Table 1 are thus the identities observed after alignment of the 1/2-cystine residues. The first line of the table presents the number of residues in the individual species of alpha subunit in parentheses. In Table 2, these same data are presented as percent identity (the shorter of the two subunits compared is taken as 100%).

In Tables 3 and 4 are presented comparable tabular comparisons for the beta subunits of the glycoprotein hormones. These tables are more complex than the alpha subunits, since the beta subunits have three and potentially four beta subunits represented because the beta subunits are the hormone-specific subunits that combine with the common subunit (alpha)

Table 1
ALPHA SUBUNIT HOMOLOGY MATRIX (IDENTICAL RESIDUES)

	Bovine (96)	Ovine (96)	Porcine (90)	Human (92)	Equine (96)	Mouse (96)	Rat (96)	Rabbit (96)	Whale (96)
Bovine		93	87	69	76	89	86	91	89
	Ovine		86	67	74	86	84	88	89
		Porcine		67	72	88	86	88	88
			Human		66	69	70	70	66
				Equine		75	73	78	75
					Mouse		93	91	88
						Rat		88	86
							Rabbit		90
								Whale	

Note: Total amino acid residues for each α-subunit is given in parentheses.

Table 2
ALPHA SUBUNIT HOMOLOGY MATRIX (PERCENT IDENTITY)

	Bovine (96)	Ovine (96)	Porcine (90)	Human (92)	Equine (96)	Mouse (96)	Rat (96)	Rabbit (96)	Whale (96)
Bovine		97	97	75	79	93	90	95	93
	Ovine		96	73	77	90	88	92	93
		Porcine		74	80	98	96	98	98
			Human		72	75	76	76	72
				Equine		78	76	81	78
					Mouse		97	95	92
						Rat		92	93
							Rabbit		94
								Whale	

Note: Total amino acid residues for each α-subunit is given in parentheses.

to produce TSH, LH, or FSH in the pituitary, or CG in the placenta of those species that have this form of glycoprotein hormone. The placenta has also been reported to have other putative glycoprotein hormones, e.g., placental TSH[18,19] but these have not been characterized chemically, nor their physiology defined. Thus, only the chorionic gonadotropins will be considered for our present purpose.

In order to arrive at the comparisons presented in Tables 3 and 4, we have again resorted to alignment of the 1/2-cystine residues as the most constant structural feature throughout the series. In order to place these 1/2-cystines in register there are certain adjustments imposed. These are (1) starting the N-terminal residue numbering with "missing" residues in the FSH and TSH series, (2) putting an additional two residues on the rabbit sequence, designated 1' and 2' to distinguish these residues from the usual leader sequences, which are usually numbered in series −1, −2, −3, etc. (The leader sequence for rabbit [l for lagomorph] LH beta has not yet been studied), and (3) allowing for a two-residue insertion in the thyrotropin beta sequences at positions 52 and 54, as suggested by the gaps in Figure 2, LH/CG and FSH subunits sequences.

The foregoing comparative adjustments have some implications for the consideration of heterogeneity in the glycoprotein hormones. The N-terminal adjustments — item (1) above — indicate that the TSH and FSH beta subunits are exposed to either natural proteolytic processing to produce shorter molecules than in the LH series, or (and this is more likely) the LH and CG subunits N-terminal area is more tightly bound to the alpha subunits, thus

Table 3
BETA SUBUNIT HOMOLOGY MATRIX (IDENTICAL RESIDUES)

	eCGβ (149)	hCGβ (145)	eLHβ (149)	hLHβ (121)	oLHβ (119)	bLHβ (121)	pLHβ (119)	lLHβ (119)	rLHβ (121)	wLHβ (118)	sGTHβ (119)	eFSHβ (118)	hFSHβ (111)	oFSHβ (111)	bFSHβ (110)	pFSHβ (107)	hTSHβ (112)	bTSHβ (118)	pTSHβ (112)	mTSHβ (118)	rTSHβ (118)
eCGβ	eCGβ	76	149	78	91	92	93	92	90	91	50	41	41	39	38	34	41	39	39	41	39
hCGβ		hCGβ	76	97	73	76	72	74	74	69	52	39	38	37	37	33	43	43	43	42	43
eLHβ			eLHβ	78	91	92	93	92	90	91	50	41	41	39	38	34	41	39	39	41	39
hLHβ				hLHβ	81	86	84	85	87	77	51	41	40	39	38	34	48	48	47	48	48
oLHβ					oLHβ	115	101	95	96	94	45	42	43	41	40	35	41	38	39	41	40
bLHβ						bLHβ	101	98	100	93	46	42	43	41	40	35	41	38	39	41	40
pLHβ							pLHβ	104	104	104	46	39	40	38	37	34	43	40	40	44	42
lLHβ								lLHβ	109	104	45	38	39	37	36	33	42	40	40	42	41
rLHβ									rLHβ	97	46	39	40	38	37	34	43	41	41	45	44
wLHβ										wLHβ	44	34	35	34	32	34	39	37	37	41	40
sGTHβ											sGTHβ	38	39	38	38	33	40	40	39	40	42
eFSHβ												eFSHβ	101	98	97	83	42	42	43	43	44
hFSHβ													hFSHβ	97	100	83	45	42	45	43	44
oFSHβ														oFSHβ	101	84	44	43	45	44	45
bFSHβ															bFSHβ	83	42	41	44	42	43
pFSHβ																pFSHβ	39	40	42	41	42
hTSHβ																	hTSHβ	100	100	97	101
bTSHβ																		bTSHβ	104	98	100
pTSHβ																			pTSHβ	94	96
mTSHβ																				mTSHβ	108
rTSHβ																					rTSHβ

Note: Total amino acid residues for each β-subunit is given in parentheses.

Table 4

BETA SUBUNIT HOMOLOGY MATRIX (PERCENT IDENTITY)

	eCGβ (149)	bCGβ (145)	eLHβ (149)	hLHβ (121)	oLHβ (119)	bLHβ (121)	pLHβ (119)	lLHβ (119)	rLHβ (121)	wLHβ (118)	sGTHβ (119)	eFSHβ (118)	hFSHβ (111)	oFSHβ (111)	bFSHβ (110)	pFSHβ (107)	hTSHβ (112)	bTSHβ (118)	pTSHβ (112)	mTSHβ (118)	rTSHβ (118)
eCGβ (149)	eCGβ	52	100	64	76	76	78	77	74	77	42	35	37	35	35	32	37	33	35	35	33
bCGβ (145)		bCGβ	52	80	61	63	61	62	61	58	44	33	34	33	34	31	38	36	38	36	36
eLHβ (149)			eLHβ	64	76	76	78	77	74	77	42	35	37	35	35	32	37	33	35	35	33
hLHβ (121)				hLHβ	68	71	71	71	72	65	43	35	36	35	35	32	43	41	42	41	41
oLHβ (119)					oLHβ	97	85	80	81	80	38	36	39	37	36	33	37	32	35	35	34
bLHβ (121)						bLHβ	85	82	83	79	39	36	39	37	36	33	37	32	35	35	34
pLHβ (119)							pLHβ	87	87	88	39	33	36	34	34	32	38	34	36	37	36
lLHβ (119)								lLHβ	92	85	38	32	35	33	33	31	38	34	36	38	35
rLHβ (121)									rLHβ	82	39	33	36	34	34	32	38	35	37	35	37
wLHβ (118)										wLHβ	37	29	32	31	29	32	35	31	33	34	34
sGTHβ (119)											sGTHβ	32	35	34	35	31	36	34	35	36	36
eFSHβ (118)												eFSHβ	91	35	35	32	38	36	38	39	37
hFSHβ (111)													hFSHβ	88	88	78	41	38	41	41	40
oFSHβ (111)														oFSHβ	91	78	40	39	41	39	41
bFSHβ (110)															bFSHβ	78	38	37	40	40	39
pFSHβ (107)																pFSHβ	36	37	39	39	39
hTSHβ (112)																	hTSHβ	89	89	87	90
bTSHβ (118)																		bTSHβ	93	83	85
pTSHβ (112)																			pTSHβ	84	86
mTSHβ (118)																				mTSHβ	92
rTSHβ (118)																					rTSHβ

Note: Total amino acid residues for each β-subunit is given in parentheses.

Glycoprotein Hormone α Subunit Sequences

```
                               10          20          30          40          50
(1) Bovine   F P D G E F T M Q G C P E C K L K E N K Y F S K P D A P I Y Q C M G C C F S R A Y P T P A R S K K T
(2) Ovine    F P E G E F T M Q G C P E C K L K E N K Y F S K P D A P I Y Q C M G C C F S R A Y P T P A R S K K T
(3) Porcine              T M Q G C P E C K L K E N K Y F S K L G A P   Y Q C M G C C F S R A Y P T P A R S K K T
(4) Human    A P D V Q         D C P E C T L Q E N P F F S Q P G A P   L Q C M G C C F S R A Y P T P L R S K K T
(5) Equine   F P D G E F T T Q D C P E C K L R E N K Y F F K L G V P I Y Q C K G C C F S R A Y P T P A R S R K T
(6) Mouse    L P D G D F I I Q G C P E C K L K E N K Y F S K L G A P   Y Q C M G C C F S R A Y P T P A R S K K T
(7) Rat      L P D G D L I I Q G C P E C K L K E N K Y F S K L G A P I Y Q C M G C C F S R A Y P T P A R S K K T
(8) Rabbit   F P D G E F A M Q G C P E C K L K E N K Y F S K L G A P I Y Q C M G C C F S R A Y P T P A R S K K T
(9) Whale    F P N G E F T M Q G C P E C K L K Q N K Y F S K L G A P I Y Q C M G C C F S R A Y P T P A R S K K T

                               60          70          80          90
(1) Bovine   M L V P K N I T S E A T C C V A K A F T K A T V M G N V R V E N H T E C H C S T C Y Y H K S
(2) Ovine    M L V P K N I T S E A T C C V A K A F T K A T V M G N V R V Z N H T Z C H C S T C Y Y H K S
(3) Porcine  M L V P K N I T S E A T C C V A K A F T K A T V M G N A R V E N H T E C H C S T C Y Y H K S
(4) Human    M L V Q K N V T S E S T C C V A K S Y N R V T V M G G F K V E N H T A C H C S T C Y Y H K S
(5) Equine   M L V P K N I T S E S T C C V A K A F I R V T V M G N I K L E N H T Q C Y C S T C Y H H K I
(6) Mouse    M L V P K N I T S E A T C C V A K A F T K A T V M G N A R V E N H T E C H C S T C Y Y H K S
(7) Rat      M L V P K N I T S E A T C C V A K S F T K A T V M G N A R V E N H T D C H C S T C Y Y H K S
(8) Rabbit   M L V P K N I T S E A T C C V A K A F T K A T V M G N A K V E N H T E C H C S T C Y Y H K S
(9) Whale    M L V P K N I T S E A T C C V A K A F T K A T V M G N A R V Q N H T Z C H C S T C Y Y H K S
```

FIGURE 1. A summary of the glycoprotein hormone alpha subunit sequences, presented with the single-letter code for the amino acids. The single-letter amino acid code is simply the first letter of the amino acid — wherever possible. Where more than one amino acid has the same first letter, alternatives are used. A. Ala. B. Asx. C. Cys. D. Asp. E. Glu. F. Phe. G. Gly. H. His. I. Ile. K. Lys. L. Leu. M. Met. N. Asn. P. Pro. Q. Gln. R. Arg. S. Ser. T. Thr. V. Val. W. Trp. Y. Tyr. Z. Glx. The sequences presented are from the following sources: bovine (8), ovine (23), porcine (24), human (12,13, these studies showed that the "deletion" of the four residues [6 through 9] in human alpha subunit is the result of a corresponding deletion in the gene near an intron), equine (25), mouse (10), rat (11), rabbit (26), and whale (27,28). In those instances where the acid or amide form has not been established (B,Z), these positions have been regarded as not identical for purposes of the tabulations in Tables 1 and 2.

protecting this part of the molecule from incidental proteolytic shortening during isolation. In support of this interpretation, N-terminal heterogeneity has rarely been documented for any LH or CG beta subunit preparation. However, N-terminal heterogeneity is commonly observed for the TSH and FSH beta subunits that have been isolated. Ward et al.[20] and Glenn et al.[21] reported an additional pyroglutamyl-prolyl sequence on rLH beta subunit, as judged by the location of the first 1/2-cystine in the sequence when compared to other LH sequences. This is interesting, since this is the only LH sequence from a species that is a reflex ovulator, but no functional significance has been established.

The two-residue insert in the TSH molecules is a consistent structural feature of all the TSH beta sequences so far reported. Some of these sequences have been deduced from the gene sequences, so it is a feature that is attributable to the corresponding gene insertion (or deletion, if one considers the case for FSH and LH beta subunits). It has been proposed that the glycoprotein hormones all evolved from a common primordial gonadotropin.[22]

The information in Tables 1 through 4 summarizes all the glycoprotein hormone sequences for which we have complete amino acid sequences at this time. We have tried to use the most accurate sequence proposals available (in our judgement). The reader should appreciate that the sequencing methodology has undergone a steady improvement with time. Current methods are usually more accurate and certainly more sensitive than those available to some of the earlier investigators. Moreover, recent reexaminations of a sequence also have the benefit of previous reports for their design and improvement. From the summaries in these tables it is at once apparent that we have a substantial body of sequence information for comparison of the glycoprotein hormones' protein structures. From these tables we can gain an appreciation of the degree of homology that exists within this group of hormones.

In Figure 1 we have summarized the proposed specific sequences of the alpha subunits that have been reported. Again we have made use of the 1/2-cystine placements to put these comparisons in register throughout the series. In the original reports the reader should appreciate that the species indicated would be numbered with the residue in the N-terminal position as number 1. As a convenience to our comparative presentation we will utilize the numbering system of the ovine/bovine alpha subunit as a basis for all the reported sequences, and this must be taken into account in reference to the individual sequence reports in the literature. To conserve space we have used the single letter code for the amino acids. For those readers unfamiliar with this code it is summarized in the legend to the figure in terms of the three-letter abbreviations. In this figure the location of areas of identity among all the species studied is immediately apparent. The most significant aspect of this presentation is the high degree of homology of the subunit among the species. If we use the average of the tabulations in Table 2, this homology may be quantitated as 86.6%.

In Figure 2 we present a similar summary for the beta subunits that have been reported, using the same conventions as put forward for Figure 1 above. Here we see a greater diversity in the location of areas of identity. The sequences have been grouped to combine the LH/CG series, FSH series, and TSH series together. For our comparative numbering we have taken the longest beta subunit sequences, those of equine CG and LH, as a reference for our comparative numbering. The same precaution applies in reference to numbering in the original publications of those subunits with "foreshortened" N-terminal sequences. The important comparison to be obtained from this presentation is the high degree of identical homologies observed within each series — LH/CG, FSH, and TSH. The average percentages, respectively, are 75.2, 85.0, and 87.8%. But the three series differ much more when compared one with another. Here the comparisons, averaged from Table 4, are LH/CG vs. FSH, 34.1% homology; LH/CG vs. TSH, 35.9%; and FSH vs. TSH, 38.8%.

Interestingly, the salmon (s) gonadotropic hormone beta subunit, the only beta sequence known outside the mammalian series, does not show a high homology with any of the three mammalian groups, but shares about the same average homology with all of them (average 37.2%).

There was at the time of the early structure studies[7,23,24,48] a tendency to attribute hetero-geneity, particularly charge heterogeneity, to loss of amide nitrogen, e.g., conversions of the type Gln → Glu or Asn → Asp, with the implied production of a negative charge (at physiological pH) where there was no charge previously. This was a "ready" explanation for obvious charge differences. However, we have searched for well-documented examples of this occurrence among the glycoprotein hormones. We are not aware of such an example. In fact, we postulated such an occurrence to explain the increased FSH-like activity of eCG (pregnant mare serum gonadotropin, PMSG) over that observed in eCG from the endometrial cups. This consideration led us to the proposal of the "determinant loop" hypothesis.[51,52] This postulate presumed a Gln-94β conversion to Glu with an increased negative charge to account for the FSH activity. Our recent extensive study of this area of the sequence of eCGβ[29] has failed to provide any evidence of the acid form of the residue in this position. Thus, although an amide to acid conversion remains a potential source of protein hetero-geneity, we regard it as an improbable source in the case of the glycoprotein hormones unless these molecules have been subjected to rather drastic treatments in the course of their handling.

Finally, in the consideration of heterogeneity, the most obvious source of heterogeneity is from contamination with foreign proteins or other substances. Although this type of heterogeneity often influences research conclusions (e.g., Reference 53), it can only be addressed on an ad hoc basis for individual samples. Thus, in many of the discussions that follow in this and other chapters there will be an implicit assumption that this type of heterogeneity has been eliminated. But the reader should always bear in mind that in any given situation this may not indeed be true.

Microheterogeneity of Glycoprotein Hormones

Glycoprotein Hormone ß Subunit Sequences

```
                    10          20          30          40          50
 (1) eCGß    SRGPLRPLCRPINATLAAEKEACP CITFTTS CAGYCPSMVRVMPAA  P
 (2) hCGß    SKEPLRPRCRPINATLAVEKEGCPVC ITVNTT CAGYCP MTRVLGGVI P

 (3) eLHß    SRGPLRPLCRPINATLAAEKEACPIC ITFTTS ICAGYCPSMVRVMPAAL P
 (4) hLHß    SREPLRPWCHPINAILAVEKEGCPVC ITVNTTICAGYCP MMRV OAVL P
 (5) oLHß    SRGPLRPLCQPINATLAAEKEACPVC ITFTTSICAGYCPSMKRV PVIL P
 (6) bLHß    SRGPLRPLCQPINATLAAEKEACPVC ITFTTSICAGYCPSMKRV PVIL P
 (7) pLHß    SRGPLRPLCRPINATLAAEDEACPVC ITFTTS CAGYCPSMRRVLPAAL P
 (8) lLHß    pEPARGPLRPLCRPVNATLAAENEACPVC TFTTSICAGYCPSMVRVLPAAL P
 (9) rLHß    SRGPLRPLCRPVNATLAAENEFCPVC ITFTTSICAGYCPSMVRVLPAAL P
(10) wLHß    PRGPLRPLCRPINATLAAQNZACPVC ITFTTSICAGYCPSMVRVLPAAL P
(11) sGTHß   SLMQ  P CQPINQTVSLEKEGCPTCLVIRAPICSGHCVTKEPVFKSP  S

(12) eFSHß        NSCELTNITIAVEKEGCRFC IT INTTWCAGYCYTRDLVYKDPA R
(13) hFSHß        NSCELTNITIAIEKEECRFC IS INTTWCAGYCYTRDLVYKDPA R
(14) oFSHß        SCELTNITITVEKEECSFC IS INTTWCAGYCYTRDLVYKBPA R
(15) bFSHß        SCELTNITITVEKEECGFC IS INTTWCAGYCYTRD LVYRDPA R
(16) pFSHß          CELTNITITVEKEECTFC IS INTTWCAGYCYTRDLVYKBPA R

(17) hTSHß        FCIPTEYMTHIERRECAYCLTINTTICAGYCMTRD INGKLFL P
(18) bTSHß        FC PTEYMMHVERKECAYCLTINTTVCAGYCMTRDVNGK FL P
(19) pTSHß        FCIPTEYMMHVERKECAYC YVNSTICAGYCMTRDFDGKLFL P
(20) mTSHß       FCIPTEYTMYVDRRECAYC LT NTT CAGYCMTRD INGKLFL P
(21) rTSHß       FCIPTEYMMYVDRRECAYCLTINTTICAGYCMTRD INGKLFL P


                    60          70          80          90         100
 (1) eCGß A  IP QPVCTYRELRFAS IRLPGCPPGVDPMVSFPVA SCHCSPCQIK T DC
 (2) hCGß A  LP GVVCNYRDVRFESIRLPGCPRGVNPVVSYAVA SQQCALCRRSTIDC

 (3) eLHß A  IP QPVCTYRELRFAS IRLPGCPPGVDPMVSFPVALSCHCGPCGIKTIDC
 (4) hLHß P  LP QVVCTYRDVRFESIRLPGCPRGVDPVVSFPVALSCRCGPCRRSISDC
 (5) oLHß P  MP QRVCTYHELRFASVRLPGCPPGVDPMVSFPVA SCHCGPCRLSSTDC
 (6) bLHß P  MP QRVCTYHELRFASVRLPGCPPGVDPMVSFPVA SCHCGPCRLSSTDC
 (7) pLHß P  VP QPVCTYRELIFASIRLPGCPPGVDPVVSFPVALSCHCGPCRLSSSDC
 (8) lLHß P  VP QPVCTYRELRFAS IRLPGCPPGVDPEVSFPVALSCRCGPCRLSSSDC
 (9) rLHß P  VP QPVCTYRELRFASVRLPGCPPGVDPIVSFPVALSCRCGPCRLSSSDC
(10) wLHß P  VP ZPVCTYRGLRFAS RLPGCPPGVNPMVSFPVALSCHCGPCRLSSSDC
(11) sGTHß T  VT QHVCTYRDVRYEMIRIPDCPPWSEPHVTYPVALSCDCSLCNMDTSDC

(12) eFSHß P  NI QKTCTFKELVYETVKVPGCAHHADSLYTYPVATZCHCGKCBSDSTBC
(13) hFSHß P  KI QKTCTFKELVYETVRVPGCAHHADSLYTYPVATQCHCGKCDSDSTDC
(14) oFSHß P  BI QKTCTFKELVYETVKVPGCAHHADSLYTYPVATECHCGKCDSDSTDC
(15) bFSHß P  NI QKTCTFKELVYETVKVPGCAHHADSLYTYPVATECHCSKCDSDSTDC
(16) pFSHß P  BI ZKTCTYRZLVYZTVKVPGCAHHABSLYTYPVATZCHCGKCBSBSTBC

(17) hTSHß KYALSQDVCTYRDFIYRTVEIPGCPLHVAPYISYPVALSCKCGKCDTDYSDC
(18) bTSHß KYALSQDVCTYRDFMYKTAE PGCPRHVTPVFSYPVA SCKCGKCNTDYSDC
(19) pTSHß KYALSQDVCTYRDFMYKTVEIENRECPHHVTPVTSYPVALSCKCGKCDTDYSDC
(20) mTSHß KYALSQDVCTYRDFIYRTVEIPGCPHHVTPVFSFPVAVSCKCGKCNTDNSDC
(21) rTSHß KYALSQDVCTYRDFTYRTVEIPGCPHHVAPVFSYPVALSCKCGKCNTDYSDC


                   110         120         130         140
 (1) eCGß    GVFRDQPLACAPQASSSSKDPPSQPLTS TSTPTPGASRRSSHPLPIKTS
 (2) hCGß    GGPKDHPLTCDDPRFQDSSSSKAPPPSLPSPSRLPGPSDTPILPQ

 (3) eLHß    GVFRDQPLACAPQASSSSKDPPSQPLTSTSTP TPGASRRSSHPLPIKTS
 (4) hLHß    GGPKDHPLTCDHPQLSGLLFL
 (5) oLHß    GPGRTZPLACBHPPLPDIL
 (6) bLHß    GGPRTQPLACDHPP PDILFL
 (7) pLHß    GPGRAQPLACDRPPLPGLL
 (8) lLHß    GGPRAEPLACDLPHLPS
 (9) rLHß    GGPRTQPMTCDLPHLPGLLLF
(10) wLHß    GPGRAQPLACNRSPRPGL
(11) sGTHß   TIESLQPDFCITQRVLTDGDMW

(12) eFSHß   TVRGLGPSYCSFGDMKZYPVALSY
(13) hFSHß   TVRGLGPSYCSFGEMKE
(14) oFSHß   TVRGLGPSYCSFSDIERZ
(15) bFSHß   TVRGLGPSYCSFREIKE
(16) pFSHß   TVRGLGPSYCSFGE
```

FIGURE 2.   A summary of the glycoprotein hormone beta subunit sequences. See the legend for Figure 1 for a comment on the single-letter code for the amino acids. The sequences presented are from the following sources: equine-eCGß (29), human-hCGß (30), equine-eLHß (31), human-hLHß (32), ovine-oLHß (33), bovine-bLHß (34), porcine-pLHß (35), lagomorph (rabbit)-lLHß (21), rat-rLHß (36,37), whale-wLHß (28,38), salmon gonadotropin-sGTHß (39), eFSHß (40), hFSHß (41), oFSHß (42), bFSHß (43), pFSHß (44,45), hTSHß (46), bTSHß (7,47), pTSHß (48), mTSHß (49), and rTSHß (50). This list of references has been selected previously for what we believe to be the most current sequence information without attempting to acknowledge all the contributors that may have provided substantial sequence information.

```
(17) hTSHβ   I  H E A   K T N Y C T K P Q K S Y
(18) bTSHβ   I  H E A   K T N Y C T K P Q K S Y M V G F S
(19) pTSHβ   I  H E A  I K T N Y C T K P E K S Y
(20) mTSHβ   I  H E A V R T N Y C T K P Q S F Y L G G F S V
(21) rTSHβ   T H E A V K T N Y C T K P Q T F Y L G G F S G
```

FIGURE 2 continued

II. LH/CG PROTEIN HETEROGENEITY

A. N-Terminal Heterogeneity

For LH and CG, N-terminal heterogeneity is usually restricted to the α-subunit, whereas the N terminus of the β-subunit is usually intact. Figure 1 summarizes the known structures of the α-subunits. Most α-subunits consist of 96 residues. N-terminal heterogeneity has been reported for oLHα,[23,54] bLHα,[55] eLHα,[56] and whale (w) LHα.[27] Rabbit LHα was not reported to have significant N-terminal heterogeneity although the "background" when sequencing intact α-subunit was higher than when N-terminal fragments were sequenced.[26] In our laboratory eCGα did not show the N-terminal heterogeneity that characterize eLHα, eFSHα, and "free" α-subunit,[56,57] however, others have reported N-terminal heterogeneity for eCGα.[58] The α-subunit of hLH and hCG consists of only 92 amino acid residues (Figure 1). N-terminal heterogeneity was reported for hCGα,[59] but none was noted for an early report of the structure of hLHα,[60] which was (erroneously) assigned only 89 residues. Keutmann et al.[61] surveyed the N-terminal heterogeneity of the human α-subunits and confirmed this structural feature for hLHα as well as for hCGα preparations. The 92-residue structure has been determined to arise from a four-codon deletion in the human α-subunit gene.[12,13] Since a single gene exists for the α-subunits in the human[12,13] and the same promoter is used in both pituitary and placenta,[62] the same prealpha subunit is the result. Speculation has been that differential processing of the prealpha subunit results in different N termini.[61] However, it is difficult to separate this from differential exposure of the α N termini in the various hormones to proteolytic degradation during gland collection and hormone isolation. The latter phenomenon is demonstrably responsible for at least some of the heterogeneity observed.[23] Porcine (p) LHα reportedly consists of only 90 amino acid residues[24] and no N-terminal heterogeneity was indicated. This result has been examined in our laboratory. It was shown that purified pLH showed only two amino acids,[63] Phe and Ser (although with the Dns-amino acid method then available Ser and Thr were not well resolved). However, the two reports considered in retrospect, plus the sequence comparisons now available, suggest the preparation examined by Chu[63] contained Phe and possibly Thr (positions one and seven on the alpha subunit) plus Ser (position one on the beta subunit). This question should be reexamined with more recent analytical techniques. Nevertheless, the foregoing indicates that the N-terminal 4 to 6 residues are exposed in the various species of LH alpha and are apparently unnecessary for biologic activity.

N-terminal heterogeneity has been observed in only two LHβ-subunits. Glenn et al.[21] reported a two-amino acid deletion at the N terminus of rabbit LHβ occurred in 3% of the molecules; however, since there is a two-amino acid extension at the N terminus of rabbit LHβ, this merely trims a small percentage to the same-sized N terminus as all other LH or CGβ-subunits. Pankov and Karasev[38] have recently noted N-terminal heterogeneity in sperm whale LHβ.

B. C-Terminal Heterogeneity

With the exception of one unconfirmed report,[24] there is no C-terminal heterogeneity of the C termini of glycoprotein hormone alpha subunits. Furthermore, the structures after the last 1/2-Cys are identical in all species except the horse (Figure 1). The horse α-subunit has some unique properties and it is tempting to attribute them to this portion of the molecule although the carbohydrate moieties must also be considered. There appears to be a reason for conservation of structure at the C terminus: Cheng et al.[64] demonstrated in 1973 that the integrity of the α-subunit C terminus was important for biologic activity of TSH and LH. Although deletion of the α C terminus prevented recombination with TSHβ, LHβ was able to recombine with the α-subunit derivatives. The importance of the α-subunit C terminus has been shown by Merz for hCGα.[65] The opposite situation exists for the LHβ-subunit where the C terminus is the least-conserved portion of the molecule and complete removal has no effect on either ability to recombine with α-subunit or on biologic activity.[66] C-terminal heterogeneity has been reported for hLHβ,[67-70] oLHβ,[33] bLHβ,[35] and wLHβ.[28,38] Neither hCGβ[30,71,72] nor eCGβ[29] has been shown to have any C-terminal "fraying". This might be attributed to the O-linked oligosaccharide on the C-terminal extension protecting the C terminus from protease. However, eLHβ, which also possesses a glycosylated C-terminal extension,[31] has been found to be "frayed" in that 20 to 30% of the preparation lacks the last two residues at the C terminus. This is in spite of the O-linked oligosaccharide at either Thr-148 or Ser-149, which makes the C terminus resistant to digestion by carboxypeptidase. The major difference between LH and CG preparations is that the latter are isolated from physiologic fluids of living organisms whereas LH is isolated postmortem. Some heterogeneity of hCG α-subunits has been reported, and fragments of hCGβ have recently been isolated and characterized.[73] These arise from the hormone fraction that has been cleared from the serum, therefore it is perhaps not unexpected to find some degradative products present. For eCG, the circulating form of the hormone is isolated (its heavy glycosylation prevents its clearance through the kidney) and that may account for its pristine condition. It has been demonstrated that endometrial cup-extracted eCG has N-terminal heterogeneity.[74] These data fit with the general consensus that the condition of N termini and C termini of glycoprotein hormone subunits depends largely upon the condition of the starting tissue and precautions taken to limit proteolysis during isolation.[2]

C. Intrachain Proteolytic Nicking

Three examples of intrachain proteolytic nicking have been reported in oLHα, hLHβ, and bovine α-subunit.[75-77] During an examination of methods of preparing oLHα-subunits it was noted that preparations isolated at low pH gave poor recovery of biological activity and upon reduction, SDS polyacrylamide gel patterns showed three bands instead of the expected single band.[76] Free N-terminal lysine was also noted in N-terminal analysis of these preparations. This was determined to be the result of cleavage of the peptide bond between lysine residues at α-48 and α-49. Initially it was believed that low pH was the cause of the nicking. However, subsequently it was determined to result from a highly specific protease from *Alkaligenes odorans*, a microorganism common in the water supply that apparently can become trapped in the beads of Sephadex® columns.[75] We have also observed this nick in the α-subunits of eLH and eFSH at the analogous site (Arg[48]–Lys[49]).[78] There is a nick which occurs in the "free α-like" material that has been isolated from bovine pituitaries.[77] Even when fresh-frozen glands were isolated in the presence of protease inhibitors evidence of some nicking was evident on SDS polyacrylamide gels following reduction of the disulfide bonds.[79]

The best-characterized example of intrachain nicking is the nick observed in hLHβ. In one of the early hLHβ sequence determinations, Shome and Parlow[68] observed nicks in the sequence Ala[46]–Val–Leu–Pro[49] resulting in the appearance of Val and Leu in N-terminal determinations. Subsequently, Sairam and Li[69] reported nicking of hLHβ in a similar region:

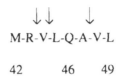

$$\downarrow\downarrow \qquad \downarrow$$
$$\text{M-R-V-L-Q-A-V-L}$$

$$42 \qquad 46 \qquad 49$$

Keutmann et al.[70] also observed nicking in the area of residues 44 to 48. Recently, our laboratory undertook the analysis of two preparations of hLH.[75] One was predominantly intact as judged by SDS polyacrylamide gel analysis under reducing conditions. The other was nicked in the β-subunit. Analysis of the nicks by automated Edman degradation indicated the following pattern of nicks:

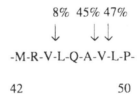

$$8\% \ 45\% \ 47\%$$
$$\downarrow \quad \downarrow\downarrow$$
$$\text{-M-R-V-L-Q-A-V-L-P-}$$

$$42 \qquad\qquad 50$$

Comparison of the relative biological activities indicated that the nicked hLH had only 53% of the receptor-binding activity of the largely intact hLH preparation.[75] Hartree et al.[80] analyzed 14 fractions of hLH that were prepared from one hLH preparation by fast protein liquid chromatography (FPLC) and found that some of the fractions were enriched for β nicks. They concluded that the nicks described above contribute to the heterogeneity observed in hLH preparations.

D. Allelic Variation

Another potential source of heterogeneity in glycoprotein hormones at the polypeptide level is allelic variation. However, many of the reported or suggested genetic alleles (i.e., phenotypic expression of two forms of the gene product detected by two different residues at the same position in otherwise identical peptides) did not rigorously exclude simple sequencing artifacts, and the same could be true of contemporary recombinant DNA studies. Among the early glycoprotein hormone sequence reports there were several reports of evidence for allelic variation in hLHβ,[81] pLHβ,[35] and bLHα.[55] Most of the evidence for sequence variations appears to have arisen from the state of the chemistry then employed in sequence determinations. Because of the obvious clinical interest, the "variations" in the hLHβ sequence drew considerable attention and these discrepancies were resolved first. The variations in the sequence of bLHα were not seen upon subsequent examination when more modern sequencing methods were applied, i.e., sequencing in a vapor-phase sequencer.[82] The pLHβ sequence has never been rechecked. Cloning of glycoprotein hormone genes promises to help resolve some of the difficulties to establish some of the sequences that remain uncertain at the present time. Sequencing cDNA or genomic clones affords an independent check of such amino acid residues as Thr, Ser, Trp, Lys, Gln, Glu, Asn, or Asp which are difficult to accurately quantitate with the protein sequencer (particularly with early methodology). Because of the amplification involved in the selection procedure, the chance of uncovering allelic variations in gonadotropins is increased. However, the most convenient method of screening for such variants, restriction mapping, may be too insensitive for the single amino acid substitutions that might exist within a population. By restriction mapping of individuals, the human glycoprotein hormone alpha subunit appears to be the product of a single gene,[12,13] and yet, as described below, there may be another gene for α-subunit which has been missed by restriction map studies.[83] Comparisons of the sequences of LHβ and α determined from cloned cDNA with cloned genomic DNA in the cow have revealed differences in the nucleotide sequences, some of which predict amino acid substitutions.[9,84] These do not correlate with the reported sequence variations based on the sequence studies by protein analysis.[55] More work will be required to determine whether the differences between the cDNA sequences and the genomic DNA sequences represent true genetic variability or whether they are simply cloning artifacts (see Reference 9).

Restriction enzyme digestion analysis of DNA derived from normal and tumor-derived human tissues has produced a consensus that there is a single gene for the glycoprotein hormone α-subunits in the human (vide supra). Recently, ectopic alpha subunit was purified from the urine of a cancer patient and sequenced.[83] A single amino acid substitution of Ala for Glu[56] was noted. In humans, ectopic hCG or its subunits are frequently produced by tumors.[85] Analysis of α-subunits from these sources suggests an apparent higher molecular weight than that of placental hCGα based on gel permeation studies.[85] The higher molecular weight has been attributed to O-linked glycosylation of Thr,[43] as has been demonstrated by Parsons et al. for free α-subunit isolated from bovine pituitaries.[77,79] Nishimura et al.[83] detected Thr at position 43 while sequencing ectopic α and concluded that glycosylation of this residue had not occurred. Since the yield of Thr at that degradative cycle was not reported, it is difficult to decide whether or not this is a conclusive result. In our own experience with the O-linked glycosylated Thr and Ser residues on the C terminus of eLHβ, we were able to detect Thr or Ser at suspected glycosylation sites, although at greatly reduced yields compared with the yields of nonglycosylated Thr and Ser.[29] In contrast to this we observed no recovery of Asn (or even Asp) at positions of N-linked glycosylation. Perhaps this happens since N-linked glycosylation occurs cotranslationally while O-linked glycosylation is a posttranslational event. In any case, Nishimura et al.[83] presented a generalized hypothesis that the Glu→Ala substitution resulted in the hydrophilic segment –Thr[54]–Ser[55]–Glu[56]–Ser[57]–Thr[58]– (predicted by the method of Kyte and Doolittle[86]) becoming an extension of the preceding hydrophobic section when Glu[56] is replaced with Ala. Although no effect on conformation was apparent by circular dichroism (CD), the technique is insensitive enough that the data are often ignored in order to hypothesize that a subtle shift in conformation resulting from the Glu→Ala substitution can account for the inability of ectopic α-subunit to recombine with hCGβ[83] and the higher apparent molecular weight of ectopic alpha subunit. The latter might be the consequence of either aggregation or altered glycosylation due to the more extensive hydrophobic region. Parsons and Pierce[79] reported that the free α that they characterized was identical in amino acid sequence to bovine α-subunit where they had determined the sequence. The only difference that they encountered was the glycosylation of Thr[43] and removal of this oligosaccharide moiety permitted recombination with β-subunit to take place. We have isolated significant quantities of free α-subunit from horse pituitaries.[57] In agreement with Pierce's laboratory, we find this material to be identical in amino acid sequence to the equine α-subunits, including having Glu at position 56.[57] Since we had no control over collection of the glands (horse gonadotropinologists are grateful to get what they can get), the free α-subunit is 90% nicked (estimated from recoveries of reduced, carboxymethylated fragments following Sephacryl® S-200 chromatography). This thus provides the fortuitous access to an enriched N terminal 48 to 50 residue fraction for further analysis. Although the exact attachment site remains to be determined, the N-terminal peptide is glycosylated as indicated by the presence of amino sugars in the amino acid analysis of this fragment. Admittedly, what Pierce has described in the cow and what we have extended to the horse may only hold for pituitary free α-

Table 5

AMINO ACID COMPOSITIONS OF hCGβ C-TERMINAL PEPTIDES

Amino acid	CTP-1	CTP-2	Theor.
Asp	1.5	1.6	1
Thr	—	0.9	1
Ser	3.2	3.5	8
Glu	0.7	1.6	2
Pro	5.7	6.9	10
Gly	1.1	1.1	1
Ala	1.2	1.1	1
$^1/_2$-Cys	—	—	—
Val	?	—	?
Met	—	—	—
Ile	—	0.9	1
Leu	1.9	2.5	3
Tyr	—	—	—
Phe	0.9	0.8	1
Lys	1.0	1.0	1
His	—	—	—
Arg	1.6	1.6	2

Note: Probable sequence suggested:
CTP-1 PRFQDSSSSKAPPPSLPSPSRLPGPSD
CTP-2 PRFQDSSSSKAPPPSLPSPSRLPGPSDTPILPQ

subunit. Further work will be required to determine if Nishimura et al.[83] are correct in their hypothesis regarding ectopic α-subunit. The consequences of their prediction are of some interest. If correct, they have described a potential allelic variant (in an individual) that they suspect has general significance for the structure of ectopically produced alpha subunit. However, if this single amino acid substitution has the consequences that they predict, then there is selective transcription of one of two (implied) forms of the alpha gene. In view of the importance of hCG in human gestation, the correct alpha gene must be present for normal hCG to be produced in order to maintain the early pregnancy. If the higher apparent molecular weight of other ectopic α-subunits arises from this Glu→Ala substitution, this allele must be fairly common in the general population. Since the amino acid substitution predicts loss of a restriction enzyme site, presumably screening of individuals for this variant should be feasible. On the other hand, these results could mean that there is a second gene that is expressed (only) in tumors. Either way, the developments of this line of investigation should be interesting. This represents possible protein heterogeneity with its source at the gene level.

The organization of the hCGβ genes is unique for those glycoprotein hormone genes that have yet been described in that there are six or seven (the number keeps dropping) along with the hLHβ gene.[87-92] Talmadge et al.[91] found evidence that only two of these genes were expressed. They did not have sufficient data (or actually the gene itself since their isolate was foreshortened on the 3′ end) to decide for CGβ1 expression. Policastro et al.[92] have determined that all six CGβ and the LHβ genes are linked and that CGβ1 is probably a pseudo gene, therefore, only two genes are expressed in placenta (CGβ3 and CGβ5).

We recently investigated the question of which genes might be expressed by analyzing hCGβ. We exploited the fact that there is a single Asp–Pro bond in hCGβ just outside the disulfide bonded core. A single break in the polypeptide chain can be made by a 30 min incubation in 0.013 N HCl at 110°C. The C-terminal peptide (residues 113 to 145) were partially purified by Sephacryl® S-200 chromatography[66] and then purified by reverse phase HPLC. Two C-terminal fragments were recovered. The amino acid compositions are given in Table 5. The results suggested some C-terminal heterogeneity such as has been observed

Table 6
AMINO ACID SEQUENCE DETERMINATION FOR THE TWO C-TERMINAL PEPTIDES DERIVED FROM hCGβ

#	CTP-1				CTP-2	
	PTH-aa	Yield	PTH-aa	Yield	PTH-aa	Yield
1	Pro	7.0			Pro	9.0
2	Arg	5.0			Arg	5.4
3	Phe	6.5			Phe	10.0
4	Gln	3.2[a]			Gln	6.0[a]
5	Asp	1.5	Val	3.1	Asp	4.8
6	Ser	0.9	Ala	3.1	Ser	0.4
7	Ser	0.7	Pro	2.7	Ser	0.5
8	Ser	0.6	Pro	3.1	Ser	0.3
9	Ser	[b]	Pro	2.8		
10	Lys	0.9	?			
11	Ala	3.6	Gly	1.1		

[a]   Yield of PTH-Gln + PTH-Glu.
[b]   Integrator did not register.

Table 7
AMINO ACID SEQUENCE OF hCGβ C TERMINUS

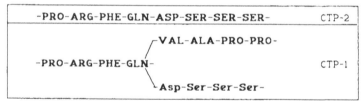

```
                115                      120
-ASP-PRO-ARG-PHE-GLN-ASP-SER-SER-SER-SER-LYS-ALA-PRO-
```

for other glycoprotein hormone beta subunits. Since the molar ratios were essentially integer values, the results suggested a clear C terminal. These findings were intriguingly close to the predicted composition that would result if the two-base insertion had not occurred which gives rise to the last eight residues of the full-length hCG sequence, as indicated by the reduction in Pro and Leu and by the absence of Thr and Ile.[6,93] On the other hand, the presence of as much Asp as in the full-length C-terminal peptide suggests mere C-terminal heterogeneity. Nevertheless, we sequenced both peptides to see whether there was any evidence for the expression of CGβ6 which has an Ala substituted for Asp.[117] The results of both sequencing experiments are shown in Table 6. CTP-2 was sequenced first and gave the same sequence as hCGβ (Figure 2). There was no indication of Ala at position 117, in accord with the conclusion from the recombinant DNA studies that this gene is not expressed. This may be even more likely since Policastro et al.[92] think that this gene does not exist. The sequencing of CTP-1 started out the same way as CTP-2. However, at step 5 we began to see two sequences. The interpretation of the data is summarized in Table 7. There appears to be a five-amino acid deletion in roughly two thirds of the peptides present in this fraction (obviously a mixture of two peptides) with a single Val substituted for it. Why we did not detect the expected valine in the amino acid analysis (Table 5) is a problem we will have to resolve by further characterization of these fragments once more material is obtained. At this time the source of this heterogeneity is unexplained.

III. FSH PROTEIN HETEROGENEITY

For practical reasons the available information about both FSH and TSH is less extensive than is the case for LH and CG. The reasons relate principally to the relative abundance of these hormones. The pituitary contains about one tenth (or less) FSH and TSH than it does LH (on a molar basis). It should be noted that the units of FSH and TSH measured in a pituitary may seem much larger than the units of LH, but this is a consequence of the fortuitous definition of the unit for each individual hormone. Thus, FSH and TSH have been considerably more difficult to purify and to obtain in sufficient quantities for detailed biochemical studies (for a review see Liu and Ward[94]). The problem has been alleviated as methodology becomes more sensitive and more studies on TSH and FSH are beginning to appear. For similar practical considerations, biochemical studies of hCG are by far the most numerous. It is readily obtained from pregnancy urine.

The heterogeneity of the alpha subunits as described in the preceding section can generally be presumed to apply to the alpha subunits of FSH and TSH although detailed studies comparable to the LH/CG are seldom available. It was noted in the original sequence studies of hFSHβ[85,96] that there was some N-terminal and C-terminal heterogeneity. This also appears to be the case for equine,[40] ovine,[42] and porcine[44] FSH.

The review by Chappel et al.[97] of the biosynthesis and secretion of FSH includes an excellent discussion of FSH microheterogeneity. The majority of the pleiomorphic forms examined in that review were related to variation in the carbohydrate moieties. These will be further considered in Chapters 6 and 7. At least six forms of FSH were detected by isoelectric focusing. The relative amount of the different forms (isohormones) was influenced by the physiological state of the donor animal. Moreover, the specific activity (measured by in vitro assays of receptor binding and release of plasminogen activator) correlated with the isoelectric point of the isohormone — the more basic forms being the more potent. Wide[98,99] studied the charge heterogeneity of FSH, LH, and TSH in human pituitary extracts as a function of sex and age of the subjects at autopsy. Both FSH and LH showed a strong tendency for increased charge (mobility) as a function of age, while TSH charge remained essentially level as a function of age. We presume (as will be discussed in subsequent chapters) that these differences are largely attributable to the carbohydrate processing on these hormones.

Another type of heterogeneity has been documented for a particular preparation of hFSH by Shome et al.[41] In that study a proteolytic nick of the beta subunit was shown between residues 38 and 39 (i.e., 44 and 45 in Figure 2, allowing for the displacement due to the "missing" residues in hFSHβ sequence to bring the 1/2-cystine residues in register). This nick appears to be due to a protease in the pituitary with a specificity akin to thermolysin. This form of heterogeneity was suggested to be related to the type of storage and processing the glands received at autopsy and prior to fractionation.[41] Thus, the susceptibility of the beta subunit of hFSH to pituitary proteases is similar to that studied in some detail for the hLHβ.[75] This suggests a comparable surface exposure to pituitary proteases, i.e., the approximate midpoint of the beta subunits is comprised of a surface loop susceptible to this type of attack.

IV. TSH PROTEIN HETEROGENEITY

A. N-Terminal Heterogeneity

Prior to the publication of the amino acid sequence for the alpha and beta subunits of bovine TSH, and before it had been proven (although it had been suspected) that TSH was composed of two subunits, Phe (70 to 80%), Asp (15 to 20%), Thr (7 to 9%), and Glu (4 to 5%) were the reported N termini (Shome et al.,[100] Liao et al.[14]). We now know that the

N-terminal sequences for the alpha and beta subunits of bTSH are F–P–D–G–E–F–T–M– . . . and F–C–I–P–T–E–Y–M– . . . , respectively (Figures 1 and 2). Since Phe is the N-terminal amino acid for both subunits of TSH in the bovine, it is of interest that Asp represents such a large N-terminal component. This N-terminal Asp is of alpha subunit origin since (1) the N-terminal 32 amino acids of the β-subunit contains no Asp, and (2) the leader peptide sequence for bTSHβ deduced from cDNA sequence analysis has no Asp or Asn (Mauer et al.[47]). This indicates that approximately 30 to 40% of this particular subunit preparation[100] of bTSH contains Asp as the N terminus; that is, they are obtained without the N-terminal F–P dipeptide.

The reported N terminal for the alpha and beta subunit of human TSH are Ala and Phe, respectively, (see Figures 1 and 2). However, Shome et al.[100] reported that 10 to 20% of the N terminus for TSH was Asp. Whereas the ovine/bovine alpha subunit sequence N terminus is F–P–D–G–E–T–M–Q–G–C . . . , human N-terminal sequence is A–P–D–V–Q–D–C– . . . The same P–D sequence is present in both subunits at the same distance from the N-terminal amino acid, albeit at different relative positions to the first 1/2-cystine residue. It is conceivable, then, that this P–D peptide bond is susceptible to cleavage in both the human and ovine/bovine pituitary. This P–D cleavage indicates a susceptible peptide bond, the presence of the requisite protease for cleavage during purification or pituitary handling following extraction, and the availability of this part of the molecule to proteolytic attack. The latter implies the N-terminal amino acids are not firmly bound to the beta subunit in the α-β complex.

B. C-Terminal Heterogeneity

C-terminal heterogeneity has been reported in the human and bovine TSH preparations (Shome et al.[100]). Met and Leu were detected in small amounts with the principal C-terminal sequence of –K–S–Y. Liao and Pierce[7] reported the C terminus of bovine TSHβ as –K–S–Y–M. Porcine and human TSHβ have the same C terminus of –K–S–Y. The 10 to 20% L or M reported by Shome et al.[100] is probably due to a Met and/or Leu extension on the carboxyl side of the C-terminal Tyr. Recombinant DNA sequencing has revealed the Y-117 is invariant among species so far examined for TSHβ; however, the next amino acid is Leu in mouse (Gurr et al.[49]) and rat (Croyle and Maurer[50]), and Met in bovine (Maurer et al.[47]). This suggests that a Leu or Met extension C terminal to Y-117 probably exists for the procine and human TSHβ as well. The lack of L or M in some preparations may result from chymotryptic-like activity which may be encountered during purification.

The recombinant DNA studies for bovine TSHβ also reveal that following the Met or the Leu at position 118, there is a C-terminal pentapeptide extension prior to the stop codon (Maurer et al.[47]). Similar pentapeptide extensions have been deduced from cDNA sequencing studies from the mouse[49] and rat[50] for which there are no conventional protein sequencing studies reported. Since conventional protein sequencing in those species studied has not revealed evidence for these pentapeptides, it has been postulated that this may be a terminal processing peptide[50] analogous to the N-terminal leader sequence that usually commences with Met. Alternatively, it is possible that upon processing of pituitaries during the extraction and purification steps, efficient enzymatic removal of this peptide takes place.

C. Internal Heterogeneity

Evidence for internal heterogeneity in the amino acid sequence is lacking for TSH. For example, there is no evidence for expression of more than one TSH α or β gene, unlike hCGβ where evidence exists for the expression of more than one of the six or seven genes for hCGβ (vide supra).

Recombinant DNA sequencing techniques have, however, revealed an interesting variation in the amino acid sequence of the β-subunit of TSH in the so-called determinant loop sequence from positions 93 to 100 (vide supra). Protein sequencing studies place either 3 Asp residues at positions 94, 96, and 99, as in the porcine, or one Asp at position 99 and Asx (amide or carboxyl not certain) for positions 94 and 96 as in the ovine/bovine. cDNA sequencing of TSHβ clones reveals that Asx-94 is Asn, and Asx-96 is Asp in the bovine, rat, and mouse.[47,49,50] Moreover, Gurr et al.[49] place an additional Asn at position 97 in mouse TSHβ. All previous reports have a Tyr at this position. There is a one-base difference between the three-base codon for Asp, Asn, and Tyr suggesting a simple evolutionary change in this respect has occurred in the mouse, compared to other species. Prior to this, the determinant loop hypothesis[51,52] had been formulated with the suggstion that a Tyr residue in this position of the loop was characteristic of a thyrotropin beta subunit. The report of Gurr et al.[49] indicates the hydrophobic character of mouse TSHβ is substantially different from the other TSH beta subunits so far examined. This provides one further bit of evidence that the ''determinant loop'' hypothesis was too simplistic in its formulation, and that the major utility of the hypothesis has been to focus attention on this important structural area of the beta subunits.

ACKNOWLEDGMENTS

Original research from our laboratory cited in this chapter was supported in part by the following research grants: AM-09801, HD-18210, NIH; and The Robert A. Welch Foundation Grant G-147.

REFERENCES

1. **Ward, D. N.,** Chemical approaches to the structure-function relationships of luteinizing hormone (Lutropin), in *Structure and Function of the Gonadotropins,* McKerns, K. W., Ed., Plenum Press, New York, 1978, 31.
2. **Pierce, J. G. and Parsons, T. F.,** Glycoprotein hormones: structure and function, *Annu. Rev. Biochem.,* 50, 465, 1981.
3. **Gordon, W. L. and Ward, D. N.,** Structural aspects of luteinizing hormone actions, in *Luteinizing Hormone Action and Receptors,* Ascoli, M., Ed., CRC Press, Boca Raton, Fla., 1985, 173.
4. **Strickland, T. W., Parsons, T. F., and Pierce, J. G.,** Structure of LH and hCG, in *Luteinizing Hormone Action and Receptors,* Ascoli, M., Ed., CRC Press, Boca Raton, Fla., 1985, 1.
5. **Bahl, O. P. and Wagh, P. V.,** Characterization of glycoproteins: carbohydrate structures of glycoprotein hormones, *Adv. Exp. Med. Biol.,* 206, 1, 1986 (Review).
6. **Fiddes, J. C. and Talmadge, K.,** Structure, expression, and evolution of the genes for the human glycoprotein hormones, *Recent Prog. Horm. Res.,* 40, 43, 1984 (Review).
7. **Liao, T.-H. and Pierce, J. G.,** The primary structure of bovine thyrotropin. II. The amino acid sequences of the reduced S-carboxymethylated α and β chains, *J. Biol. Chem.,* 246, 850, 1971.
8. **Nilson, J. H., Thomason, A. R., Cserbak, M. T., Moncman, C. L., and Woychik, R. P.,** Nucleotide sequence of a cDNA for the common alpha subunit of the bovine pituitary glycoprotein hormones. Conservation of nucleotides in the 3'-untranslated region of bovine and human pre-alpha subunit mRNA's, *J. Biol. Chem.,* 258, 4679, 1983.
9. **Erwin, C. R., Croyle, M. L., Donelson, J. E., and Maurer, R. A.,** Nucleotide sequence of cloned complementary deoxynucleic acid for the alpha subunit of bovine pituitary glycoprotein hormones, *Biochemistry,* 22, 4856, 1983.
10. **Chin, W. W., Kronenberg, H. M., Dee, P. C., Maloof, F., and Habener, J. F.,** Nucleotide sequence of the mRNA encoding the pre-α-subunit of mouse thyrotropin, *Proc. Natl. Acad. Sci. U.S.A.,* 78, 5329, 1981.
11. **Godine, J. E., Chin, W. W., and Habener, J. F.,** α Subunit of rat glycoprotein hormones. Primary structure of the precursor determined from the nucleotide sequence of cloned cDNAs, *J. Biol. Chem.,* 257, 8368, 1982.
12. **Fiddes, J. C. and Goodman, H. M.,** Isolation, cloning and sequence analysis of the cDNA for the α-subunit of human chorionic gonadotropin, *Nature,* 281, 351, 1979.

13. **Boothby, M., Ruddon, R. W., Anderson, C., McWilliams, D., and Boime, I.,** A single gonadotropin alpha subunit gene in normal tissue and tumor-derived cell lines, *J. Biol. Chem.,* 256, 5121, 1981.
14. **Liao, T.-H. and Pierce, J. G.,** The presence of a common type of subunit in bovine thyroid-stimulating and luteinizing hormones, *J. Biol. Chem.,* 245, 3275, 1970.
15. **Pierce, J. G., Liao, T.-H., Carlsen, R. B., and Riemo, T.,** Comparisons between the α chain of bovine thyrotropin and the CI chain of luteinizing hormones, *J. Biol. Chem.,* 246, 866, 1971.
16. **Bousfield, G. R., Liu, W.-K., and Ward, D. N.,** Hybrids from equine LH: alpha enhances, beta diminishes activity, *Mol. Cell. Endocrinol.,* 40, 69, 1985.
17. **Glenn, S. D., Liu, W.-K., and Ward, D. N.,** Characteristics of hybrids of ovine LH and human glycoprotein hormone subunits in rat and chicken *in vitro* test systems, *Biol. Reprod.,* 25, 1027, 1981.
18. **Hennen, G.,** Detection and study of a human-chorionic-thyroid-stimulating factor, *Arch. Int. Physiol. Biochim.,* 73, 689, 1965.
19. **Hennen, G., Pierce, J. G., and Freychet, P.,** Human chorionic thyrotropin: further characterization and study of its secretion during pregnancy, *J. Clin. Endocrinol. Metab.,* 29, 581, 1969.
20. **Ward, D. N., Desjardins, C., Moore, W. T., Jr., and Nahm, H. S.,** Rabbit lutropin: preparation, characterization of the hormone, its subunits and radioimmunoassay, *Int. J. Pept. Prot. Res.,* 13, 62, 1979.
21. **Glenn, S. D., Nahm, H. S., and Ward, D. N.,** The amino acid sequence of the rabbit lutropin beta subunit, *J. Prot. Chem.,* 3, 259, 1984.
22. **Fontaine, Y.-A. and Burzawa-Gerard, E.,** Esquisse de l'evolution des hormones gonadotropes et thyreotropes des vertebres, *Gen. Comp. Endocrinol.,* 32, 341, 1977.
23. **Liu, W.-K., Nahm, H. S., Sweeney, C. M., Lamkin, W. M., Baker, N., and Ward, D. N.,** The primary structure of ovine luteinizing hormone. I. The amino acid sequence of the reduced and S-aminoethylated S-subunit (LHα), *J. Biol. Chem.,* 247, 4351, 1972.
24. **Maghuin-Rogister, G., Combarnous, Y., and Hennen, G.,** Porcine luteinizing hormone. The amino acid sequence of the α-subunit, *FEBS Lett.,* 25, 57, 1972.
25. **Ward, D. N., Moore, W. T., Jr., and Burleigh, B. D.,** Structural studies on equine chorionic gonadotropin, *J. Prot. Chem.,* 1, 263, 1982.
26. **Glenn, S. D., Nahm, H. S., and Ward, D. N.,** The amino acid sequence of the rabbit glycoprotein hormone alpha subunit, *J. Prot. Chem.,* 3, 143, 1984.
27. **Pankov, Y. A. and Karasev, V. S.,** Luteinizing hormone of the sperm whale. Isolation, separation into subunits, and study of the amino acid sequence of the α-subunit, *Biokhimiya,* 49, 111, 1984.
28. **Pankov, Y. A. and Karasev, V. S.,** Primary structure of sperm whale luteinizing hormone, *Int. Pept. Prot. Res.,* 28, 124, 1986.
29. **Sugino, H., Bousfield, G. R., Moore, W. T., Jr., and Ward, D. N.,** Structural studies on equine glycoprotein hormones: amino acid sequence of equine chorionic gonadotropin β-subunit, *J. Biol. Chem.,* 262, 8603, 1987.
30. **Morgan, F. J., Birken, S., and Canfield, R. E.,** The amino acid sequence of human chorionic gonadotropin. The alpha and beta subunit, *J. Biol. Chem.,* 250, 5247, 1975.
31. **Bousfield, G. R., Liu, W.-K., Sugino, H., and Ward, D. N.,** Structural studies on equine glycoprotein hormones: amino acid sequence of equine lutropin β subunit, *J. Biol. Chem.,* 262, 8610, 1987.
32. **Talmadge, K., Vamvakopoulos, N. C., and Fiddes, J. C.,** Evolution of the genes for the β subunits of human chorionic gonadotropin and luteinizing hormone, *Nature,* 307, 37, 1984.
33. **Liu, W.-K., Nahm, H. S., Sweeney, C. M., Holcomb, G., and Ward, D. N.,** The primary structure of ovine luteinizing hormone. II. The structure of the reduced, S-carboxymethylated A-subunit (LH-β), *J. Biol. Chem.,* 247, 4365, 1972.
34. **Virgin, J. A., Silver, B. J., Thomason, A. R., and Nilson, J. H.,** The gene for the β subunit of bovine luteinizing hormone encodes a gonadotropin mRNA with an unusually short 5'-untranslated region, *J. Biol. Chem.,* 260, 7072, 1985.
35. **Maghuin-Rogister, G. and Hennen, G.,** Luteinizing hormone. The primary structure of the β-subunit from bovine and porcine species, *Eur. J. Biochem.,* 39, 235, 1973.
36. **Ward, D. N., Reichert, L. E., Jr., Fitak, B. A., Nahm, H. S., Sweeney, C. M., and Neill, J. D.,** Isolation and properties of subunits of rat pituitary luteinizing hormone, *Biochemistry,* 10, 1796, 1971.
37. **Chin, W. W., Godine, J. E., Klein, D. R., Chang, A. S., Tan, L. K., and Habener, J. F.,** Nucleotide sequence of the cDNA encoding the precursor of the β subunit of rat lutropin, *Proc. Natl. Acad. Sci. U.S.A.,* 80, 4649, 1983.
38. **Pankov, Y. A. and Karasev, V. S.,** Amino acid sequence of reduced and carboxymethylated β-subunit of sperm whale luteinizing hormone, *Biokhimiya,* 49, 1004, 1984.
39. **Trinh, K.-T., Wang, N. C., Hew, C. L., and Crim, L. W.,** Molecular cloning and sequencing of salmon gonadotropin β subunit, *Eur. J. Biochem.,* 159, 619, 1986.
40. **Fujiki, Y., Rathnam, P., and Saxena, B. B.,** Amino acid sequence of the β-subunit of the follicle-stimulating hormone from equine pituitary glands, *J. Biol. Chem.,* 253, 5363, 1978.

41. Shome, B., Parlow, A. F., Liu, W.-K., Nahm, H. S., Wen, T., and Ward, D. N., A reevaluation of the amino acid sequence of human follitropin beta subunit, manuscript in preparation.
42. Sairam, M. R., Seidah, N. G., and Chretien, M., Primary structure of the ovine pituitary follitropin β-subunit, *Biochem. J.*, 197, 541, 1981.
43. Esch, F. S., Mason, A. J., Cooksey, K., Mercado, M., and Shimasaki, S., Cloning and DNA sequence analysis of the cDNA for the precursor of the β chain of bovine follicle stimulating hormone, *Proc. Natl. Acad. Sci. U.S.A.*, 83, 6618, 1986.
44. Closset, J., Maghuin-Rogister, G. M., Hennen, G., and Strosberg, A. D., Porcine follitropin: the amino-acid sequence of the β subunit, *Eur. J. Biochem.*, 86, 115, 1978.
45. Sugino, H. and Ward, D. N., A reevaluation of the amino acid sequence of porcine follitropin beta subunit, manuscript in preparation.
46. Hayashizaki, Y., Miyai, K., Kato, K., and Matsubara, K., Molecular cloning of the human thyrotropin-beta subunit gene, *FEBS Lett.*, 188, 394, 1985.
47. Maurer, R. A., Croyle, M. L., and Donelson, J. E., The sequence of a cloned cDNA for the β subunit of bovine thyrotropin predicts a protein containing both NH_2- and COOH-terminal extensions, *J. Biol. Chem.*, 259, 5024, 1984.
48. Maghuin-Rogister, G., Hennen, G., Closset, J., and Kopeyan, C., Porcine thyrotropin. The amino acid sequence of the alpha and beta subunits, *Eur. J. Biochem.*, 61, 157, 1976.
49. Gurr, J. A., Catterall, J. F., and Kourides, I. A., Cloning of cDNA encoding the pre-β subunit of mouse thyrotropin, *Proc. Natl. Acad. Sci. U.S.A.*, 80, 2122, 1983.
50. Croyle, M. L. and Maurer, R. A., Thyroid hormone decreases thyrotropin subunit mRNA levels in rat anterior pituitary, *DNA*, 3, 231, 1984.
51. Ward, D. N. and Moore, W. T., Jr., Comparative study of mammalian glycoprotein hormones, in *Animal Models for Research on Fertility and Contraception*, Alexander, N. J., Ed., Harper & Row, Hagerstown, Md., 1979, 151.
52. Moore, W. T., Jr., Burleigh, B. D., and Ward, D. N., Chorionic gonadotropins: comparative studies and comments on relationships to other glycoprotein hormones, in *Chorionic Gonadotropin*, Segal, S. J., Ed., Plenum Press, New York, 1980, 89.
53. Parlow, A. F. and Shome, B., The immunoreactive hTSH contamination of "Immunochemical Grade" hFSH preparation, *J. Clin. Endocrinol. Metab.*, 41, 189, 1975.
54. Sairam, M. R., Papkoff, H., and Li, C.-H., The primary structure of ovine interstitial cell-stimulating hormone. I. The α-subunit, *Arch. Biochem. Biophys.*, 153, 551, 1972.
55. Maghuin-Rogister, G., Closset, J., and Hennen, G., The carboxy-terminal primary structure of the α subunit from bovine and porcine luteinizing hormone, *FEBS Lett.*, 13, 301, 1971.
56. Bousfield, G. R. and Ward, D. N., Purification of lutropin and follitropin in high yield from horse pituitary glands, *J. Biol. Chem.*, 259, 1911, 1984.
57. Bousfield, G. R. and Ward, D. N., unpublished data.
58. Papkoff, H., Chemical and biological properties of the subunits of pregnant mare serum gonadotropin, *Biochem. Biophys. Res. Comm.*, 58, 397, 1974.
59. Morgan, F. J., Birken, S., and Canfield, R. E., The amino acid sequence of human chorionic gonadotropin, the α and β subunit, *J. Biol. Chem.*, 250, 5247, 1975.
60. Sairam, M. R., Papkoff, H., and Li, C.-H., Human pituitary interstitial cell stimulating hormone: primary structure of the α subunit, *Biochem. Biophys. Res. Comm.*, 48, 530, 1972.
61. Keutmann, H. T., Dawson, B., Bishop, W. T., and Ryan, R. J., Structure of human luteinizing hormone alpha subunit, *Endocrinol. Res. Common.*, 5, 57, 1978.
62. Jameson, J. L., Lindell, C. M., and Habener, J. F., Evolution of different transcriptional start sites in the human luteinizing hormone and chorionic gonadotropin β-subunit genes, *DNA*, 5, 227, 1986.
63. Chu, W. P.-C., The Physical and Chemical Properties and the C- and N-Terminal Amino Acids Determination of Porcine Luteinizing Hormone and its Subunits, M. S. thesis, The University of Texas Graduate School of Biomedical Sciences at Houston, 1972, 61.
64. Cheng, K.-W., Glazer, A. N., and Pierce, J. G., The effects of modification of the COOH-terminal regions of bovine thyrotropin and its subunits, *J. Biol. Chem.*, 248, 7930, 1973.
65. Merz, W. E., Studies of the specific role of the subunits of choriogonadotropin for biological, immunological and physical properties of the hormone. Digestion of the α-subunit with carboxypeptidase A, *Eur. J. Biochem.*, 101, 541, 1979.
66. Bousfield, G. R. and Ward, D. N., Biologic activity of eLH and hCG following removal of the C-terminal glycopeptide, *Endocrinology*, 118 (Suppl.), 530, 1986.
67. Closset, J., Hennen, G., and Lequin, R. M., Human luteinizing hormone: the amino acid sequence of the β subunit, *FEBS Lett.*, 29, 97, 1973.
68. Shome, B. and Parlow, A. F., The primary structure of the hormone-specific, beta subunit of human pituitary luteinizing hormone (hLH), *J. Clin. Endocrinol. Metab.*, 36, 618, 1973.

69. **Sairam, M. R. and Li, C.-H.,** Human pituitary lutropin: isolation, properties, and the complete amino acid sequence of the β-subunit, *Biochim. Biophys. Acta,* 412, 70, 1975.

70. **Keutmann, H. T., Williams, R., and Ryan, R. J.,** Structure of human luteinizing hormone beta subunit: evidence for a related carboxyl-terminal sequence among certain peptide hormones, *Biochem. Biophys. Res. Comm.,* 90, 842, 1979.

71. **Carlsen, R. B., Bahl, O. P., and Swaminathan, N.,** Human chorionic gonadotropin: linear amino acid sequence of the β subunit, *J. Biol. Chem.,* 248, 6810, 1973.

72. **Keutmann, H. T. and Williams, R. M.,** Human chorionic gonadotropin: amino acid sequence of the hormone-specific COOH-terminal region, *J. Biol. Chem.,* 252, 5393, 1977.

73. **Birken, S., Kolks, M. A. G., Agosto, G. M., Armstrong, E. G., Krichevsky, A., and Canfield, R. E.,** 68th Annu. Meet. Endocrine Society, Abstr. #514, 1986.

74. **Papkoff, H., Farmer, S. W., and Cole, H. H.,** Isolation of a gonadotropin (PMEG) from pregnant mare endometrial cups: comparison with PMSG, *Proc. Soc. Exp. Biol. Med.,* 158, 373, 1978.

75. **Ward, D. N., Glenn, S. D., Nahm, H. S., and Wen, T.,** Characterization of cleavage products in selected human lutropin preparations, *Int. J. Pept. Prot. Res.,* 27, 70, 1986.

76. **Liu, W.-K., Ascoli, M., and Ward, D. N.,** Ovine lutropin subunit isolation: comparison of salt precipitation and countercurrent distribution procedures, *J. Biol. Chem.,* 252, 5274, 1977.

77. **Parsons, T. F., Bloomfield, G. A., and Pierce, J. G.,** Purification of an alternate form of the α subunit of the glycoprotein hormones from bovine pituitaries and identification of its O-linked oligosaccharide, *J. Biol. Chem.,* 258, 240, 1983.

78. **Bousfield, G. R. and Ward, D. N.,** unpublished experiments.

79. **Parsons, T. F. and Pierce, J. G.,** Free α-like material from bovine pituitaries. Removal of its O-linked oligosaccharide permits combination with lutropin-β, *J. Biol. Chem.,* 259, 2662, 1984.

80. **Hartree, A. S., Lester, J. B., and Shownkeen, R. C.,** Studies of the heterogeneity of human pituitary LH by fast protein liquid chromatography, *J. Endocrinol.,* 105, 405, 1985.

81. **Ward, D. N., Reichert, L. E., Jr., Liu, W.-K., Nahm, H. S., Hsia, J., Lamkin, W. M., and Jones, N. S.,** Chemical studies of luteinizing hormone from human and ovine pituitaries, *Rec. Prog. Horm. Res.,* 29, 533, 1973.

82. **Sharp, S. B. and Hunkapiller, M. W.,** Bovine luteinizing hormone iodinated at αTyr^{21}, αTyr^{92}, or αTyr^{93} retains specific binding activity, *Endocrinology,* 115, 2183, 1984.

83. **Nishimura, R., Shin, J., Ji, I., Middaugh, C. R., Kruggel, W., Lewis, R. V., and Ji, T. H.,** A single amino acid substitution in an ectopic α subunit of a human carcinoma choriogonadotropin, *J. Biol. Chem.,* 261, 10475, 1986.

84. **Maurer, R. A.,** Analysis of several bovine lutropin β subunit cDNA's reveals heterogeneity in nucleotide sequence, *J. Biol. Chem.,* 260, 4684, 1985.

85. **Franchimont, P., Reuter, A., and Gospard, U.,** Ectopic production of human chorionic gonadotropin and its alpha- and beta- subunits, *Curr. Top. Exp. Endocrinol.,* 3, 201, 1978 (Review).

86. **Kyte, J. and Doolittle, R. F.,** A simple method for displaying the hydropathic character of a protein, *J. Mol. Biol.,* 157, 105, 1982.

87. **Boorstein, W. R., Vamvakopolous, N. C., and Fiddes, J. C.,** Human chorionic gonadotropin β-subunit is encoded by at least eight genes arranged in tandem and inverted pairs, *Nature,* 300, 419, 1982.

88. **Policastro, P., Ovitt, C. E., Hoshima, M., Fukuoka, H., Boothby, M. R., and Boime, I.,** The β subunit of human chorionic gonadotropin is encoded by multiple genes, *J. Biol. Chem.,* 258, 11492, 1983.

89. **Talmadge, K., Boorstein, W. R., and Fiddes, J. C.,** The human genome contains seven genes for the β-subunit of chorionic gonadotropin but only one gene for the β-subunit of luteinizing hormone, *DNA,* 2, 281, 1983.

90. **Chin, W. W., Maizel, J. V., Jr., and Habener, J. F.,** Differences in sizes of human compared to murine α-subunits of the glycoprotein hormones arises by a four-codon deletion or insertion, *Endocrinology,* 112, 482, 1983.

91. **Talmadge, K., Boorstein, W. R., Vamvakopolous, N. C., Gething, M.-J., and Fiddes, J. C.,** Only three of the seven human chorionic gonadotropin beta subunit genes can be expressed in the placenta, *Nucleic Acids Res.,* 12, 8414, 1984.

92. **Policastro, P. F., Daniels-McQueen, S., Carle, G., and Boime, I.,** A map of the hCGβ-LHβ gene cluster, *J. Biol. Chem.,* 261, 5907, 1986.

93. **Fiddes, J. C. and Goodman, H. M.,** The cDNA for the β-subunit of human chorionic gonadotropin suggests evolution of a gene by read-through into the 3'-untranslated region, *Nature,* 286, 684, 1980.

94. **Liu, W.-K. and Ward, D. N.,** The purification and chemistry of pituitary glycoprotein hormones, *Pharmacol. Ther. B,* 1, 545, 1975.

95. **Shome, B. and Parlow, A. F.,** Human follicle stimulating hormone: first proposal for the amino acid sequence of the hormone-specific, β subunit (hFSHβ), *J. Clin. Endocrinol. Metab.,* 39, 203, 1974.

96. **Saxena, B. B. and Rathnam, P.,** Amino acid sequence of the β subunit of follicle-stimulating hormone from human pituitary glands, *J. Biol. Chem.,* 251, 993, 1976.

97. **Chappel, S. C., Ulloa-Aguirre, A., and Coutifaris, C.,** Biosynthesis and secretion of follicle-stimulating hormone, *Endocr. Rev.,* 4, 179, 1983.
98. **Wide, L.,** Median charge and charge heterogeneity of human pituitary FSH, LH and TSH. I. Zone electrophoresis in agarose suspension, *Acta Endocrinol.,* 109, 181, 1985.
99. **Wide, L.,** Median Charge and charge heterogeneity of human pituitary FSH, LH and TSH. II. Relationship to sex and age, *Acta Endocrinol.,* 109, 190, 1985.
100. **Shome, B., Brown, D. M., Howard, S. M., and Pierce, J. G.,** Bovine, human and porcine thyrotropins: molecular weights, amino- and carboxyl-terminal studies, *Arch. Biochem. Biophys.,* 126, 456, 1968.

Chapter 2

OLIGOSACCHARIDE STRUCTURES OF THE ANTERIOR PITUITARY AND PLACENTAL GLYCOPROTEIN HORMONES

H. Edward Grotjan, Jr.

TABLE OF CONTENTS

I. INTRODUCTION

Classically, the family of glycoprotein hormones includes luteinizing hormone (LH; lutropin), follicle-stimulating hormone (FSH; follitropin), and thyroid-stimulating hormone (TSH; thyrotropin) produced by the anterior pituitary as well as the chorionic gonadotropins (CG; choriogonadotropin) produced by the placenta. These four hormones have rather complex chemical structures because (1) they are heterodimers, (2) they contain numerous intrachain disulfide bonds, and (3) they are glycoproteins. The significance of glycoprotein hormone subunits as well as the structure of the peptide chains will be considered in detail in other chapters of this book (see Chapters 1 and 3). The structures of the oligosaccharides associated with these four glycoprotein hormones will be considered herein. Additional information may be found in previous reviews which discuss this topic.[1-8]

Beginning in the late 1950s significant progress was made in the purification and structural analysis of the glycoprotein hormones (reviewed by Liu and Ward;[1] Parsons and Pierce[3]). However, only in the last decade has significant progress been made in elucidating the fine structure of their carbohydrate moieties. The oligosaccharides of certain glycoprotein hormones (e.g., human CG) have been described in detail. Information is extremely limited in other cases, particularly with regard to FSH. When considering a given glycoprotein hormone (e.g., LH), variations in oligosaccharide structure occur between species. Moreover, any given glycoprotein hormone may possess oligosaccharides which vary slightly to markedly in their structures, even at a single glycosylation site. The observation that the common alpha subunit, assumed to be the product of a single gene, appears to possess distinct oligosaccharide structures when associated with different beta subunits[9,10] is quite intriguing. Even more fascinating is the possibility that a single cell type, the gonadotrope, may produce two glycoprotein hormones with distinctly different oligosaccharide structures.[9,10] Assuming that the carbohydrate moieties are involved in the biological functions of hormones, it is desirable to have more complete information regarding their oligosaccharide structures, particularly as glycoprotein hormones produced by recombinant DNA technology (e.g.,[11,12]) become available.

II. STRUCTURAL ANALYSIS OF OLIGOSACCHARIDES

Although it is not the purpose of this chapter to serve as a treatise on oligosaccharide structural analysis, certain comments are necessary to facilitate understanding of the information presented. Detailed information on the techniques involved in oligosaccharide structural analysis may be found in Volumes 8, 28, 50, and 83 in the series *Methods in Enzymology* as well as certain review articles.[8,13-16]

In order to completely and correctly delineate the structure of a given oligosaccharide one must identify the protein sugar linkage, the constituent sugars, and the sequence of the sugar residues including their linkages complete with anomeric configurations.[8] The two classes of oligosaccharides most commonly found in glycoproteins are (1) N-linked in which N-acetyl-D-glucosamine (GlcNAc) is N-glycosidically linked to the β-amide nitrogen of asparagine (Asn) and (2) O-linked in which N-acetyl-D-galactosamine (GalNAc) is O-glycosidically linked to the hydroxyl group of serine (Ser) or threonine (Thr).

The first step in structural analysis is normally liberation of the sugar chains. This may be preceded by the preparation of glycopeptides[15,16] representing distinct glycosylation sites if the glycoprotein under consideration contains multiple sugar chains and one desires to ascertain the oligosaccharide structures associated with particular attachment points. In some cases it is desirable to remove as many amino acids as possible with proteolytic enzymes such as pronase.[15,16] If desired, the glycopeptide may be labeled by acetylating the free amino groups of the peptide with radioactive acetic anhydride (e.g.,[17]) to facilitate identification during subsequent purification.

Liberation of sugar chains may be achieved chemically or enzymatically. O-linked oligosaccharides may be selectively released by β-elimination with weak alkali[18] while N-linked oligosaccharides may be released by strong alkali[19] or hydrazinolysis.[20] Such reactions are routinely performed in conjunction with or followed by borohydride reduction which converts the attachment residue to a sugar alcohol (N-acetylglucosaminitol or N-acetylgalactosaminitol in N- and O-linked oligosaccharides, respectively). Thus, the residue at the "reducing terminus" can be used to identify the type of linkage. Tritiated borohydride may be included in the reduction reaction to facilitate subsequent identification of the oligosaccharides. Alternatively, liberation of sugar chains may be achieved with endoglycosidases which cleave oligosaccharides at or near the attachment site.[8,14,16] Susceptibility to cleavage by specific endoglycosidases usually infers certain structural features.

After cleavage, it is desirable to purify the liberated oligosaccharides into homogeneous components by one or more chromatographic techniques (reviewed by Bahl and Wagh;[8] Kobata[16]). Structural analysis of each component typically involves a series of chemical reactions, some of which fragment oligosaccharides in a specific manner, as well as enzymatic digestions both of which reveal specific features. Some of the chemical reactions used in the structural analysis of glycoproteins include: (1) permethylation[21] which converts the hydroxyl groups of sugars to their O-methylated derivatives; points of linkage are then deduced from potential sites *not* methylated (for example see[10]), (2) periodate oxidation[22] (also called Smith degradation) which destroys monosaccharides possessing vicinal hydroxyl groups, (3) acetolysis[23] which preferentially cleaves α1–6 glycosidic linkages, and (4) nitrous acid deamination[19] which fragments oligosaccharides at the N-acetylhexosamines converting each GlcNAc to 2,5-anhydromannitol and each GalNAc to 2,5-anhydrotalitol.[8] These chemical reactions are generally applied in conjunction with exoglycosidase and endoglycosidase digestions[13,14,16] which reveal the anomeric configurations of the linkages and, if applied systematically, may be useful in deducing the sugar sequences. Information may also be gleaned from lectin affinity studies[24] which separate classes of oligosaccharides (e.g., sialylated vs. nonsialylated or biantennary vs. triantennary vs. bisecting GlcNAc). However, these typically do not yield precise structural information.

N-linked oligosaccharides typically possess a common pentasaccharide core which contains two GlcNAc and three D-mannose (Man) residues: $Man_3GlcNAc_2$ or more precisely Manα1–3(Manα1–6)Manβ1–4GlcNAcβ1–4GlcNAc. They can be further classified according to the sugars found on their outer branches (reviewed by Bahl and Wagh;[8] Snider;[25] Kornfeld and Kornfeld[26]). High mannose-type oligosaccharides possess an additional two to six Man residue attached to the pentasaccharide core. Complex-type oligosaccharides typically possess two or more sugar chains attached to core Man residues via GlcNAc and may possess an L-fucose (Fuc) residue attached to the innermost (attachment) GlcNAc. Hybrid-type oligosaccharides possess one chain of Man residues and one or more sugars chains attached to Man via GlcNAc. Hybrid and complex oligosaccharides are thought to arise from high mannose oligosaccharides as the result of an intricate series of processing reactions (reviewed by Bahl and Kalyan;[4] Snider;[25] Kornfeld and Kornfeld[26]).

The glycoprotein hormones generally possess complex-type N-linked oligosaccharides. Besides GlcNAc, Man, and Fuc, their oligosaccharides may contain D-galactose (Gal) and N-acetylneuraminic acid (NeuAc; also known by the more general term — sialic acid). The complex-type N-linked oligosaccharides of LH, TSH, and FSH are somewhat unique because they usually contain some GalNAc. Furthermore, the N-linked oligosaccharides of some of the glycoprotein hormones have O-sulfated N-acetylhexosamines at nonreducing termini. The beta subunits of hCG and equine (e) CG as well as a portion of the uncombined common alpha subunits possess O-linked oligosaccharides which generally contain GalNAc, Gal, and NeuAc, but may also contain GlcNAc. The sugars in glycoproteins are assumed to exist in the pyranose form, an observation which has been repeatedly confirmed by methylation analysis.

All of the aforementioned sugars are electrochemically neutral except for NeuAc. Some oligosaccharides have anions such as sulfate or phosphate attached to them. Oligosaccharides containing NeuAc, sulfate, or phosphate are negatively charged and hence acidic rather than neutral. Oligosaccharides containing sulfate or phosphate can frequently be identified by biosynthetic radiolabeling (e.g.,[9,27-29]) providing it is demonstrated that the sulfate or phosphate is incorporated into the carbohydrate portion of the particular glycoprotein under consideration. Sulfate and phosphate associated with oligosaccharides can be quantitated colorimetrically[30,31] but this usually requires relatively large amounts of materials. Ion chromatographic approaches have recently been introduced.[32,33] These appear to hold great promise, particularly the method of Ward et al.,[33] because both sulfate and phosphate can simultaneously be determined which reduces the quantity of hormone required for analysis.

Glycoproteins typically exhibit heterogeneous oligosaccharide structures ("microheterogeneity"), even at a single glycosylation site.[8,16] For N-linked oligosaccharides which possess the typical pentasaccharide core, microheterogeneity can be expressed as: (1) the presence or absence of Fuc linked to the attachment GlcNAc, (2) variations in the number and arrangement of sugars in the peripheral chains as well as in their linkages and degree of branching, and (3) the presence or absence of anionic substituents such as sulfate and phosphate. Oligosaccharide microheterogeneity is thought to primarily be a product of biosynthesis rather than resulting from degradation.[8] In some studies, care was taken to isolate the oligosaccharides associated with a single glycosylation site. However, homogeneity of oligosaccharides at the particular glycosylation site was not considered. In fact, this represented the state of the art until recently. Studies which employ such an approach usually yield an average or general structure rather than the precise structures which are present.[8,16] The most precise and reliable information is gleaned from studies in which the oligosaccharides have been purified to homogeneity with respect to both size and charge prior to detailed structural analysis.

III. ANTERIOR PITUITARY GLYCOPROTEIN HORMONES

LH, FSH, and TSH were recognized to be glycoproteins many years ago (reviewed by Liu and Ward[1]). In recent years other anterior pituitary hormones, not classically included in the family of glycoprotein hormones, have been shown to exist in molecular forms which contain carbohydrate. For example, certain molecular forms of the precursor of adrenocorticotropic hormone (ACTH), proopiomelanocortin, are glycosylated (reviewed by Eipper and Mains[34]). Prolactin, classically considered to be a protein hormone, is now known to exist in forms which are glycoproteins (see Chapter 5) and glycosylated variants of growth hormone (GH) have recently been identified.[35] Thus, the list of anterior pituitary glycoprotein hormones has recently been expanded. Nonetheless, this discussion will focus on the three classic glycoprotein hormones listed above and will also be limited to mammalians because of the limited amount of information available in lower species.

As noted previously, the anterior pituitary glycoprotein hormones contain approximately one residue GalNAc per oligosaccharide. They were some of the first glycoproteins in which GalNAc was observed to be present in N-linked rather than O-linked sugar chains. Similarly, these were some of the first glycoproteins in which sulfated *N*-acetylhexosamines were observed and perhaps are the most thoroughly characterized glycoproteins of this type.

Complete, detailed oligosaccharide structures have only been proposed for a portion of the anterior pituitary glycoprotein hormones from three species: ovine (o), bovine (b), and human (h). Among these species, a majority of the work has been directed towards LH. Thus, information from these three species will be examined first.

A. Bovine, Ovine, and Human

In terms of composition, the oligosaccharides of LH, FSH, and TSH from these three

Table 1

CARBOHYDRATE COMPOSITION[a] OF bLH, oLH, AND bTSH

Hormone	GlcNAc	Fuc	Man	Gal	NeuAc	GalNAc	SO$_4$	Ref.
bLH	3.0	0.3	3.0	≈0	≈0	1.1	1.9	36
bLHα	2.7	0.3	3.0	≈0	≈0	1.0	1.0—2.7	36
oLH	4.2	0.6	3.0	0.5	≈0	1.2	≈1	19
oLHα	3.3	0.3	3.0	0.2	≈0	0.8	N.R.[b]	19
oLHβ	3.8	0.6	3.0	0.3	≈0	1.6	N.R.	19
oLH	4.3	0.5	3.0	0.4	<0.1	1.4	0.8	37
bTSH	3.4	0.4	3.0	0.1	≈0	1.4	N.R.	36, 38
bTSHα	3.3	0.2	3.0	0.1	≈0	1.2	2.5	36, 38
bTSHβ	3.7	1.1	3.0	≈0	≈0	1.8	N.R.	36, 38

[a] Values have been normalized such that Man = 3.0.

[b] N.R. = not reported.

species can be divided into two general groups: (1) those which essentially lack NeuAc, and for the most part, Gal and (2) those which contain significant quantities of NeuAc. Among the former group, bovine and ovine LH have been the most thoroughly investigated. Information regarding bTSH and oTSH oligosaccharides will be considered in conjunction with LH because TSH from these species is presumed to have similar oligosaccharide structures. Among the latter group, the human anterior pituitary glycoprotein hormones have been the most thoroughly characterized. Information regarding bovine and ovine FSH will be considered subsequent to the human hormones because their oligosaccharides appear to fall in that group.

1. N-Linked Oligosaccharides Lacking NeuAc

a. Bovine and Ovine LH

These hormones typically contain 14 to 20% carbohydrate by weight in three N-linked oligosaccharides, two of which reside on the alpha subunit (Asn56 and Asn82) and one of which is found on the beta subunit (Asn13) (reviewed by Liu and Ward;[1] Pierce and Parsons[3]). Their constituent sugars as analyzed by different laboratories have been reviewed in detail by Liu and Ward.[1] Representative compositions are presented in Table 1 emphasizing recent work which includes sulfate quantitation.

Early attempts to delineate the carbohydrate structures of oLH and bLH were hampered becaused their oligosaccharides are resistant to many glycosidases.[36,39] However, in the early 1980s two laboratories[19,36,40] proposed general structures with only slight differences. Both laboratories noted that Fuc was located in a nonreducing terminal position and that approximately one residue of GlcNAc and one residue of GalNAc were located in positions peripheral to periodate-sensitive Man residues.[19,36,40]

Parsons and Pierce[36] suggested that approximately two sulfate residues were present on each oligosaccharide of bLH. Under the assumptions that each *N*-acetylhexosamine was a terminal sugar and that each contained one sulfate residue which would render it resistant to periodate oxidation and exoglycosidase cleavage, the structure illustrated in Figure 1A was proposed.[36] They also speculated that Fuc was most likely linked in the core region.

Studies published by Bahl and co-workers[19,40] were more complete in that methylation analyses, acetolysis, and deamination studies as well as exoglycosidase digestions were performed. Fuc, when present, was shown to be bound to the attachment GlcNAc via an α1–6 linkage. Both GlcNAc and GalNAc appeared to be linked to Man residues and Gal was found to occur in a terminal nonreducing position in oLH but was absent from bLH. The major evidence that GalNAc (actually GalNAc-sulfate) was bound directly to Man resulted from the appearance of *N*-acetylgalactosaminyl glyceraldehyde after periodate ox-

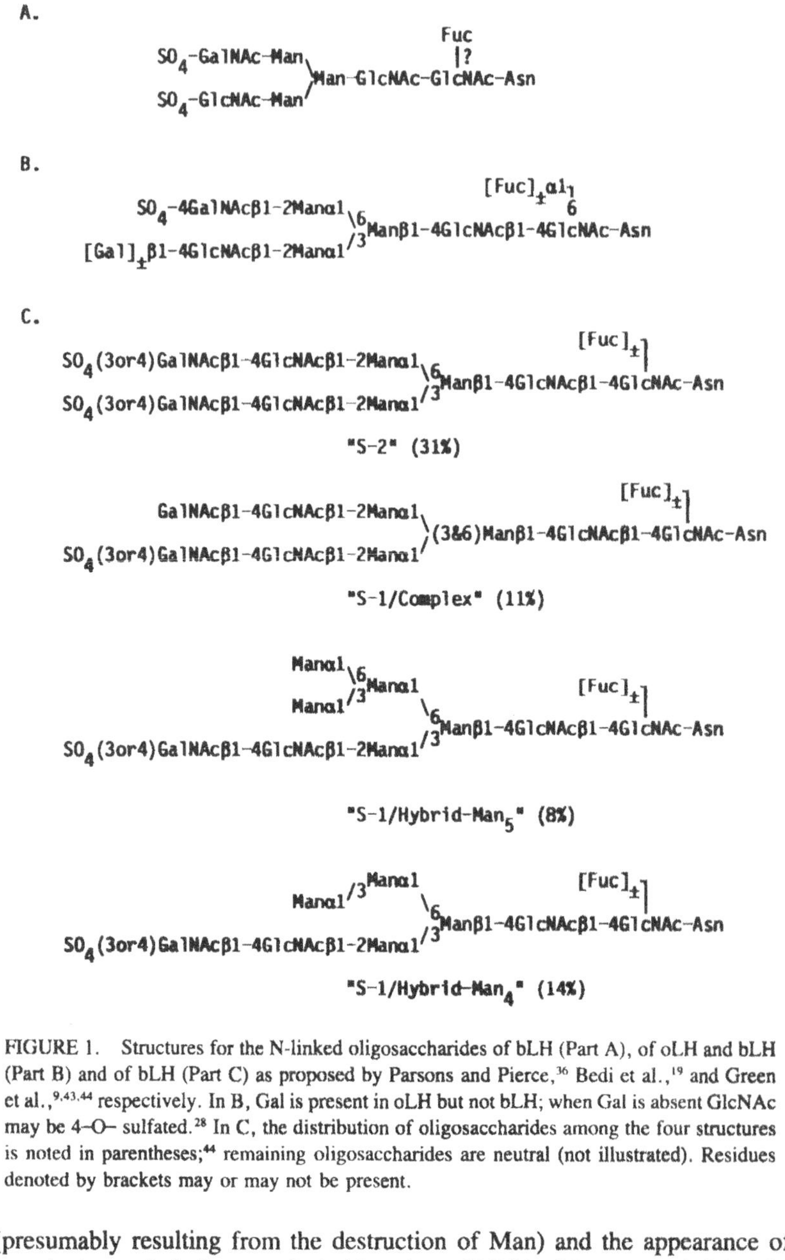

A.

```
SO4-GalNAc-Man                         Fuc
             \                          |?
              Man-GlcNAc-GlcNAc-Asn
             /
SO4-GlcNAc-Man
```

B.

```
                                        [Fuc]±α1
SO4-4GalNAcβ1-2Manα1                           6
                   \6
                    Manβ1-4GlcNAcβ1-4GlcNAc-Asn
                   /3
[Gal]±β1-4GlcNAcβ1-2Manα1
```

C.

```
                                               [Fuc]±
SO4(3or4)GalNAcβ1-4GlcNAcβ1-2Manα1
                                 \6
                                  Manβ1-4GlcNAcβ1-4GlcNAc-Asn
                                 /3
SO4(3or4)GalNAcβ1-4GlcNAcβ1-2Manα1
```

"S-2" (31%)

```
                                          [Fuc]±
GalNAcβ1-4GlcNAcβ1-2Manα1
                         \
                          (3&6)Manβ1-4GlcNAcβ1-4GlcNAc-Asn
                         /
SO4(3or4)GalNAcβ1-4GlcNAcβ1-2Manα1
```

"S-1/Complex" (11%)

```
Manα1
     \6
      Manα1
     /3
Manα1     \6                               [Fuc]±
           Manβ1-4GlcNAcβ1-4GlcNAc-Asn
          /3
SO4(3or4)GalNAcβ1-4GlcNAcβ1-2Manα1
```

"S-1/Hybrid-Man5" (8%)

```
       /3Manα1
Manα1          \6                          [Fuc]±
                Manβ1-4GlcNAcβ1-4GlcNAc-Asn
               /3
SO4(3or4)GalNAcβ1-4GlcNAcβ1-2Manα1
```

"S-1/Hybrid-Man4" (14%)

FIGURE 1.   Structures for the N-linked oligosaccharides of bLH (Part A), of oLH and bLH (Part B) and of bLH (Part C) as proposed by Parsons and Pierce,[36] Bedi et al.,[19] and Green et al.,[9,43,44] respectively. In B, Gal is present in oLH but not bLH; when Gal is absent GlcNAc may be 4-O- sulfated.[28] In C, the distribution of oligosaccharides among the four structures is noted in parentheses;[44] remaining oligosaccharides are neutral (not illustrated). Residues denoted by brackets may or may not be present.

idation (presumably resulting from the destruction of Man) and the appearance of *N*-acetylgalactosaminyl mannitol after acetolysis (denoting a GalNAc–Man sugar chain). Methylation analysis suggested a nonreducing terminal position for GalNAc which was assigned to the α1–6 antennae on the basis of the products resulting from acetolysis. β-*N*-acetylhexosaminidase cleaved approximately one residue of GlcNAc after removal of Gal on oLH with β-galactosidase suggesting that GlcNAc was terminal if Gal was absent. Only 0.1 residue GalNAc per oligosaccharide was liberated by β-*N*-acetylhexosaminidase under the conditions utilized (i.e., without prior removal of sulfate). In the first study[40] GalNAc was observed to be substituted with an acid labile group at C-4 which was subsequently demonstrated to be sulfate.[19] On the basis of these results, the biantennary structure illustrated in Figure 1B was proposed.

Thus, the two laboratories proposed similar albeit not precisely identical structures for oLH and bLH oligosaccharides. The primary difference between these two structures is the number of sulfates and whether sulfate occurs exclusively on GalNAc or may also occur on GlcNAc. In a subsequent study, Anumula and Bahl[28] published data which suggest that sulfate can occur on *both* GalNAc and GlcNAc. Nonetheless, both proposed structures (Figure 1A and 1B) are derived from studies where the entire population of oligosaccharides on the hormones were implicitly assumed to be homogeneous on the basis of similarity of oligosaccharide compositions between the intact hormones and glycopeptides which represented distinct glycosylation sites. Careful examination of the data from both laboratories suggests oligosaccharide microheterogeneity (slight deviations from the proposed general structures). This is perhaps best illustrated by the presence of 0.5 and 0.6 residues of 2,3,4,6-tetra-*O*-methyl mannose (indicative of a terminal position) in oLH and bLH, respectively, following permethylation.[40]

The Galβ1–4GlcNAc sugar chains of oLH offer a potential sialylation site. Some laboratories[41,42] have found a small amount of NeuAc (0.2 to 0.3 residue per oligosaccharide) associated with oLH. Thus, it is possible that some preparations of this hormone possess oligosaccharides with terminal NeuAc residues.

In recent studies, Green et al.[43,44] utilized a combination of biosynthetic radiolabeling and chemical approaches to reveal salient structural features of bLH oligosaccharides. They also utilized techniques which appear to be valuable in delineating certain structural features of these types of oligosaccharides. First, they established that methanolysis (0.5 N HCl in anhydrous methanol) could liberate sulfate without significantly altering the remainder of the oligosaccharide. Second, they demonstrated that β-*N*-acetylhexosaminidase could cleave both GlcNAc and GalNAc when peripherally located if the oligosaccharides were previously desulfated. Third, they employed two recently described enzymes which cleave complex N-linked oligosaccharides: endo-β-*N*-acetylglucosaminidase F (endoF) which cleaves between the two GlcNAc residues of the pentasaccharide core[45,46] and peptide: *N*-glycanase (PNGase; sometimes called peptide: *N*-glycanase F or peptide: *N*-glycosidase F) which cleaves oligosaccharides between the attachment GlcNAc and Asn.[47,48]

Using these techniques as well as high performance liquid chromatography (HPLC) to establish the homogeneity of oligosaccharides with respect to size and charge, it was demonstrated that oligosaccharides which contained two ("S-2"), one ("S-1"), or zero sulfate residues (neutral) were present.[43,44] Effort was initially directed towards determining the detailed structure of disulfated oligosaccharides.[43] The structure of the core region was established by lectin affinity chromatography utilizing concanavalin A (ConA) as well as sensitivity to endoF. (ConA preferentially binds α-linked Man or glucose residues not substituted at C-3, C-4, and C-6 as well as high mannose, hybrid, and biantennary complex oligosaccharides whose α-linked Man residues are not substituted at positions other than C-2.[24,49] EndoF cleaves high mannose, hybrid, and complex *biantennary* N-linked oligosaccharides between the two GlcNAc residues of the di-*N*-acetylchitobiose core.[47]) After desulfation GalNAc, but not GlcNAc, was sensitive to periodate oxidation suggesting that GalNAc, but not GlcNAc, was located in a terminal position. Digestion of the asulfo oligosaccharides with β-*N*-acetylhexosaminidase released four residues, two GalNAc and two GlcNAc. On the basis of these and other observations including methylation analysis, the disulfated structure illustrated in Figure 1C was proposed.[43] The primary difference between this structure and those presented earlier is that only GalNAc is sulfated (not GlcNAc) and GalNAc is bound to GlcNAc rather than directly to Man. Because care was taken to purify oligosaccharides to homogeneity, the disulfated biantennary oligosaccharide illustrated in Figure 1C is likely to be one of the structures found on bLH.

In companion studies Green et al.[9,44] noted that a portion of the oligosaccharides on bLH were sensitive to endo-β-*N*-acetylglucosaminidase H (endoH; endoH cleaves high mannose and hybrid oligosaccharides[14]). The endoH-sensitive sugar chains were found to be of the hybrid type and contained either four or five Man residues (Figure 1C).[44] The remaining monosulfated oligosaccharides were subsequently found to be simple derivatives of the disulfated structure (Figure 1C).[44] Thus, other oligosaccharide structures which likely occur on bLH are two monosulfated hybrids and two monosulfated complex oligosaccharides depending on which antennae is sulfated (assuming that *either* antennae can be sulfated). Green et al.[44] also noted that approximately 35% of the oligosaccharides on bLH were neutral (asulfo). The neutral oligosaccharides could be fractionated into at least eight distinct components, most of which were neither of the hybrid or high mannose types. Some of these appeared to be abbreviations of complex structures noted above. None of the neutral oligosaccharides were characterized in detail because of the limited quantities available.[44] Fuc was found to be a variable constituent in that it could be present or absent from any of the above structures including the monosulfated hybrids.[44]

Studies from the author's laboratory[37,50] suggest that oLH may possess oligosaccharides similar to those of bLH as proposed by Green et al.[43,44] First, oLH oligosaccharides were quantitatively (or nearly quantitatively) cleaved by endoF. The composition suggested four GlcNAc residues per oligosaccharide rather than three (Table 1; note that our oLH compositional analyses are nearly identical to those of Bedi et al.[19]). EndoF cleaved three of the four GlcNAc residues but did not liberate any Fuc which is consistent with linkage to the attachment GlcNAc.

The number, and perhaps location, of sulfated hexosamines on bLH and oLH are of interest because they presumably affect the isoelectric point of the hormones. Parsons and Pierce[36] originally suggested that bLH contained approximately two sulfates per oligosaccharide. However, other studies[19,33,37] have found an average of approximately one sulfate per oLH and bLH oligosaccharide. Beta subunit sugar chains are thought to contain greater amounts of sulfate than alpha subunit sugar chains.[9,33] Green et al.[9,43,44] maintain that β1–4-linked GalNAc serves as *the* sulfate acceptor site while Anumula and Bahl[28] have observed GlcNAc-sulfate on oLH.

Thus, there is general agreement that oLH and bLH possess N-linked oligosaccharides with Manα1–3(Manα1–6)Manβ1–4GlcNAcβ1–4(Fucα1–6)±GlcNAc cores and that the majority of structures are of the biantennary complex type. There appears to be slight disagreement regarding the sugar chains present on the antennae and whether or not GlcNAc is sulfated. However, one must consider that the general structures initially proposed[19,36,40] could be slightly in error because a mixture of oligosaccharides was analyzed (as recently discussed by Bahl[8]). It is evident from the elegant work of Green et al.[43,44] that more than 13 distinct oligosaccharide structures could be present on bLH. Assuming two unique S-1/complex structures depending on which antennae is sulfated and that Fuc may be present or absent from each form, the ten major oligosaccharide structures associated with bLH can be derived from those illustrated in Figure 1C. oLH is likely to possess some of the same oligosaccharide structures as bLH but probably possesses additional structures containing Gal and perhaps NeuAc, GalNAc-Man, and GlcNAc-sulfate.

b. Bovine and Ovine TSH

Three N-linked oligosaccharides are found on bTSH, one on TSHβ (Asn23), and two on TSHα (Asn56 and Asn82) (reviewed by Liu and Ward[1]; Pierce and Parsons[3]). Their composition is similar to those of bLH (Table 1). The oligosaccharide structures of bTSH are presumed to be similar to those found on bLH (biantennary complex with sulfated termini) because the sugar chains: (1) contain sulfate,[9,36] (2) bind to ConA,[9] and (3) can be cleaved with endoF.[9,51] Disulfated oligosaccharides are thought to occur with a greater frequency on bTSH than bLH.[9]

As evident from the reviews by Liu and Ward[1] as well as Condliffe and Weintraub,[52] there is surprisingly little information available regarding the chemistry of ovine TSH even though purified preparations are available from the National Institutes of Health (NIH). Of

Table 2

CARBOHYDRATE COMPOSITION[a] OF hLH, hFSH, hTSH, oFSH,
AND bFSH

Hormone	GlcNAc	Fuc	Man	Gal	NeuAc	GalNAc	SO₄	Ref.
hLH	3.1	0.5	3.0	1.4	0.9	1.0	N.R.[b]	54
hLHα	3.6	0.2	3.0	1.3	1.0	1.1	N.R.	54
hLHβ	4.0	1.1	3.0	1.7	1.2	1.0	N.R.	54
hLHα	2.7	0.1	3.0	2.2	1.7	0.8	N.R.	55[c]
hLH	4.0	0.8	3.0	1.2	N.R.	0.7	1.4	36
hLH	3.5	0.6	3.0	1.6	2.1	0.5	N.R.	37
hLHα	4.7	0.4	3.0	1.3	1.1	0.5	N.R.	10
hFSH	2.2	0.3	3.0	2.1	1.6	0.2	N.R.	56
hFSHα	3.5	0.4	3.0	2.2	2.7	0.7	N.R.	57
hFSHβ	3.7	0.6	3.0	2.9	2.7	0.7	N.R.	57
hFSHα	6.2	0.4	3.0	2.6	2.2	0.2	N.R.	10
hTSH	3.5	0.5	3.0	0.9	0.9	1.1	N.R.	58
hTSHα	3.8	0.1	3.0	1.0	1.0	1.1	N.R.	58
hTSHβ	3.2	0.8	3.0	0.6	0.5	1.1	N.R.	58
hTSHα	2.8	Trace	3.0	2.1	Trace	0.6	N.R.	55[c]
hTSHα	4.2	0.3	3.0	1.1	0.8	0.8	N.R.	10
oFSH	4.3	0.5	3.0	1.5	1.9	0.7	N.R.	59
oFSHα	3.7	0.2	3.0	0.9	1.1	0.7	N.R.	59
oFSHβ	5.5	1.0	3.0	2.3	3.1	0.7	N.R.	59
bFSH	4.4	1.1	3.0	1.9	1.0	1.5	N.R.	60

[a] Values have been normalized such that Man = 3.0.
[b] N.R. = not reported.
[c] hLHα = LGP-II; hTSHα = TGP-II.

the papers dealing with oTSH, one which provides some information regarding its carbohydrate composition states that oTSH is devoid of sialic acids and Gal.[53] Thus, oTSH appears
to fall within the same class as bLH, oLH, and bTSH.

2. N-Linked Oligosaccharides Containing NeuAc
a. Human LH, FSH, and TSH

Human LH, FSH, and TSH are thought to have oligosaccharide structures distinct from
ovine and bovine LH because they contain significant quantities of NeuAc and Gal (Table
2). A portion of their oligosaccharides are susceptible to exoglycosidase digestion[36] (also,
see below). The common alpha subunit possesses two N-linked oligosaccharides (Asn52
and Asn78) while hLHβ possesses one (Asn30), hTSHβ possesses one (Asn23), and hFSHβ
possesses two (Asn7 and Asn24).[1,3] The constituent sugars for these hormones as reported
by a variety of laboratories are presented in Table 2.

Studies relating to the carbohydrate chains of hFSH were conducted by Kennedy, Chaplin,
and colleagues in the late 1960s and early 1970s. They utilized periodate oxidation,[61]
glycosidase digestions,[62] and methylation analyses[63] to reveal certain structural features and
to study structure-function relationships. Based on these and later experiments, it was proposed that some terminal chains of hFSH were NeuAc2–2Gal1–6GlcNAc1–6Man.[64]

In 1978, Hara et al.[55] proposed a general structure for the oligosaccharides of the human
pituitary glycoprotein hormones. Using an approach which emphasized exoglycosidase digestions of hLH, hTSH, and hFSH alpha subunit glycopeptides (Asn52 for hLH and hTSH;
Asn78 for hFSH), they established that some outer chains had the sequence Neu-Ac
–α–Gal–β–GlcNAc–β– and that some sugar chains were attached to α-linked Man. NeuAc,
Gal, and Fuc were located in terminal positions as judged by susceptibility to periodate
oxidation. Fuc was assigned a position of attachment to the core region based on the ability

$$NeuAc-Gal\beta-GlcNAc\beta1-2Man\alpha1$$
$$[Fuc]_{\pm}]$$
$$\diagdown(3\&6)Man\beta1-4GlcNAc\beta1-4GlcNAc-Asn$$
$$SO_4-GalNAc\beta-GlcNAc\beta1-2Man\alpha1\diagup$$

FIGURE 2.   General structure for the monosulfated, monosialylated N-linked oligosaccharides of hLH and hTSH as proposed by Green et al.[9] Residues denoted by brackets may or may not be present. Oligosaccharides illustrated in Figure 1C are thought to also occur on hLH and hTSH except that no hybrid structures were observed on hLH.[9]

of exoglycosidases other than α-fucosidase to remove peripheral sugars. Thus, these workers elucidated certain important features which have subsequently been substantiated (see below). However, they also suggested that the oligosaccharides of hLH, hFSH, and hTSH possessed a unique core region: Man–β–GalNAc–GlcNAc–Asn. The general structure proposed from these studies[55] consisted of three NeuAc–α–Gal–β–GlcNAc–β–Man chains attached to a –α–Man–β–GalNAc–GlcNAc core with Fuc linked to a residue in the core.

In 1982, Tolvo et al.[57] published additional studies where they examined the oligosaccharide structures associated with hFSHα (Asn52) and hFSHβ (Asn7). Further evidence presented to support the location of GalNAc in the core included a lack of sensitivity of GalNAc to cleavage by exo-α-N-acetylgalactosaminidase and of the core to cleavage by endoH as well as endoglycosidase D (endoD).

The proposal that GalNAc occupies a position in the core region of N-linked oligosaccharides[55,57] is unique and controversial. It has yet to be confirmed for the human glycoprotein hormones or any other glycoprotein. Moreover, an oligosaccharide core containing GalNAc would require a biosynthetic mechanism distinct from the standard dolichol-high mannose pathway.[4,25,26]

Evidence that the human pituitary glycoprotein hormones possess typical Manα1-3(Manα1-6)Manβ1-4GlcNAcβ1-4GlcNAc pentasaccharide cores, with or without Fuc attached to the innermost GlcNAc, comes from several recent studies.[9,10,37,50] Most notably the oligosaccharides of hLH, hTSH, and hFSH are susceptible to cleavage by endoF;[9,37,50] Fuc associated with hLH oligosaccharides is not liberated by endoF.[37] Methylation studies of alpha subunits derived from native hormones suggest that GalNAc is located peripherally rather than internally.[10]

Green et al.[9] suggested that hLH and hTSH possess a mixture of disulfated ("S-2"), monosulfated ("S-1"), and monosulfated plus monosialylated ("S-N") oligosaccharides. hLH oligosaccharides possessing sialylated termini were susceptible to sequential neuraminidase and β-galactosidase digestion. A small percentage of hTSH oligosaccharides was cleaved by endoH suggesting hybrid structures but these were not present on hLH.[9] hLH and hTSH oligosaccharides quantitatively bound to ConA suggesting biantennary structures.[9] Based on these considerations, the general structure illustrated in Figure 2 was proposed. This structure, its abbreviated derivatives, the structures illustrated in Figure 1C, and possibly other monosulfated and/or monosialylated oligosaccharides and unidentified neutral glycans were postulated to occur on hLH and hTSH. As in the bovine, hTSH was found to possess a higher percentage of disulfated structures than hLH.[9] Consistent with these observations, hLH has been reported to contain approximately 0.3 residues sulfate per oligosaccharide.[33]

Green et al.[9] reported that hFSH did not possess endoH-sensitive oligosaccharides. Approximately 40% of the oligosaccharides of hFSH did not bind to ConA suggesting that this hormone contains some structures more complex than those found on hLH and hTSH.[9] Furthermore, hFSH did not routinely serve as a sulfate acceptor unless desialylated suggesting a high percentage of sialylated termini.

Recent methylation studies[10] of the alpha subunits of the human anterior pituitary glycoprotein hormones are consistent with the results of Green et al.[9] Both hLHα and hTSHα appeared to possess biantennary structures with Fuc linked to the core and considerable heterogeneity in their outer antennae.[10] GalNAc was located at some nonreducing termini and some were 4–O– substituted. In hTSHα there were approximately equal numbers of chains terminating in NeuAc and GlcNAc with some GlcNAc being 4–O– substituted. hLHα exhibited greater heterogeneity with Man, Gal, GlcNAc, and NeuAc residues present as nonreducing termini.[10] The remaining Gal was both 3–O– and 6–O– substituted (perhaps to NeuAc) while GlcNAc was 4–O– or 6–O– monosubstituted or 4,6–O– disubstituted.

The oligosaccharides of hFSHα were even more complex in that a portion of the chains appeared to be triantennary and some bisecting GlcNAc structures (GlcNAcβ1–4 linked to the innermost Man of the pentasaccharide core; see Figure 6B for an example of a bisecting GlcNAc oligosaccharide) were present. Some Gal was terminal; when not terminal, Gal was either 3–O– or 6–O– substituted. GalNAc was terminal and a percentage was 4–O– substituted. GlcNAc was also terminally located, 4–O– monosubstituted or 3,4–O– as well as 4,6–O– disubstituted. Some Fuc appeared to be attached in the core region while the remainder appeared to be attached to GlcNAc in peripheral chains as Fuc1–3GlcNAc.

These data suggest that additional studies will be required to precisely define the oligosaccharide structures present on the human anterior pituitary glycoprotein hormones. The best available evidence suggests that they possess the common $Man_3GlcNAc_2$ pentasaccharide core. hLH and hTSH appear to possess biantennary sugar chains which terminate in sulfate or NeuAc but both exhibit considerable heterogeneity. hFSH appears to possess similar chains as well as additional structures which are triantennary or of the bisecting GlcNAc type.

b. Bovine and Ovine FSH

oFSH contains four N-linked oligosaccharides, two on each subunit.[65,66] Those on the oFSHα[65] are located at the same positions as oLHα while those on oFSHβ are located at Asn6 and Asn23.[66] FSH from these two species contains significant quantities of NeuAc (Table 2) and can be sulfated in vitro[9] suggesting their oligosaccharides are more similar to those found on the human glycoprotein hormones than those of oLH and bLH. Like hFSH, approximately 40% of bFSH oligosaccharides do not bind to ConA suggesting some oligosaccharides possess bisecting GlcNAc and/or more than two antennae.

B. Other Mammalian Species

The oligosaccharide structures associated with the anterior pituitary glycoprotein hormones from other species remain to be determined. However, information from available composition studies and other experiments which suggest certain characteristics for the oligosaccharides of these hormones is reviewed below. If β1–4-linked GalNAc indeed serves as the exclusive sulfate acceptor,[43,44] the concentration of GalNAc in N-linked oligosaccharides could be indicative of their sulfation.

1. Porcine

The composition of porcine (p) LH, FSH, and TSH oligosaccharides (Table 3) is quite comparable to those of the respective hormones from the bovine and ovine. This author could find no data in the literature regarding the NeuAc content of pLH. However, it has been reported that pTSH contains "insignificant" amounts of NeuAc.[75] pFSH contains fairly large amounts of NeuAc and relatively low amounts of GalNAc (Table 3).

2. Equine

The anterior pituitary glycoprotein hormones of the equine appear to be quite interesting. The amino acid sequences of eLH and eCG are thought to be identical raising the possibility that eLH may possess O-linked oligosaccharides on the beta subunit. The high GalNAc and NeuAc concentrations of eLH and eLHβ (Table 3) are consistent with this possibility. The

Table 3

CARBOHYDRATE COMPOSITION[a] OF LH, TSH, AND FSH FROM A VARIETY OF SPECIES

Hormone	GlcNAc	Fuc	Man	Gal	NeuAc	GalNAc	SO$_4$	Ref.
pLH	2.9	0.5	3.0	0.4	N.R.[b]	1.1	N.R.	67
pLHα	1.8	0.4	3.0	0.4	N.R.	0.7	N.R.	67
pLHβ	3.4	0.9	3.0	0.2	N.R.	1.9	N.R.	67
pTSH	4.1	0.7	3.0	0.3	N.R.	1.3	N.R.	68
pFSH	3.6	0.9	3.0	2.3	4.8	0.4	N.R.	69
pFSHα	2.4	0.4	3.0	1.4	8.1	0.4	N.R.	69
pFSHβ	4.6	1.1	3.0	2.9	6.5	0.4	N.R.	69
eLH[a]	7.2	1.4	4.1	9.9	6.1	6.4	1.8	33, 70
eLH[a]	9.4	1.3	8.5	6.5	8.8	5.8	N.R.	71
eLHα	3.2	0.2	3.0	1.0	1.6	0.9	N.R.	71
eLHβ[a]	6.2	0.9	3.3	4.7	7.0	2.1	N.R.	71
eFSH	2.8	0.6	3.0	1.6	2.1	1.2	N.R.	72
eFSHα	3.7	0.4	3.0	1.7	2.2	0.8	N.R.	72
eFSHβ	3.6	0.8	3.0	1.7	2.4	0.8	N.R.	72
rLH	6.3	0.8	3.0	0.7	N.R.	1.5	N.R.	73
rLHα	7.8	0.4	3.0	1.5	N.R.	1.4	N.R.	73
rLHβ	5.2	1.2	3.0	0.9	N.R.	1.3	N.R.	73
lLH	4.9	0.6	3.0	0.4	0.5	2.0	N.R.	74

[a]   Values have been normalized such that Man = 3.0 except for eLH and eLHβ where the values are mole per mole of hormone because eLHβ may possess O-linked oligosaccharides.[70]

[b]   N.R. = not reported.

equine appears to be one species where LH is more heavily sialylated than FSH. Nonetheless, a portion of the oligosaccharides of eLH are sulfated (1.8 mole sulfate per mole hormone).[33] eFSH also contains significant quantities of both NeuAc (Table 3) and sulfate (2.4 mol/mol hormone).[33] eTSH appears to contain fairly low concentrations of NeuAc (1.4 g/100 g hormone).[76]

3. Rodents

Although rat (r) LH, FSH, and TSH are available from NIH, there is little information available regarding their carbohydrate compositions. The carbohydrate composition of rLH, as reported by Ward et al.,[73] indicates oligosaccharides rich in GalNAc, GlcNAc, and Gal (relative to oLH and bLH) (Table 3). There appears to be some confusion as to whether rLH oligosaccharides are sulfated and/or sialylated. In the original carbohydrate composition studies, NeuAc was simply not quantitated (D. N. Ward, personal communication). It has been reported that the charge heterogeneity of rLH is related to terminal NeuAc residues.[77] In contrast, Boime and co-workers have demonstrated that rLH oligosaccharides can be sulfated in vitro.[27,78] Additional studies will be required to clarify these issues.

The charge heterogeneity of rFSH can, to a large extent, be ablated by neuraminidase digestion[79,80] (Chapter 7) suggesting a high percentage of sialylated termini. rFSH does not quantitatively bind to ConA[81] (Chapter 7). Although there are difficulties in drawing conclusions from lectin affinity studies for molecules with multiple oligosaccharides, the latter observation is consistent with the possibility that rFSH possesses some multiantennary sugar chains. Similar data are available with regard to hamster FSH.[82]

4. Rabbit

With regard to the rabbit (denoted by "l" for lagomorph), only the carbohydrate composition of lLH is available.[74] lLH oligosaccharides contain fairly low concentrations of NeuAc and Gal but relatively high concentrations of GalNAc (Table 3).

Table 4

CARBOHYDRATE COMPOSITION (% BY WEIGHT) OF hCG AND
eCG

Hormone	GlcNAc	Fuc	Man	Gal	NeuAc	GalNAc	Total	Ref.
hCG	8.9	0.5	5.6	5.2	11.8	1.6	33.5	86
hCGα	11.7	Trace	7.7	4.2	9.0	Trace	32.6	86
hCGβ	7.8	0.8	4.5	5.7	13.0	3.3	35.1	86
eCG	9.2	0.8	2.3	11.6	14.5	3.3	41.7	87
eCGα	5.1	0.5	2.5	8.5	5.0	0.0	21.6	87
eCGβ	9.9	0.3	1.6	13.5	18.0	2.8	46.1	87

5. Nonhuman Primates

Digestion of rhesus monkey FSH, LH, and TSH with neuraminidase reduces their apparent molecular size during gel filtration[83] suggesting each possesses oligosaccharides which terminate in NeuAc. Based on the change in molecular size after neuraminidase digestion, the relative content of NeuAc in the three hormones is FSH > LH > TSH.[83] Similarly, the charge heterogeneity of FSH from cynomolgus monkeys appears to be related to terminal NeuAc[84] suggesting a high percentage of sialylated termini.

IV. CHORIONIC GONADOTROPINS

Placental gonadotropins have been identified in a variety of species (reviewed by Moore et al.[85]). However, detailed information regarding the chemistry of placental gonadotropins is available for only two species: human and equine. The structure of the glycans associated with hCG have been thoroughly investigated. Recently, information concerning the sugar chains of eCG has become available. eCG is unique in that its beta subunits contain a particularly high percentage of sugars most of which reside in O-linked oligosaccharides.

A. Human Chorionic Gonadotropin

hCG contains approximately 33% carbohydrate (reviewed in Bahl and Shah[2]). Furthermore, each subunit is approximately one third carbohydrate by weight (Table 4). Two aspects regarding its carbohydrate composition are worthy of note. First, almost all of the Fuc is associated with hCGβ. Second, GalNAc is present but only associated with the beta subunit of hCG. hCG possesses four N-linked oligosaccharides, two on each subunit (Asn52 and Asn78 of hCGα; Asn13 and Asn30 of hCGβ)[86,88,89] and four O-linked oligosaccharides all of which reside at the carboxy terminal region of hCGβ (Ser121, Ser127, Ser132, and Ser138).[18,89] In contrast to the anterior pituitary glycoprotein hormones, hCG oligosaccharides contain no sulfate.[33,36] hCG can be sulfated in vitro but the sulfate is incorporated into the protein chains rather than the sugars.[9,29]

1. N-Linked Oligosaccharides

The first attempts to elucidate the structure of hCG oligosaccharides were performed in the late 1960s. Using a series of exoglycosidases (neuraminidase, β-galactosidase, β-N-acetylglucosaminidase, α-mannosidase, and α-fucosidase), Bahl[90] established that the outer chains consisted of NeuAc–α–Gal–β–GlcNAc–β– and that these chains were attached to Man residues in the core region, some of which were α linked. Based on the observation that Fuc was located at nonreducing termini, it was suggested that Fuc could replace NeuAc in some of the outer chains. A putative structure for hCG oligosaccharides was proposed.[2,90,91]

In 1973, Kennedy and Chaplin[64] confirmed that Fuc and GalNAc were essentially absent from hCGα and reported that a small percentage (≈7%) of the sialic acid was present as

N-glycol rather than N-acetylneuraminic acid. After preparing hCGα glycopeptides, methylation analysis was performed. Based on the data obtained by methylation analysis, as well as earlier periodate degradation studies,[56] structures for the oligosaccharide at each glycosylation site on hCGα were proposed.[64] The structures, which will not be reiterated here, generally consisted of linear chains of the sugars with minimal branching. Nonetheless, Kennedy and Chaplin[64] did observe some nonreducing terminal Man, which has been confirmed in subsequent studies.

The fine, detailed structure of the N-linked oligosaccharides of hCG was determined in the late 1970s by two independent laboratories.[86,88,92] Each group used a slightly different approach which yielded complementary information.

The general approach used by Bahl and co-workers was to separate hCG derived from urinary sources into subunits, prepare glycopeptides from each subunit (representing distinct glycosylation sites), and to purify the glycopeptides to homogeneity.[86,88] An implicit assumption in this approach is that the oligosaccharides on a given glycosylation site are homogeneous. After a series of chemical fragmentation reactions, methylation analyses to establish linkages as well as a series of glycosidase digestions to establish anomeric configurations, the biantennary structure illustrated in Figure 3A was proposed. It should be noted that this oligosaccharide represents a general or average structure because the oligosaccharides at each glycosylation site were not resolved into homogeneous components prior to structural analysis. In fact, Bahl et al.[88] acknowledged that there might be slight deviations from this general structure due to the inherent "microheterogeneity" of glycoproteins.

Kobata and co-workers[92] used a slightly different approach which involved cleavage of oligosaccharides from native hCG (the α-β dimer; also of urinary origin) followed by purification of the oligosaccharides to homogeneity with respect to both size and charge prior to detailed structural analysis. This approach has the limitation that specific oligosaccharide structures cannot be associated with specific glycosylation sites. It was observed that greater than 90% of the oligosaccharides were acidic because of nonreducing terminal NeuAc. A detailed structure for each of the acidic hCG oligosaccharides was proposed; these are illustrated as the acidic oligosaccharides (denoted with the letter A) along with their corresponding neutral counterparts (denoted with the letter N) in Figure 3B.

A detailed distribution of hCG oligosaccharides among these structures was not presented but it was stated that only 25% were fucosylated raising the possibility that Fuc might be found on a single glycosylation site of hCGβ. However, the carbohydrate compositions of hCGβ glycopeptides[86,88] suggests that Fuc is associated with both hCGβ glycosylation sites. Of the acidic oligosaccharides observed,[92] approximately 56% were disialylated (structures A-1 and A-3 in Figure 3B), approximately 44% were monosialylated (structures A-2, A-4, and A-5 in Figure 3B) and approximately 13% possessed a nonreducing terminal Man (structure A-5 in Figure 3B).

All of the specific oligosaccharide structures proposed by Endo et al.[92] were consistent with the general structure proposed by Kessler et al.[86] Both studies noted that linkages between terminal NeuAc and Gal were α2–3 rather than α2–6 which was thought to be somewhat unique at the time. Recent methylation studies of hCGα[10] are consistent with the structures proposed by Kessler et al.[86] and Endo et al.[92] However, some terminal GlcNAc was observed raising the possibility that other oligosaccharide structures may be present on hCG.

In a subsequent study, Mizuochi and Kobata[93] examined the distribution of specific oligosaccharide structures on the subunits of hCG as well as compared hCG derived from the placenta to that present in urine. Urinary hCG possessed a higher percentage of acidic oligosaccharides than placentally derived hCG consistent with the possibility that asialo forms of hCG are cleared from the circulation more rapidly (see Chapters 3 and 11). Nonetheless, placentally derived hCGβ oligosaccharides were approximately 90% acidic

A.

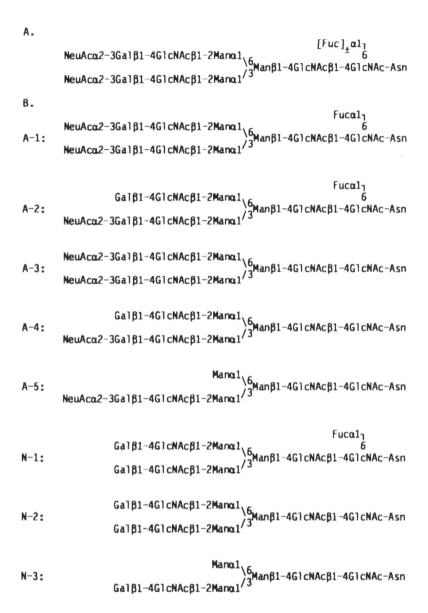

FIGURE 3. Detailed structures proposed for the N-linked oligosaccharides of hCG as proposed by Kessler et al.[86] (Part A) and by Mizuochi and Kobata[93] (Part B). In A, the brackets denote a residue which may or may not be present. In B, acidic oligosaccharides are denoted by "A-" and neutral oligosaccharides are denoted by "N-". These structures are thought to occur on hCG derived from the placenta and urine. A higher percentage of acidic oligosaccharides are found on urinary vs. placental hCG.

and approximately equally distributed between monosialylated and disialylated forms.[93] Placental hCGα contained 20 to 30% neutral chains and a greater abundance of monosialylated than disialylated oligosaccharides. An extremely important observation was that specific oligosaccharide structures were somewhat systematically distributed between the two hCG subunits. To facilitate structural analysis, terminal NeuAc residues were removed by exhaustive neuraminidase digestion. The resulting N-2 (Figure 3B) oligosaccharides (and presumably their acidic counterparts which in this case are A-3 and A-4) were found on both subunits but N-3 oligosaccharides (which correspond with A-5) were found almost exclusively on hCGα and N-1 oligosaccharides (which correspond with A-1 and A-2) were found only on hCGβ. Additional details regarding the distribution of specific structures were subsequently published.[94] One aspect is particularly worthy of note. Oligosaccharides with nonreducing terminal Man (A-5 and N-3) appear to constitute approximately 25% of the structures present on urinary hCG suggesting that they are likely to appear on a single glycosylation site of hCGα. Thus, specific structures appear to reside at particular glycosylation sites on the peptide chains of hCG[93] yielding a somewhat systematic distribition of oligosaccharides.

It is evident that there is agreement regarding the general structure of hCG oligosaccharides. The small differences in the completeness of the antennae presumably result from the inherent microheterogeneity of glycoproteins.[8,16] These relatively minor changes result in the presence of at least eight distinct oligosaccharide structures on hCG (Figure 3B). The distribution of these eight distinct structures among the four glycosylation sites of hCG is not random but appears to be somewhat systematic.

More recently, structures for the oligosaccharides on hCG derived from hydatidiform moles and choriocarcinomas have been proposed.[94,95] Hydatidiform molar hCG excreted in the urine had identical oligosaccharide structures (Figure 3B) and distributions to those observed for normal pregnancy (urinary) hCG.[94] In contrast, hCG derived from choriocarcinomas contained approximately 50% fucosylated oligosaccharides (Figure 4). In the earlier study,[95] only 3% acidic oligosaccharides were observed and all of these were monosialylated. In a subsequent study which included data from four individual patients[94] it was observed that the degree of sialylation could vary from 3 to 100% (observed values were 3, 6, 92, and 100%). Initially it was proposed that choriocarcinoma hCG primarily possessed eight distinct oligosaccharides[95] which are illustrated as neutral structures in Figure 4. Only three of these are found on urinary or placental hCG (E, F, and H of Figure 4). Two of the eight oligosaccharides of choriocarcinoma hCG were triantennary as the result of three Galβ1–4GlcNAc chains attached to the pentasaccharide core (A and B of Figure 4). Two others were biantennary with Man residues at one nonreducing terminal (C and D in Figure 4). These four oligosaccharides are somewhat unique relative to placental hCG because of extra Galβ1–4GalNAc chains β1–4 linked to α-linked Man. In contrast to hCG derived from normal tissues, a fucosylated monoantennary structure (G in Figure 4) was observed on hCG produced by this tumor.[95] It was subsequently demonstrated that some choriocarcinoma patients also produced a multitude of sialylated variants of these structures[94] presumably with α2–3-linked NeuAc on the various antennae. Thus, it appears that hCG produced by hydatidiform moles is similar to pregnancy hCG but that choriocarcinoma hCG has a distinct and more complex pattern of N-linked oligosaccharides.

2. O-Linked Oligosaccharides

Bahl[90] was the first to recognize that hCG possesses both N- and O-linked oligosaccharides. Based on the sugars present, as well as their susceptibility to exoglycosidases and β-elimination, hCG was initially thought to possess three O-linked oligosaccharides whose general structure was NeuAc–Gal–GalNAc–Ser.[2,90,91] In subsequent studies it was recognized that hCGβ possessed four, rather than three, O-linked oligosaccharides.[88,89]

Detailed structural studies were performed by first preparing glycopeptides from urinary hCG, purifying the glycopeptides to homogeneity and subjecting their oligosaccharides to chemical fragmentation, exoglycosidase digestions as well as methylation analyses.[18,88] Based on the observation that NeuAc, GalNAc, and Gal were present in the ratio of 1.84:0.97:1.18, it was assumed that each oligosaccharide contained one GalNAc, one Gal, and two NeuAc residues. A small amount of GlcNAc was also observed (0.1 residue per oligosaccharide) but this was considered to be contaminating N-linked oligosaccharides.[18] As a result of these studies, the general structure illustrated in Figure 5A was proposed.

However, in biosynthetic studies, radioactive glucosamine was incorporated into hCG

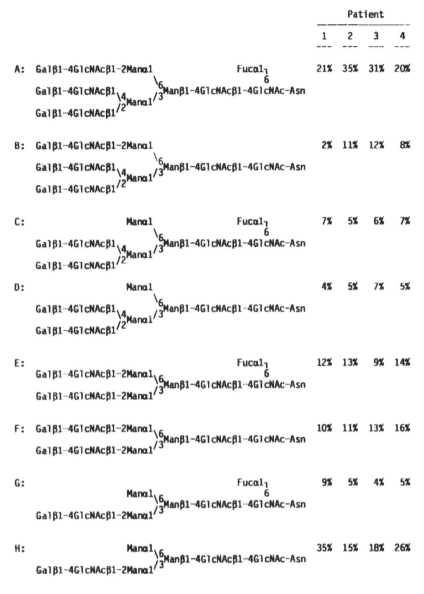

FIGURE 4.   Detailed structures proposed for the N-linked oligosaccharides of hCG derived from choriocarcinomas as proposed by Mizuochi et al.[94,95] All of the above are illustrated as neutral (asialo) structures but considerable quantities of sialylated oligosaccharides were observed in some patients (up to 100%).[94] The distribution of oligosaccharides in four individual patients as reported by Mizuochi et al.[94] is noted.

oligosaccharides of various sizes (3 to 7 sugar residues) which were susceptible to β-elimination suggesting that alternative O-linked structures might be present.[96] Cole et al.[97-99] subsequently observed that the O-linked oligosaccharides of hCG were a mixture of di-, tri-, tetra-, and hexasaccharides. After demonstrating that each oligosaccharide was homogeneous with respect to size and charge, the detailed structures were delineated utilizing an approach which emphasized glycosidases, periodate oxidation and lectin affinity chromatography.[97-99] The fine structure of each of these six oligosaccharides as well as their distribution is illustrated in Figure 5B. Purified urinary hCG (CR121) possessed four sialylated structures (O-1, O-2, O-3, and O-4 in Figure 5B).[97,99] Three of the four glycans (O-2, O-3, and O-4) were consistent with the general structure proposed by Kessler et al.[18,88]

A.

$$\begin{array}{c} \text{NeuAc}\alpha2 \\ \backslash 6 \\ \text{NeuAc}\alpha2-3\text{Gal}\beta1 / 3 \end{array}\text{GalNAc-Ser}$$

B.

	Urine	Placenta		Choriocarcinoma	
	hCG	hCG	free-α	hCG	free-α
O-1: NeuAcα2-3Galβ1-4GlcNAcβ1\6 NeuAcα2-3Galβ1/3 GalNAc-Ser	13%	14%	18%	51%	55%
O-2: NeuAcα2\6 NeuAcα2-3Galβ1/3 GalNAc-Ser	34%	73%	71%	29%	25%
O-3: NeuAcα2-3Galβ1-3GalNAc-Ser	43%	<1%	3%	10%	9%
O-4: NeuAcα2-6GalNAc-Ser	10%	5%	1%	2%	<1%
O-5: Galβ1-4GlcNAcβ1\6 Galβ1/3 GalNAc-Ser	≈0%	<1%	<1%	1%	2%
O-6: Galβ1-3GalNAc-Ser	≈0%	7%	6%	7%	8%

FIGURE 5. Detailed structures of the O-linked oligosaccharides of hCG as proposed by Kessler et al.[18] (Part A) and of hCG as well as free-hCGα as proposed by Cole[98,99] (Part B). The distribution of structures on urinary hCG (CR121), on hCG, and free-hCGα produced second trimester placental explants and on hCG and free-hCGα produced by choriocarcinoma (JAr) cells as reported by Cole[98,99] are also presented.

but the fourth was unique (O-1) and contained GlcNAc. These four structures were demonstrated to be randomly distributed among the four glycosylation sites.[97,99] It is interesting to note at the O-linked hexasaccharide (O-1) occurred with a frequency of 13% on urinary hCG which agrees favorably with the 0.1 residue of GlcNAc per O-linked oligosaccharide reported by Kessler et al.[88]

It was subsequently observed that cultured placental explants and choriocarcinoma (JAr) cells produced hCG with these same O-linked structures as well as a small percentage (≤10%) of their asialo variants (Figure 5B). Interestingly, the tetrasaccharide (O-2) is the predominate form produced by placental explants while low amounts of an asialo derivative (O-3; a major componet of urinary hCG) are produced. Perhaps the tetrasaccharide is desialylated yielding the trisaccharide which appears in urine. In choriocarcinoma cells, the hexasaccharide was clearly the predominate form produced with the tetrasaccharide also being made in substantial quantities. Thus, it appears that the O-linked oligosaccharides of hCG are heterogeneous and are present in at least four sialylated structures in urinary hCG as well as two additional asialo variants in placental tissues. The distribution of oligosaccharide structures appears to vary with source of the hormone and to be altered markedly in the malignant trophoblast. Nonetheless, the observed structures appear to be randomly distributed among the four O-glycosylation sites.

B. Equine Chorionic Gonadotropin

eCG contains approximately 40% carbohydrate (Table 4; reviewed by Moore et al.[85]). The sugars of eCGα, which constitute approximately 22% by weight, are present as two N-linked oligosaccharides located at Asn56 and Asn82.[100] eCGβ is almost 50% carbohydrate by weight (Table 4) and has one N- and six O-linked oligosaccharides.[8] Among the glycoprotein hormones, the oligosaccharides of eCG are the most complex identified to date because of their high degree of branching and because some contain up to 50 sugar residues. Furthermore, a portion of both the O- and N-linked oligosaccharides of eCG contain repeating N-acetyllactosamine disaccharides (Galβ1–4GlcNAcβ1–3).[8] eCG may also be an example of a sulfated placental glycoprotein hormone because it has recently been reported to contain approximately one sulfate residue per mole hormone.[33] However, it is not evident from the compositional data whether the sulfate is associated with oligosaccharides.

1. N-Linked Oligosaccharides

Preliminary information on the structure of the N-linked oligosaccharides of eCG has recently been published by Bahl and Wagh.[8] After preparation of glycopeptides, oligosaccharides were released with endoF and purified with respect to size. Structural analysis then was achieved using a combination of chemical methods (methylation analysis, periodate oxidation, and deamination) and enzymatic digestions. The enzymatic digestions emphasized the use of two exoglycosidases (β-galactosidase and β-N-acetylglycosaminidase) and one endoglycosidase (endo-β-galactosidase). Endo-β-galactosidase hydrolyzes the internal β-galactosidic linkages of R-GlcNAc(or GalNAc)β1–3Galβ1–4GlcNAc(or Glc)[101] and appears to be particularly useful in structural analysis of oligosaccharides containing repeating N-acetyllactosamine sequences.

The N-linked sugars chains on eCGβ and on one glycosylation site of eCGα ("αGP-I") appeared to be similar.[8] After exhaustive desialylation, oligosaccharides of four major sizes were present. All of the desialylated oligosaccharides could be reduced to pentasaccharide core structures (with or without Fuc) by the application of β-galactosidase, β-N-acetylglucosaminidase, and endo-β-galactosidase either sequentially or in combination. Based on the number of sugar residues released after the application of each enzyme it was determined that three of these four oligosaccharides were biantennary and one was triantennary. One of the biantennary sugar chains contained a bisecting GlcNAc residue while another contained a repeating N-acetyllactosamine sequence. The general structures for these oligosaccharides are illustrated as A, B, C, and D in Figure 6. The four oligosaccharides were found with a frequency of 51, 24, 9, and 16%, respectively, on eCGβ while their frequency on αGP-I was 60, 18, 10, and 12%, respectively.

The oligosaccharides on the remaining glycosylation site of eCGα were even more complex. Using the same general approach as outlined above it was determined that some oligosaccharides were biantennary, some were biantennary with bisecting GlcNAc, some were triantennary, some were tetraantennary, and some were even pentaantennary.[8] All of the oligosaccharides contained from three to nine N-acetyllactosamine sequences. Proposed general structures for these oligosaccharides are illustrated as E through J in Figure 6.

Thus, it would appear that the N-linked oligosaccharides of eCG possess typical Man$_3$GlcNAc$_2$ cores but exhibit variable degrees of branching. From the general structures proposed thus far, literally hundreds to thousands of specific structures can be envisioned. Nonetheless, the distribution of structures among the glycosylation sites does not appear to be completely random because the simpler structures are preferentially found on eCGβ and one glycosylation site of eCGα.

2. O-Linked Oligosaccharides

The O-linked oligosaccharides of eCG are found on six glycosylation sites in the beta subunit and constitute approximately 75% of its carbohydrate. General structures have recently been proposed by Bahl and Wagh.[8] Structural analysis proceeded as outlined above for the N-linked oligosaccharides except that the O-linked oligosaccharides were released by β-elimination and desialylated by mild acid hydrolysis. After separation on the basis of size it was observed that 3% of the desialylated oligosaccharides were disaccharides con-

```
A:  NeuAcα2-3(6)Galβ1-4GlcNAcβ1-2Manα1
                                        \6
                                          Man-GG
                                        /3
    NeuAcα2-3(6)Galβ1-4GlcNAcβ1-2Manα1


B:  NeuAcα2-3(6)Galβ1-4GlcNAcβ1-2Manα1
                                        \6
                 GlcNAcβ1-4Man-GG
                                        /3
    NeuAcα2-3(6)Galβ1-4GlcNAcβ1-2Manα1


C:    NeuAcα2-3(6)Galβ1-4GlcNAcβ1
                                  \2
                                    Manα1┐
    NeuAcα2-3(6)Galβ1-4GlcNAcβ1/6(4)      (3&6)Man-GG
                                    Manα1┘
      NeuAcα2-3(6)Galβ1-4GlcNAcβ1-2Manα1


D:  NeuAcα2-3(6)Galβ1-4GlcNAcβ1-3Galβ1-4GlcNAcβ1-2Manα1┐
                                                        (3&6)Man-GG
          NeuAcα2-3(6)Galβ1-4GlcNAcβ1-2Manα1┘


E:          [NeuAc]±α2-3(6)Galβ1-4GlcNAcβ1-2Manα1┐
                                                  (3&6)Man-GG
    NeuAcα2-3(6)Galβ1-4GlcNAcβ1-3Galβ1-4GlcNAcβ1-2Manα1┘


F:          [NeuAc]±α2-3(6)Galβ1-4GlcNAcβ1-2Manα1┐
                                                  (3or6)
                              GlcNAcβ1-4Man-GG
                                                  (3or6)
    NeuAcα2-3(6)Galβ1-4GlcNAcβ1-3Galβ1-4GlcNAcβ1-2Manα1┘


G:          [NeuAc]±α2-3(6)Galβ1-4    ┌  GlcNAcβ1-2Manα1┐
                                      │                 (3or6)
                NeuAcα2-3(6)Galβ1-4   │                 Man-GG
                                      │  GlcNAcβ1        (3or6)
    NeuAcα2-3(6){Galβ1-4GlcNAcβ1-3}1-2Galβ1-4  │          \4(6)Manα1┘
                                      │  GlcNAcβ1/2
                                      └


H:          [NeuAc]±α2-3(6)Galβ1-4    ┌  GlcNAcβ1
                                      │          \6Manα1┐
                NeuAcα2-3(6)Galβ1-4   │  GlcNAcβ1/2     (3or6)
                                      │                 Man-GG
                NeuAcα2-3(6)Galβ1-4   │  GlcNAcβ1        (3or6)
    NeuAcα2-3(6){Galβ1-4GlcNAcβ1-3}1-2Galβ1-4  │          \4Manα1┘
                                      │  GlcNAcβ1/2
                                      └
```

FIGURE 6.   Structures proposed by Bahl and Wagh[8] for the N-linked oligosaccharides of eCG. Residues denoted by brackets may or may not be present. Sugar chains enclosed in pointed brackets ({}) may be repeated multiple times with the range subscripted. The location of the various antennae on the more complex oligosaccharides (G through J) remain to be assigned (denoted by vertical lines). Man–GG denotes Manβ1-4GlcNAcβ1-4[Fucα1-6]. GlcNAc–Asn.

sisting of Galβ1–3GalNAc (structure A in Figure 7). Approximately one fourth had structures similar to B in Figure 7. The remaining oligosaccharides generally had Galβ1–4GlcNAcβ1 –6(Galβ1–3)GalNAc cores and were distributed among several more complex structures (C through G in Figure 7). In general they were extended on the β1–6 linked GlcNAc residues by varying numbers of repeating *N*-acetyllactosamine units. Methylation analysis revealed that some Gal was linked at C-1, C-3, and C-6 giving rise to oligosaccharides with several branches.[8]

```
I:  NeuAcα2-3(6)Galβ1-4GlcNAcβ1                    GlcNAcβ1-2Manα1
                              \6Galβ1-4                      (3or6)
    NeuAcα2-3(6)Galβ1-4GlcNAcβ1/3               GlcNAcβ1        Man-GG
                                               \4(6)  (3or6)
    NeuAcα2-3(6){Galβ1-4GlcNAcβ1-3}1-2Galβ1-4  GlcNAcβ1/2Manα1
                                                  /
    NeuAcα2-3(6){Galβ1-4GlcNAcβ1-3}1-2Galβ1-4  GlcNAcβ1

                          ** or **

                         NeuAcα2-3(6)Galβ1-4         GlcNAcβ1-2Manα1
                                                              (3or6)
                         NeuAcα2-3(6)Galβ1-4                  Man-GG
                                                  GlcNAcβ1\4(6)(3or6)
    NeuAcα2-3(6){Galβ1-4GlcNAcβ1-3}1-2Galβ1-4  GlcNAcβ1    Manα1
                              \6Galβ1-4         GlcNAcβ1/2
    NeuAcα2-3(6){Galβ1-4GlcNAcβ1-3}1-2Galβ1-4  GlcNAcβ1/3


J:  NeuAcα2-3(6)Galβ1-4GlcNAcβ1                 GlcNAcβ1
                              \6Galβ1-4                 \6Manα1
    NeuAcα2-3(6)Galβ1-4GlcNAcβ1/3               GlcNAcβ1/2
                                                              (3or6)
    NeuAcα2-3(6)Galβ1-4GlcNAcβ1-3Galβ1-4                      Man-GG
                                                             (3or6)
    NeuAcα2-3(6)Galβ1-4GlcNAcβ1-3Galβ1-4        GlcNAcβ1\4Manα1
                                                GlcNAcβ1/2
    NeuAcα2-3(6)Galβ1-4GlcNAcβ1-3Galβ1-4

                          ** or **

                    NeuAcα2-3(6)Galβ1-4             GlcNAcβ1
                                                           \6Manα1
                    NeuAcα2-3(6)Galβ1-4             GlcNAcβ1/2
                                                                 (3or6)
            NeuAcα2-3(6)Galβ1-4GlcNAcβ1-3Galβ1-4               Man-GG
                                                              (3or6)
    NeuAcα2-3(6)Galβ1-4GlcNAcβ1-3Galβ1-4GlcNAcβ1  GlcNAcβ1\4Manα1
                                               \6Galβ1-4     /2
    NeuAcα2-3(6)Galβ1-4GlcNAcβ1-3Galβ1-4GlcNAcβ1/3 GlcNAcβ1
```

FIGURE 6 continued

Thus, the O-linked oligosaccharides of eCGβ appear to possess Galβ1–4GlcNAcβ1–6(Galβ1–3)GalNAc cores and the most prevalent structures are the sialylated variants of this glycan. However, some of the O-linked oligosaccharides of eCG are much more complex and contain up to 50 sugars in various branches attached to this core.

C. Other Mammalian Species

Aggarwal et al.[102] has isolated and partially characterized donkey CG. This hormone contained approximately 32% carbohydrate by weight, of which approximately 7% was NeuAc.

V. UNCOMBINED ("FREE") ALPHA SUBUNITS

Although native bLH contains only N-linked oligosaccharides, Parsons et al.[103] identified a molecular form of uncombined bLHα which possesses an O-linked oligosaccharide at Thr43. The O-linked oligosaccharides of free-bLHα are apparently devoid of sulfate[104] and contain NeuAc, Gal, and GalNAc in the approximate molar ratios of 1.7:0.8:0.9.[103] Similarly, Cole et al.[106] identified a molecular form of free-hCGα which possesses an O-linked oligosaccharide at Thr39 and contains NeuAc, Gal, GalNAc, and GlcNAc. At present only the detailed structures for O-linked oligosaccharides of free-hCGα are available. Recently, Cole[98,99] (Chapter 3) reported that the O-linked oligosaccharides of free-hCGα are similar in structure to those which occur on hCGβ (Figure 5B). The distribution of structures from normal and neoplastic sources closely corresponds to that found on hCGβ, i.e., the tetrasaccharide is the major form on free-hCGα produced by the placenta whereas the hexasaccharide is the predominate form produced by choriocarcinomas.[98]

ConA lectin affinity studies suggest that some of the N-linked oligosaccharides on free-

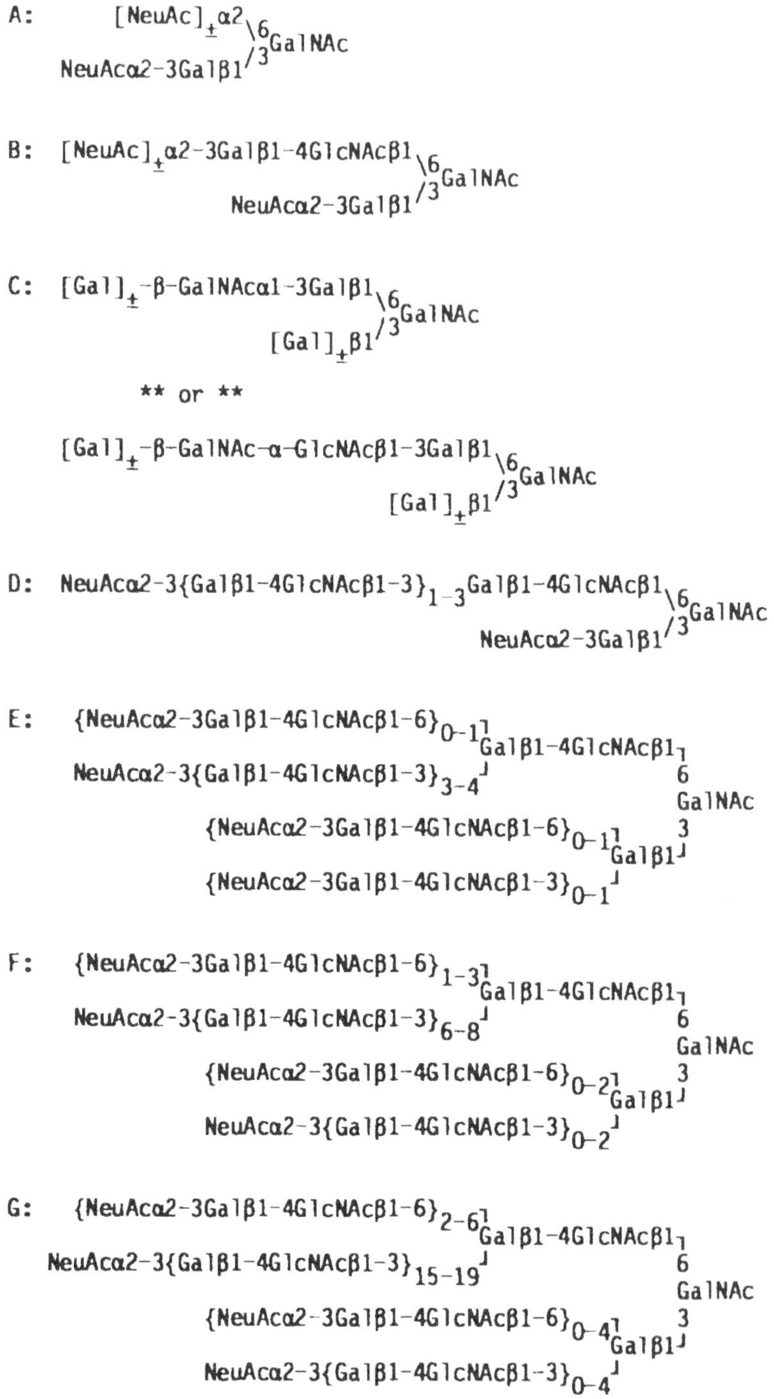

A: [NeuAc]$_{\pm}\alpha 2\backslash 6$GalNAc
      NeuAcα2-3Galβ1/3

B: [NeuAc]$_{\pm}\alpha$2-3Galβ1-4GlcNAcβ1$\backslash 6$GalNAc
      NeuAcα2-3Galβ1/3

C: [Gal]$_{\pm}$-β-GalNAcα1-3Galβ1$\backslash 6$GalNAc
      [Gal]$_{\pm}\beta$1/3

      ** or **

      [Gal]$_{\pm}$-β-GalNAc-α-GlcNAcβ1-3Galβ1$\backslash 6$GalNAc
      [Gal]$_{\pm}\beta$1/3

D: NeuAcα2-3{Galβ1-4GlcNAcβ1-3}$_{1-3}$Galβ1-4GlcNAcβ1$\backslash 6$GalNAc
      NeuAcα2-3Galβ1/3

E: {NeuAcα2-3Galβ1-4GlcNAcβ1-6}$_{0-1}$⌉
      {NeuAcα2-3{Galβ1-4GlcNAcβ1-3}$_{3-4}$⌋ Galβ1-4GlcNAcβ1⌉
                                                              6
                                                            GalNAc
      {NeuAcα2-3Galβ1-4GlcNAcβ1-6}$_{0-1}$⌉ 3
                                                       Galβ1⌋
      {NeuAcα2-3Galβ1-4GlcNAcβ1-3}$_{0-1}$⌋

F: {NeuAcα2-3Galβ1-4GlcNAcβ1-6}$_{1-3}$⌉
      NeuAcα2-3{Galβ1-4GlcNAcβ1-3}$_{6-8}$⌋ Galβ1-4GlcNAcβ1⌉
                                                              6
                                                            GalNAc
      {NeuAcα2-3Galβ1-4GlcNAcβ1-6}$_{0-2}$⌉ 3
                                                       Galβ1⌋
      NeuAcα2-3{Galβ1-4GlcNAcβ1-3}$_{0-2}$⌋

G: {NeuAcα2-3Galβ1-4GlcNAcβ1-6}$_{2-6}$⌉
      NeuAcα2-3{Galβ1-4GlcNAcβ1-3}$_{15-19}$⌋ Galβ1-4GlcNAcβ1⌉
                                                                6
                                                              GalNAc
      {NeuAcα2-3Galβ1-4GlcNAcβ1-6}$_{0-4}$⌉ 3
                                                       Galβ1⌋
      NeuAcα2-3{Galβ1-4GlcNAcβ1-3}$_{0-4}$⌋

FIGURE 7.   Structures proposed by Bahl and Wagh[8] for the O-linked oligosaccharides of eCG. Residues denoted by brackets may or may not be present. Sugar chains enclosed in pointed brackets ({}) may be repeated multiple times with the range subscripted.

hCGα are different in structure than those found on native hCG.[105] Interestingly, the O-linked oligosaccharide[103,104,106,107] and perhaps the unique N-linked structures[105] on "free" alpha subunits are thought to render them incapable of forming α-β dimers.

VI. BIOSYNTHESIS OF OLIGOSACCHARIDES

In recent years it has been unequivocally established that the genetic information for each glycoprotein hormone subunit is encoded in a separate gene and that the subunits of the glycoprotein hormones are synthesized individually as presubunits (subunits extended on the N termini by leader sequences). The leader sequences are thought to be cleaved during translation (reviewed by Parsons and Pierce;[3] Weintraub et al.[108]).

The initial events in the biosynthesis of N-linked oligosaccharides occur on the isoprenol lipid dolichol-pyrophosphate where a high mannose oligosaccharide is assembled (reviewed in Snider;[25] Kornfeld and Kornfeld[26]). In the anterior pituitary,[23] placenta,[109,110] and other tissues[25,26] this high mannose oligosaccharide has the following structure:

$$
\begin{array}{l}
\text{Man}\alpha1\text{-}2\text{Man}\alpha1 \\
\qquad\qquad\qquad \diagdown6 \\
\qquad\qquad\qquad\quad \text{Man}\alpha1 \\
\qquad\qquad\qquad \diagup3 \\
\text{Man}\alpha1\text{-}2\text{Man}\alpha1 \\
\qquad\qquad\qquad\qquad\qquad\qquad \diagdown6 \\
\qquad\qquad\qquad\qquad\qquad\qquad\quad \text{Man}\beta1\text{-}4\text{GlcNAc}\beta1\text{-}4\text{GlcNAc} \\
\qquad\qquad\qquad\qquad\qquad\qquad \diagup3 \\
\text{Glc}\alpha1\text{-}2\text{Glc}\alpha1\text{-}3\text{Glc}\alpha1\text{-}3\text{Man}\alpha1\text{-}2\text{Man}\alpha1\text{-}2\text{Man}\alpha1
\end{array}
$$

As the peptide chains are translated, $Glc_3Man_9GlcNAc_2$ sugar chains are transferred *en bloc* from the lipid to appropriate Asn residues. Appropriate acceptors generally occur in the sequence Asn–X–Ser/Thr where X can be any amino acid except possibly proline and aspartic acid. Not all chains which fit this sequence serve as glycosylation sites (reviewed in Snider;[25] Kornfeld and Kornfeld[26]). The glucose residues are trimmed rather rapidly (perhaps during the transfer reaction itself) by glucosidases I and II, enzymes which are located in the rough endoplasmic reticulum. The four α1–2-linked Man residues are then removed by α-mannosidases located in the endoplasmic reticulum and Golgi apparatus (reviewed by Snider;[25] Kornfeld and Kornfeld[26]). Studies of the hormones produced by the pituitary,[23,111,112] placenta,[109,110,113-116] and mouse thyrotropic tumors[108,112,117-121] are consistent with these concepts. In the placenta, the four α1–2-linked Man residues appear to be removed in a specific sequence.[4,109] The resulting $Man_5GlcNAc_2$ oligosaccharides represent a pivotal point in oligosaccharide processing.

With specific regard to the glycoprotein hormones, the addition of high mannose oligosaccharides appears to aid in the proper folding of the subunits[119,122,123] and may protect them from intracellular degradation.[119] Furthermore, the subunits of the glycoprotein hormones appear to combine early in the biosynthetic process while their oligosaccharides are still of the high mannose type.[78,124,125]

The remaining steps in the processing of N-linked oligosaccharides occur in the Golgi apparatus. The next two steps (which may occur interchangeably) in the processing of $Man_5GlcNAc_2$ oligosaccharides are (1) the cleavage of two additional Man residues, one α1–3-linked and one α1–6-linked, by α-mannosidase II and (2) the addition of GlcNAc to the α1–3-linked Man antennae of the di-N-acetylchitobiose trimannosyl core by N-acetyl-glycosaminyltransferase I. Remaining sugar residues in peripheral chains (GlcNAc, GalNAc, Gal, NeuAc, and Fuc) are added by specific glycosyltransferases located in the Golgi apparatus (reviewed by Snider;[25] Kornfeld and Kornfeld[26]). A description of all of the specific reactions involved, some of which control branching of oligosaccharides, is beyond the scope of this chapter. For further information on this topic, the reader is directed to several excellent reviews prepared by Schachter.[126-129] Sulfate, when present, is transferred from

5'-phosphoadenosine 5'-phosphosulfate to GalNAc (and possibly GlcNAc) by a highly specific sulfotransferase which is found in the Golgi apparatus.[27,29,78]

The biosynthesis of O-linked oligosaccharides is thought to occur posttranslationally.[130,131] The linkage GalNAc is transferred to Ser or Thr by polypeptide α-N-acetylgalactosaminyltransferase (reviewed by Sadler;[131] Schachter and Roseman[132]). Additional sugars are presumably added by the specific glycosyltransferases located in the Golgi apparatus.

Based on this general outline, a putative scheme for the processing of sialylated oligosaccharides of the type found on hCG can be envisioned.[4] The oligosaccharides are added as Glc_{1-3} $Man_9GlcNAc_2$ chains which are subsequently trimmed of Glc and four α1–2-linked Man residues. GlcNAc is transferred to the α1–3-linked Man attached to the β-linked Man in the core by N-acetylglucosaminyltransferase I either before or after two additional Man residues are removed. A second GlcNAc is transferred to the α1–6-linked Man by N-acetylglucosaminyltransferase II and the remaining sugars (Gal, NeuAc, Fuc) are added by specific transferases to complete the Asn-linked glycans. O-linked glycans are presumably added and completed in the Golgi.

A putative scheme for the processing of sulfated oligosaccharides has recently been proposed by Green et al.[44] Oligosaccharides are added and trimmed to the $Man_5GlcNAc_2$ stage as described above. GlcNAc is added by N-acetylglucosaminyltransferase I. A GalNAc added to this GlcNAc before further processing prevents removal of additional Man residues resulting in the formation of hybrid structures. An S-1/hybrid oligosaccharide will be formed if the GalNAc is sulfated. Alternatively, the Golgi α-mannosidase can cleave two α-linked Man residues after the initial GlcNAc is added. A second GlcNAc may then be added to the α1–6-linked Man antennae and the processing completed by the addition of two GalNAc and up to two sulfate residues. According to their scheme, S-N oligosaccharides as well as more highly branched structures are generated by the second pathway.[44]

VII. CONCLUDING REMARKS

In the last 10 years significant progress has been made in understanding the oligosaccharides structures associated with the glycoprotein hormones of four species: human, bovine, ovine, and equine. Both the N- and O-linked sugar chains of hCG appear to be defined and the major carbohydrate structures associated with bLH have recently been identified. With regard to N-linked oligosaccharides of these two hormones, it is evident that each possesses several distinct structures which appear to be somewhat systematically distributed. Current evidence suggests that, within a given species, the oligosaccharides of TSH are similar to those of LH but this remains to be definitively established. The sugar chains found on FSH are just beginning to be understood. Besides variations between hormones, there also appear to be significant variations between species for any given hormone.

It is now evident that a variety of Asn-linked oligosaccharide structures are present on the glycoprotein hormones. These range in increasing degrees of complexity from the biantennary disulfated complex oligosaccharides of bLH to the biantennary sulfated/sialylated chains of hLH to the biantennary disialylated complex oligosaccharides of hCG to the multiantennary sugar chains found on FSH and eCG. Similarly, the O-linked glycans of the chorionic gonadotropins range in structure from disaccharides to multibranched sugar chains with up to 50 residues. Thus, the oligosaccharides of the glycoprotein hormones exhibit considerable diversity.

In contrast to DNA, RNA, and proteins which are linear strands of molecules, the oligosaccharides of glycoproteins have the potential to be present in exceedingly complex structures because of their branching. With regard to proteins, DNA serves as a template for RNA synthesis which then serves as a template for protein synthesis. Thus, their primary structure is encoded in the genome and faithfully reproduced during the processes of transcription and translation. In contrast, the biosynthesis of oligosaccharides does not occur from a template or blueprint but results from a complex series of reactions which is capable of generating many diverse structures. The function of diverse oligosaccharide structures on any given glycoprotein is still not established. However, the observation that a single pituitary cell type, the gonadotrope, is capable of synthesizing two distinct glycoprotein hormones which have, at least in part, distinctly different oligosaccharide structures indeed suggests that the processing of oligosaccharides is not a random event. Thus, it is likely that the small changes in molecular structure of any given glycoprotein hormone resulting from subtle differences in its sugar chains may play a role in its functions which we are just beginning to understand.

ACKNOWLEDGMENTS

The author expresses his appreciation to Dr. Laurence A. Cole for reviewing this chapter and to Drs. D. N. Ward and L. A. Cole for providing access to information in press. Studies from the author's laboratory were supported by NIH Grant HD18879.

REFERENCES

1. **Liu, W.-K. and Ward, D. N.,** The purification and chemistry of pituitary glycoprotein hormones, *Pharmacol. Therap. B.*, 1, 545, 1975.
2. **Bahl, O. P. and Shah, R. H.,** Glycoenzymes and glycohormones, in *The Glycoconjugates, Vol. 1*, Horowitz, M. I. and Pigman, W., Eds., Academic Press, New York, 1977, 385.
3. **Pierce, J. G. and Parsons, T. F.,** Glycoprotein hormones: structure and function, *Annu. Rev. Biochem.*, 50, 465, 1981.
4. **Bahl, O. P. and Kalyan, N. K,** Chemistry and biology of placental choriogonadotropin and related peptides, in *Role of Peptides and Proteins in Control of Reproduction*, McCann, S. M. and Dhindsa, D. S., Eds., Elsevier, New York, 1983, 293.
5. **Sairam, M. R.,** Gonadotropic hormones: relationship between structure and function with emphasis on antagonists, in *Hormonal Proteins and Peptides*, Vol. 11, Li, C. H., Ed., Academic Press, New York, 1983, 1.
6. **Rathnam, P. and Saxena, B. B.,** Structure-function relationships of gonadotropins, in *Hormone Receptors in Growth and Reproduction*, Saxena, B. B., Catt, K. J., Birnbaumer, L., and Martini, L., Eds., Raven Press, New York, 1984, 21.
7. **Strickland, T. W., Parsons, T. F., and Pierce, J. G.,** Structure of LH and hCG, in *Luteinizing Hormone Action and Receptors*, Ascoli, M., Ed., CRC Press, Boca Baton, Fla., 1985, 1.
8. **Bahl, O. P. and Wagh, P. V.,** Characterization of glycoproteins: carbohydrate structures of glycoprotein hormones, in *Molecular and Cellular Aspects of Reproduction*, Dhindsa, D. S. and Bahl, O. P., Eds., Plenum Press, New York, 1986, 1.
9. **Green, E. D., Baenziger, J. U., and Boime, I.,** Cell-free sulfation of human and bovine pituitary hormones. Comparison of the sulfated oligosaccharides of lutropin, follitropin, and thyrotropin, *J. Biol. Chem.*, 260, 15361, 1985.
10. **Nilsson, B., Rosen, S. W., Weintraub, B. D. D., and Zopf, D. A.,** Differences in the carbohydrate moieties of the common α-subunits of human chorionic gonadotropin, luteinizing hormone, follicle-stimulating hormone, and thyrotropin: preliminary structural inferences from direct methylation analysis, *Endocrinology*, 119, 2737, 1986.
11. **Strickland, T. W., Thomason, A. R., Nilson, J. H., and Pierce, J. G.,** The common α subunit of bovine glycoprotein hormones: limited formation of native structure by the totally nonglycosylated polypeptide chain, *J. Cell. Biochem.*, 29, 225, 1985.
12. **Chappel, S., Zabrecky, J., Hyman, L., Nugent, N., Gordon, K., and Bernstine, E.,** Characterization of bovine luteinizing hormone (bLH) produced by recombinant DNA technology, *Proc. 68th Annu. Meet. U.S. Endocrine Society*, Abstr. #610, 1986.
13. **Li, Y.-T. and Li, S.-C.,** Use of enzymes in elucidation of structure, in *The Glycoconjugates*, Vol. 1, Horowitz, M. I. and Pigman, W., Eds., Academic Press, New York, 1977, 51.

14. **Kobata, A.,** Use of endo- and exoglycosidases for structural studies of glycoconjugates, *Anal. Biochem.,* 100, 1, 1979.
15. **Kornfeld, R. and Kornfeld, S.,** Structure of glycoproteins and their oligosaccharide units, in *The Biochemistry of Glycoproteins and Proteoglycans,* Lennarz, W. J., Ed., Plenum Press, New York, 1980, 1.
16. **Kobata, A.,** The carbohydrates of glycoproteins, in *Biology of Carbohydrates,* Vol. 2, Ginsburg, V. and Robbins, P. W., Eds., John Wiley & Sons, New York, 1984, 87.
17. **Koide, N. and Muramatsu, T.,** Endo-β-*N*-acetylglucosaminidase acting on carbohydrate moieties of glycoproteins. Purification and properties of the enzyme from *Diplococcus pneumoniae, J. Biol Chem.,* 249, 4897, 1974.
18. **Kessler, M. J., Mise, T., Ghai, R. D., and Bahl, O. P.,** Structure and location of the *O*-glycosidic carbohydrate units of human chorionic gonadotropin, *J. Biol. Chem.,* 254, 7909, 1979.
19. **Bedi, G. S., French, W. C., and Bahl, O. P.,** Structure of carbohydrate units of ovine luteinizing hormone, *J. Biol. Chem.,* 257, 4345, 1982.
20. **Takasaki, S., Mizuochi, T., and Kobata, A.,** Hydrazinolysis of asparagine-linked sugar chains to produce free oligosaccharides, *Methods Enzymol.,* 83, 263, 1982.
21. **Hakomori, S.-I.,** A rapid permethylation of glycolipid, and polysaccharide catalyzed by methylsulfinyl carbanion in dimethyl sulfoxide, *J. Biochem.,* 55, 205, 1964.
22. **Goldstein, I. J., Hay, G. W., Lewis, B. A., and Smith, F.,** Controlled degradation of polysaccharides by periodate oxidation, reduction, and hydrolysis, *Methods Carbohydr. Chem.,* 5, 361, 1965.
23. **Henner, J. A., Kessler, M. J., and Bahl, O. P.,** Glycoprotein biosynthesis in calf pituitary. Chemical characterization of mannose- and glucose-radiolabeled oligosaccharides, *J. Biol. Chem.,* 256, 5997, 1981.
24. **Cummings, R. D. and Kornfeld, S.,** Fractionation of asparagine-linked oligosaccharides by serial lectin-agarose affinity chromatography: a rapid, sensitive, and specific technique, *J. Biol. Chem.,* 257, 11235, 1982.
25. **Snider, M. D.,** Biosynthesis of glycoproteins: formation of N-linked oligosaccharides, in *Biology of Carbohydrates,* Vol. 2, Ginsburg, V. and Robbins, P. W., Eds., John Wiley & Sons, New York, 1984, 163.
26. **Kornfeld, R. and Kornfeld, S.,** Assembly of asparagine-linked oligosaccharides, *Annu. Rev. Biochem.,* 54, 631, 1985.
27. **Hortin, G., Natowicz, M., Pierce, J., Baenziger, J., Parsons, T., and Boime, I.,** Metabolic labeling of lutropin with [^{35}S]sulfate, *Proc. Natl. Acad. Sci. U.S.A.,* 78, 7468, 1981.
28. **Anumula, K. R. and Bahl, O. P.,** Biosynthesis of lutropin in ovine pituitary slices: incorporation of [^{35}S]sulfate in carbohydrate units, *Arch. Biochem. Biophys.,* 220, 645, 1983.
29. **Green, E. D., Gruenebaum, J., Bielinska, M., Baenziger, J. U., and Boime, I.,** Sulfation of lutropin oligosaccharides with a cell-free system, *Proc. Natl. Acad. Sci. U.S.A.,* 81, 5320, 1984.
30. **Spencer, B.,** The ultramicro determination of inorganic sulfate, *Biochem. J.,* 75, 435, 1960.
31. **Lelior, L. F. and Cardini, C. E.,** Characterization of phosphorus compounds by acid lability, *Methods Enzymol.,* 3, 840, 1957.
32. **Grotjan, H. E., Jr., Padrnos-Hicks, P. A., and Keel, B. A.,** Ion chromatographic method for the analysis of sulfate in complex carbohydrates, *J. Chromatogr.,* 367, 367, 1986.
33. **Ward, D. N., Wen, T., and Bousfield, G. R.,** Sulfate and phosphate analysis in glycoproteins and other biologic compounds using ion chromatography: application to glycoprotein hormones and sugar esters, *J. Chromatogr.,* 398, 255, 1987.
34. **Eipper, B. A. and Mains, R. E.,** Structure and biosynthesis of pro-adrenocorticotropin/endorphin and related peptides, *Endocr. Rev.,* 1, 1, 1980.
35. **Sinha, Y. N. and Lewis, U. J.,** A lectin-binding immunoassay indicates a possible glycosylated growth hormone in the human pituitary gland, *Biochem. Biophys. Res. Comm.,* 140, 491, 1986.
36. **Parsons, T. F. and Pierce, J. G.,** Oligosaccharide moieties of glycoprotein hormones: bovine lutropin resists enzymatic deglycosylation because of terminal O-sulfated *N*-acetylhexosamines, *Proc. Natl. Acad. Sci. U.S.A.,* 77, 7089, 1980.
37. **Cole, L. A., Metsch, L. S., and Grotjan, H. E., Jr.,** Significant steroidogenic activity of luteinizing hormone is maintained following enzymatic removal of oligosaccharides, *Mol. Endocrinol.,* 1, 621, 1987.
38. **Pierce, J. G., Liao, T.-H., Howard, S. M., Shome, B., and Cornell, J. S.,** Studies on the structure of thyrotropin: its relationship to luteinizing hormone, *Rec. Prog. Horm. Res.,* 27, 165, 1971.
39. **Carlsen, R. B. and Pierce, J. G.,** Purification and properties of an α-ʟ-fucosidase from rat epididymis, *J. Biol. Chem.,* 247, 23, 1972.
40. **Bahl, O. P., Reddy, M. S., and Bedi, G. S.,** A novel carbohydrate structure in bovine and ovine luteinizing hormones, *Biochem. Biophys. Res. Comm.,* 96, 1192, 1980.
41. **Chu, W. P., Liu, W.-K., and Ward, D. N.,** Carbohydrate components of the glycopeptides of ovine lutropin, *Biochim. Biophys. Acta,* 437, 377, 1976.
42. **Papkoff, H., Gospodarowicz, D., Candiotti, A., and Li, C. H.,** Preparation of ovine interstitial cell-stimulating hormone in high yield, *Arch. Biochem. Biophys.,* 111, 431, 1965.

43. **Green, E. D., van Halbeek, H., Boime, I., and Baenziger, J. U.**, Structural elucidation of the disulfated oligosaccharide from bovine lutropin, *J. Biol. Chem.*, 260, 15623, 1985.

44. **Green, E. D., Boime, I., and Baenziger, J. U.**, Biosynthesis of sulfated asparagine-linked oligosaccharides on bovine lutropin, *J. Biol. Chem.*, 261, 16309, 1986.

45. **Elder, J. H. and Alexander, S.**, Endo-β-N-acetylglycosaminidase F: endoglycosidase from *Flavobacterium meninosepticum* that cleaves both high-mannose and complex glycoproteins, *Proc. Natl. Acad. Sci. U.S.A.*, 79, 4540, 1982.

46. **Plummer, J. H., Jr., Elder, J. H., Alexander, S., Phelan, A. W., and Tarentino, A. L.**, Demonstration of peptide: N-glycosidase F activity in endo-β-N-acetylglucosaminidase F preparations, *J. Biol. Chem.*, 259, 10700, 1984.

47. **Tarentino, A. L., Gomez, C. M., and Plummer, T. H., Jr.**, Deglycosylation of asparagine-linked glycans by peptide: N-glycosidase F, *Biochemistry*, 24, 4665, 1985.

48. **Tarentino, A. L. and Plummer, T. H., Jr.**, Oligosaccharide accessibility to peptide: N-glycosidase as promoted by protein-unfolding reagents, *J. Biol. Chem.*, 257, 10776, 1982.

49. **Baenziger, J. U. and Fiete, D.**, Structural determinants of concanavalin A specificity for oligosaccharides, *J. Biol. Chem.*, 254, 2400, 1979.

50. **Grotjan, H. E., Jr. and Cole, L. A.**, Ovine and human luteinizing hormone enzymatically deglycosylated with endoglycosidase F are biologically active, *Ann. N. Y. Acad. Sci.*, in press.

51. **Lee, K.-O., Gesundheit, N., Chen, H.-C., and Weintraub, B. D.**, Enzymatic deglycosylation of thyroid-stimulating hormone with peptide N-glycosidase F and endo-β-N-acetylglycosaminidase F, *Biochem. Biophys. Res. Comm.*, 138, 230, 1986.

52. **Condliffe, P. G. and Weintraub, B. D.**, Pituitary thyroid-stimulating hormone and other thyroid-stimulating substances, in *Hormones in Blood*, Vol. 1, 3rd ed., Gray, C. H. and James, V. H. T., Eds., Academic Press, New York, 1979, 499.

53. **Pierce, J. G. and Wynston, L. K.**, On the composition of sheep thyrotropic hormone, *Biochim. Biophys. Acta*, 43, 538, 1960.

54. **Sairam, M. R. and Li, C. H.**, Human pituitary lutropin. Isolation, properties, and the complete amino acid sequence of the β-subunit, *Biochim. Biophys. Acta*, 412, 70, 1975.

55. **Hara, K., Rathnam, P., and Saxena, B. B.**, Structure of the carbohydrate moieties of α subunits of human follitropin, lutropin, and thyrotropin, *J. Biol. Chem.*, 253, 1582, 1978.

56. **Kennedy, J. F., Chaplin, M. F., and Stacey, M.**, Periodate oxidation, acid hydrolysis, and structure-activity relationships of human-pituitary, follicle-stimulating hormone and human chorionic gonadotrophin, *Carbohydr. Res.*, 36, 369, 1974.

57. **Tolvo, A., Fujiki, Y., Bhavanandan, V. P., Rathnam, P., and Saxena, B. B.**, Studies on the unique presence of an N-acetylgalactosamine residue in the carbohydrate moieties of human follicle-stimulating hormone, *Biochim. Biophys. Acta*, 719, 1, 1982.

58. **Cornell, J. S. and Pierce, J. G.**, The subunits of human pituitary thyroid-stimulating hormone. Isolation, properties, and composition, *J. Biol. Chem.*, 248, 4327, 1973.

59. **Grimek, H. J. and McShan, W. H.**, Isolation and characterization of the subunits of highly purified ovine follicle-stimulating hormone, *J. Biol. Chem.*, 249, 5725, 1974.

60. **Grimek, H. J., Gorski, J., and Wentworth, B. C.**, Purification and characterization of bovine follicle-stimulating hormone: comparison with ovine follicle-stimulating hormone, *Endocrinology*, 104, 140, 1979.

61. **Kennedy, J. F. and Butt, W. R.**, Periodate oxidation studies of human pituitary follicle-stimulating hormone, *Biochem. J.*, 115, 225, 1969.

62. **Chaplin, M. F., Gray, C. J., and Kennedy, J. F.**, Chemical studies on a pituitary FSH preparation, in *Gonadotropins and Ovarian Development*, Butt, W. R., Crooke, A. C., and Ryle, M., Eds., Livingstone, Edinburgh, 1970, 77.

63. **Kennedy, J. F. and Chaplin, M. F.**, Structural investigation of the carbohydrate element of human pituitary follicle-stimulating hormone by methylation analysis, *Biochem. J.*, 130, 417, 1972.

64. **Kennedy, J. F. and Chaplin, M. F.**, The structures of the carbohydrate moieties of the α subunit of human chorionic gonadotropin, *Biochem. J.*, 155, 303, 1976.

65. **Sairam, M. R.**, Primary structure of the ovine pituitary follitropin α-subunit, *Biochem. J.*, 197, 535, 1981.

66. **Sairam, M. R., Seidah, N. G., and Chretien, M.**, Primary structure of the ovine pituitary follitropin β-subunit, *Biochem. J.*, 197, 541, 1981.

67. **Hennen, G., Prusik, Z., and Maghuin-Rogister, G.**, Porcine luteinizing hormone and its subunits. Isolation and characterization, *Eur. J. Biochem.*, 18, 376, 1971.

68. **Kim, J. H., Shome, B., Liao, T.-H., and Pierce, J. G.**, Analysis of neutral sugars by gas-liquid chromatography of alditol acetates: application to thyrotropic hormone and other glycoproteins, *Anal. Biochem.*, 20, 258, 1967.

69. **Closset, J. and Hennen, G.**, Porcine follitropin. Isolation and characterization of the native hormone and its α and β subunits, *Eur. J. Biochem.*, 86, 105, 1978.

70. **Bousfield, G. R., Sugino, H., and Ward, D. N.,** Demonstration of a COOH-terminal extension on equine lutropin by means of a common acid-labile bond in equine lutropin and equine chorionic gonadotropin, *J. Biol. Chem.*, 260, 9531, 1985.

71. **Landefeld, T. D. and McShan, W. H.,** Equine luteinizing hormone and its subunits. Isolation and physicochemical properties, *Biochemistry*, 13, 1389, 1974.

72. **Landefeld, T. D. and McShan, W. H.,** Isolation and characterization of subunits from equine pituitary follicle-stimulating hormone, *J. Biol. Chem.*, 249, 3527, 1974.

73. **Ward, D. N., Reichert, L. E., Jr., Fitak, B. A., Nahm, H. S., Sweeney, C. M., and Neill, J. D.,** Isolation and properties of rat pituitary luteinizing hormone, *Biochemistry*, 10, 1796, 1971.

74. **Ward, D. N., Desjardins, C., Moore, W. T., Jr., and Nahm, H. S.,** Rabbit lutropin: preparation, characterization of the hormone, its subunits and radioimmunoassay, *Int. J. Pept. Protein Res.*, 13, 62, 1979.

75. **Shome, B., Parlow, A. F., Ramirez, V. D., Elrick, H., and Pierce, J. G.,** Human and porcine thyrotropins: a comparison of electrophoretic and immunological properties with the bovine hormone, *Arch. Biochem. Biophys.*, 126, 444, 1968.

76. **Papkoff, H.,** Equine thyrotropin, in *The Anterior Pituitary Gland*, Bhatnagar, A. S., Ed., Raven Press, New York, 1983, 191.

77. **Hattori, M.-A., Ozawa, K., and Wakabayashi, K.,** Sialic acid moiety is responsible for the charge heterogeneity and the biological potency of rat lutropin, *Biochem. Biophys. Res. Comm.*, 127, 501, 1985.

78. **Hoshina, H. and Boime, I.,** Combination of rat lutropin subunits occurs early in the secretory pathway, *Proc. Natl. Acad. Sci. U.S.A.*, 79, 7649, 1982.

79. **Ulloa-Aquirre, A., Miller, C., Hyland, L., and Chappel, S.,** Production of all follicle-stimulating hormone isohormones from a purified preparation by neuraminidase digestion, *Biol. Reprod.*, 30, 382, 1984.

80. **Blum, W. F. P., Riegelbauer, G., and Gupta, D.,** Heterogeneity of rat FSH by chromatofocusing: studies on in-vitro bioactivity of pituitary FSH forms and effect of neuraminidase treatment, *J. Endocrinol.*, 105, 17, 1985.

81. **Chappel, S.,** The presence of two species of follicle-stimulating hormone within hamster anterior pituitary glands as disclosed by concanavalin A chromatography, *Endocrinology*, 109, 935, 1981.

82. **Ulloa-Aquirre, A. and Chappel, S. C.,** Multiple species of follicle-stimulating hormone exist within the anterior pituitary gland of male golden hamsters, *J. Endocrinol.*, 95, 257, 1982.

83. **Peckham, W. D. and Knobil, E.,** The effects of ovariectomy, estrogen replacement, and neuraminidase treatment on the properties of the adenohypophysial glycoprotein hormones of the rhesus monkey, *Endocrinology*, 98, 1054, 1976.

84. **Chappel, S. C., Bethea, C. L., and Spies, H. G.,** Existence of multiple forms of follicle-stimulating hormone within the anterior pituitaries of cynomolgus monkeys, *Endocrinology*, 115, 452, 1984.

85. **Moore, W. T., Jr., Burleigh, B. D., and Ward, D. N.,** Chorionic gonadotropins: comparative studies and comments on relationships to other glycoprotein hormones, in *Chorionic Gonadotropin*, Segal, S. J., Ed., Plenum Press, New York, 1980, 89.

86. **Kessler, M. J., Reddy, M. S., Shah, R. H., and Bahl, O. P.,** Structures of N-glycosidic carbohydrate units of human chorionic gonadotropin, *J. Biol. Chem.*, 254, 7901, 1979.

87. **Christakos, S. and Bahl, O. P.,** Pregnant mare serum gonadotropin. Purification and physicochemical, biological, and immunological characterization, *J. Biol. Chem.*, 254, 4253, 1979.

88. **Bahl, O. P., Marz, L., and Kessler, M. J.,** Isolation and characterization of N- and O-glycosidic carbohydrate units of human chorionic gonadotropin, *Biochem. Biophys. Res. Comm.*, 84, 667, 1978.

89. **Morgan, F. J., Birken, S., and Canfield, R. E.,** The amino acid sequence of human chorionic gonadotropin. The α subunit and β subunit, *J. Biol. Chem.*, 250, 5247, 1975.

90. **Bahl, O. P.,** Human chorionic gonadotropin. II. Nature of the carbohydrate units, *J. Biol. Chem.*, 244, 575, 1969.

91. **Bahl, O. P.,** Chemistry of human chorionic gonadotropin, in *Hormonal Proteins and Peptides*, Vol. 1, Li, C. H., Ed., Academic Press, New York, 1973, 171.

92. **Endo, Y., Yamashita, K., Tachibana, Y., Tojo, S., and Kobata, A.,** Structures of the asparagine-linked sugar chains of human chorionic gonadotropin, *J. Biochem.*, 85, 669, 1979.

93. **Mizuochi, T. and Kobata, A.,** Different asparagine-linked sugar chains on the two polypeptide chains of human chorionic gonadotropin, *Biochem. Biophys. Res. Comm.*, 97, 772, 1980.

94. **Mizuochi, T., Nishimura, R., Taniguchi, T., Utsunomiya, T., Mochizuki, M., Derappe, C., and Kobata, A.,** Comparison of carbohydrate structure between human chorionic gonadotropin present in urine of patients with trophoblastic diseases and healthy individuals, *Jpn. J. Cancer Res.*, 76, 752, 1985.

95. **Mizuochi, T., Nishimura, R., Derappe, C., Taniguchi, T., Hamamoto, T., Mochizuki, M., and Kobata, A.,** Structures of the asparagine-linked sugar chains of human chorionic gonadotropin produced in choriocarcinoma. Appearance of triantennary sugar chains and unique biantennary sugar chains, *J. Biol. Chem.*, 258, 14126, 1983.

96. **Peters, B. P., Brooks, M., Hartle, R. J., Krzesicki, R. F., Perini, F., and Ruddon, R. W.,** The use of drugs to dissect the pathway for secretion of glycoprotein hormone chorionic gonadotropin by cultured human trophoblastic cells, *J. Biol. Chem.,* 258, 14505, 1983.

97. **Cole, L. A., Birken, S., and Perini, F.,** The structures of the serine-linked sugar chains on human chorionic gonadotropin, *Biochem. Biophys. Res. Comm.,* 126, 333, 1985.

98. **Cole, L. A.,** O-glycosylation of proteins in the normal and neoplastic trophoblast, *Troph. Res.,* 2, 139, 1987.

99. **Cole, L. A.,** Distribution of O-linked sugar units on hCG and its free α subunit, *Mol. Cell. Endocrinol.,* 50, 45, 1987.

100. **Ward, D. N., Moore, W. T., Jr., and Burleigh, B. D.,** Structural studies on equine chorionic gonadotropin, *J. Prot. Chem.,* 1, 263, 1982.

101. **Fukuda, M. N.,** Isolation and characterization of a new endo-β-galactosidase from *Diplococcus penumoniae, Biochemistry,* 24, 2154, 1985.

102. **Aggarwal, B. B., Farmer, S. W., Papkoff, H., Stewart, F., and Allen, W. R.,** Purification and characterization of donkey chorionic gonadotropin, *J. Endocrinol.,* 85, 449, 1980.

103. **Parsons, T. F., Bloomfield, G. A., and Pierce, J. G.,** Purification of an alternative form of the α subunit of the glycoprotein hormones from bovine pituitaries and identification of its O-linked oligosaccharide, *J. Biol. Chem.,* 258, 240, 1983.

104. **Parsons, T. F. and Pierce, J. G.,** Free α-like material from bovine pituitaries. Removal of its O-linked oligosaccharide permits combination with lutropin-β, *J. Biol. Chem.,* 259, 2662, 1984.

105. **Blithe, D. L. and Nisula, B. C.,** Variations in the oligosaccharides on free and combined α-subunits of human choriogonadotropin in pregnancy, *Endocrinology,* 117, 2218, 1985.

106. **Cole, L. A., Perini, F., Birken, S., and Ruddon, R. W.,** An oligosaccharide of the O-linked type distinguishes the free from the combined form of hCG α subunit, *Biochem. Biophys. Res. Comm.,* 122, 1260, 1984.

107. **Corless, C. and Boime, I.,** Differential secretion of O-glycosylated gonadotropin α-subunit and luteinizing hormone (LH) in the presence of LH-releasing hormone, *Endocrinology,* 117, 1699, 1985.

108. **Weintraub, B. D., Stannard, B. S., Magner, J. A., Ronin, C., Taylor, T., Joshi, L., Constant, R. B., Menezes-Ferreira, M. M., Petrick, P., and Gesundheit, N.,** Glycosylation and posttranslational processing of thyroid-stimulating hormone: clinical implications, *Rec. Prog. Horm. Res.,* 41, 577, 1985.

109. **Henner, J. A., French, W. C., and Bahl, O. P.,** Biosynthesis of glycoproteins in human placenta: processing of oligosaccharides, *Arch. Biochem. Biophys.,* 224, 601, 1983.

110. **French, W. C., Henner, J. A., and Bahl, O. P.,** Biosynthesis of glycoproteins in human placenta: differential labeling of mannose and heterogeneity of oligosaccharide lipid intermediates, *Arch. Biochem. Biophys.,* 230, 560, 1984.

111. **Landefeld, T. D. and Kepa, J.,** The cell free synthesis of bovine lutropin β subunit, *Biochem. Biophys. Res. Comm.,* 90, 1111, 1979.

112. **Ronin, C., Stannard, B. S., and Weintraub, B. D.,** Differential processing and regulation of thyroid-stimulating hormone subunit carbohydrate chains in thyrotropic tumors and in normal and hypothyroid pituitaries, *Biochemistry,* 24, 5626, 1985.

113. **Bielinska, M. and Boime, I.,** mRNA-dependent synthesis of a glycosylated subunit of human chorionic gonadotropin in cell-free extracts derived from ascites tumor cells, *Proc. Natl. Acad. Sci. U.S.A.,* 75, 1768, 1978.

114. **Bielinska, M. and Boime, I.,** Glycosylation of human chorionic gonadotropin in mRNA-dependent cell-free extracts: post-translational processing of an asparagine-linked mannose-rich oligosaccharide, *Proc. Natl. Acad. Sci. U.S.A.,* 76, 1208, 1979.

115. **Ruddon, R. W., Hanson, C. A., and Addison, N. J.,** Synthesis and processing of human chorionic gonadotropin subunits in cultured choriocarcinoma cells, *Proc. Natl. Acad. Sci. U.S.A.,* 76, 5143, 1979.

116. **Ruddon, R. W., Hartle, R. J., Peters, B. P., Anderson, C., Huot, R. I., and Stromberg, K.,** Biosynthesis and secretion of chorionic gonadotropin subunits by organ cultures of first trimester human placenta, *J. Biol. Chem.,* 256, 11389, 1981.

117. **Weintraub, B. D., Stannard, B. S., Linnekin, D., and Marshall, M.,** Relationship of glycosylation to de novo thyroid-stimulating hormone biosynthesis and secretion by mouse pituitary tumor cells, *J. Biol. Chem.,* 255, 5715, 1980.

118. **Chin, W. W. and Habener, J. F.,** Thyroid-stimulating hormone subunits: evidence from endoglycosidase-H cleavage for late presecretory glycosylation, *Endocrinology,* 108, 1628, 1981.

119. **Weintraub, B. D., Stannard, B. S., and Meyers, L.,** Glycosylation of thyroid-stimulating hormone in pituitary tumor cells: influence of high mannose oligosaccharide units on subunit aggregation, combination and intracellular degradation, *Endocrinology,* 112, 1331, 1983.

120. **Ronin, C., Stannard, B. S., Rosenbloom, I. L., Magner, J. A., and Weintraub, B. D.,** Glycosylation and processing of high-mannose oligosaccharides of thyroid-stimulating hormone subunits: comparison to nonsecretory cell glycoproteins, *Biochemistry,* 23, 4503, 1984.

121. **Magner, J. A., Ronin, C., and Weintraub, B. D.,** Carbohydrate processing of thyrotropin differs from that of free α-subunit and total glycoproteins in microsomal subfractions of mouse pituitary tumor, *Endocrinology*, 115, 1019, 1984.

122. **Strickland, T. W. and Pierce, J. G.,** The α subunit of pituitary glycoprotein hormones. Formation of three-dimensional structure during cell-free biosynthesis, *J. Biol. Chem.*, 258, 5927, 1983.

123. **Strickland, T. W. and Pierce, J. G.,** The β subunits of glycoprotein hormones. Formation of three-dimensional structure during cell-free biosynthesis of lutropin-β, *J. Biol. Chem.*, 260, 5816, 1985.

124. **Magner, J. A. and Weintraub, B. D.,** Thyroid-stimulating hormone subunit processing and combination in microsomal subfractions of mouse pituitary tumor, *J. Biol. Chem.*, 257, 6709, 1982.

125. **Peters, B. P., Krzesicki, R. F., Hartle, R. J., Perini, F., and Ruddon, R. W.,** A kinetic comparison of the processing and secretion of the αβ dimer and the uncombined α and β subunits of chorionic gonadotropin synthesized by human choriocarcinoma cells, *J. Biol. Chem.*, 259, 15123, 1984.

126. **Schachter, H., Narashimhan, S., Gleeson, P., and Vella, G.,** Control of branching during the biosynthesis of asparagine-linked oligosaccharides, *Can. J. Biochem. Cell. Biol.*, 61, 1049, 1983.

127. **Schachter, H.,** Glycoproteins: their structure, biosynthesis and possible clinical implications, *Clin. Biochem.*, 17, 3, 1984.

128. **Schachter, H.,** Biosynthetic controls that determine the branching and microheterogeneity of protein-bound oligosaccharides, *Can. J. Biochem. Cell. Biol.*, 64, 163, 1986.

129. **Schachter, H.,** Biosynthetic controls that determine the branching and microheterogeneity of protein-bound oligosaccharides, in *Molecular and Cellular Aspects of Reproduction*, Dhindsa, D. S. and Bahl, O. P., Eds., Plenum Press, New York, 1986, 53.

130. **Hanover, J. A., Elting, J., Mintz, G. R., and Lennarz, W. J.,** Temporal aspects of the N- and O-glycosylation of human chorionic gonadotropin, *J. Biol. Chem.*, 257, 10172, 1982.

131. **Sadler, J. E.,** Biosynthesis of glycoproteins: formation of O-linked oligosaccharides, in *Biology of Carbohydrates*, Vol. 2, Ginsburg, V. and Robbins, P. W., Eds., John Wiley & Sons, New York, 1984, 199.

132. **Schachter, H. and Roseman, S.,** Mammalian glycosyltransferases. Their role in the synthesis and function of complex carbohydrates and glycolipids, in *The Biochemistry of Glycoproteins and Proteoglycans*, Lennarz, W. J., Ed., Plenum Press, New York, 1980, 85.

Chapter 3

OCCURRENCE AND PROPERTIES OF GLYCOPROTEIN HORMONE FREE SUBUNITS

Laurence A. Cole

TABLE OF CONTENTS

I. INTRODUCTION

The glycoprotein hormones, pituitary lutropin (LH), follitropin (FSH), and thyrotropin (TSH), and placental choriogonadotropin (CG), are all noncovalently linked heterodimers. Each hormone is composed of a like alpha subunit, and a differing beta subunit which infers the unique biological and immunological activities.[1,2] The subunits, in addition to being heterogeneous in structure, as in isohormones, are also heterogeneous in state, existing in both an uncombined or free form, and as part of an alpha-beta dimer.

Uncombined placental or pituitary glycoprotein hormone subunits are present in tissues and culture fluids, and in normal or pregnancy sera and urine,[3-47] however, they have no, or as yet undefined, biological activities.[7,8] Oligosaccharide structural differences have been reported between combined and uncombined subunits,[9-25] and changing balances of free and combined alpha and beta have been associated with different stages of pregnancy and reproductive development.[26-32] Elevated levels of free subunits have been associated with pituitary and placental neoplasms and with thyroid and gonadal dysfunction.[33-38] Ectopic free alpha and free hCG beta subunit have been detected in the sera and tissues of subjects with a variety of nonendocrine neoplasms.[39-46] Such findings have generated much interest in glycoprotein hormone free subunits, and their possible roles as clinical markers.

Small forms of glycoprotein hormone subunits have been detected in urines.[48-53] The small form of hCG beta subunit found in pregnancy urine (beta core fragment), has been purified, and has been shown to consist of two fragments, residues 6 to 40 and 55 to 92, disulfide-linked.[48] The fragments present in urine also have potential applications as tumor markers.[50-52]

This article reviews current research on glycoprotein hormone free subunits and fragments, their production and detection, how they are generated, their physical and chemical properties, their clearance from circulation, and their possible biological functions. Since levels of circulating hCG and its subunits and fragments in normal gestation well exceed those of the pituitary hormones in the nonpregnant subject, occurrence and properties of placental glycoprotein hormone subunits and fragments is best documented. As such, this article will concentrate on hCG subunits and fragments, although wherever possible, analogies with data published on free TSH, FSH, and LH subunits and fragments will be discussed.

II. OCCURRENCE OF FREE GLYCOPROTEIN HORMONE SUBUNITS

The free common alpha subunit of the glycoprotein hormones has been detected in extracts and cultures of pituitary and placental tissues, and in sera or urines of humans,[1,2,17-20,24-33,41,42] rodents,[4,5,12,13,23] and cattle.[10,15] Lesser amounts of the four free beta subunits have also been detected in cultures of pituitary and placental tissues, and in the sera of humans,[24,25,27,28,32,33,35,37,43,45,46] and rodents.[4,5]

Figures 1 and 2 show the results of recent studies in the author's laboratory in which hCG dimer and free subunit levels were individually quantitated in 196 serum samples from 81 subjects at different stages of pregnancy. hCG was measured using the Hybritech "Tandem" assay and free beta and free alpha subunit by radioimmunoassay using free subunit-specific antibodies.[47] Figure 1 shows levels in serum from participants in the in vitro fertilization program at Yale-New Haven Hospital 0 to 30 days after embryo transfer (2 days postovulation). As shown, exogenous hCG administered 3 days before embryo transfer to induce ovulation is cleared from the circulation by day 10. Endogenous hCG is first detected (>0.01 nmol/ℓ) 1 to 2 days afterwards. hCG levels rise sharply in the following weeks, doubling approximately every 2 days. Free beta is not detected (>0.02 nmol/ℓ) in the circulation until a week or more after the appearance of exogenous hCG. In our laboratory, serum samples were examined from 36 woman achieving pregnancy by in vitro fertilization. The average time of first detection of free beta was 21 days after embryo transfer (range: day 15 to 28). Free alpha cannot be detected

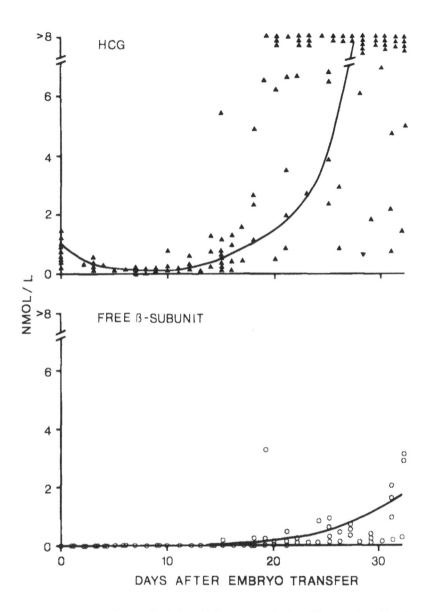

FIGURE 1. hCG and free beta levels in multiple serum samples from 36 women that achieved pregnancy following in vitro fertilization and embryo transfer. To accommodate individual variation, lines were drawn from 9-day (±4 days) averages. hCG was measured using the Hybritech "Tandem" immunoradiometric assay. Free alpha was quantitated by radioimmunoassay using antisera H7 (generated against choriocarcinoma free alpha, gift from R.O. Hussa of the Medical College of Wisconsin) and dissociated ^{125}I-hCG tracer (hCG cross-reactivity <0.5%). Free beta was measured in the radioimmunoassay using the monoclonal antibody 1E5 (gift from Hybritech) and ^{125}I-dissociated hCG beta subunit tracer (hCG cross-reactivity <0.2%).

(<0.02 nmol/ℓ) in early pregnancy serum (0 to 30 days after embryo transfer). Figure 2 shows levels of hCG and free subunits in serum from regular pregnancies, 17 to 280 days after ovulation. hCG and free beta levels rise to a peak at 45 to 60 days after ovulation, and diminish thereafter. Free beta concentration averages 2.6% of hormone level (nmol/nmol) up to the time of the hCG peak, but average just 0.5% of hormone level in the months thereafter. Free alpha is first detected in serum (<0.04 nmol/ℓ) at the time of the hCG peak. Free alpha levels rise

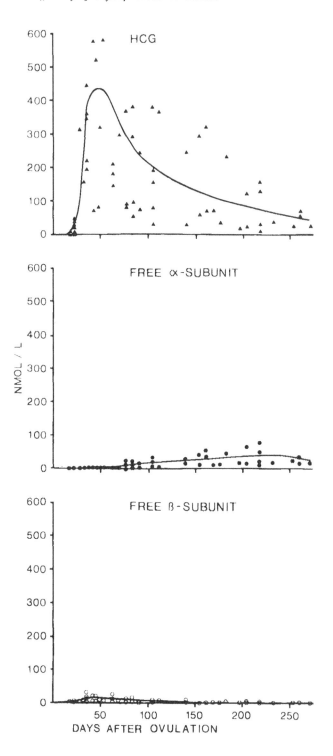

FIGURE 2. hCG, free alpha, and free beta levels in multiple serum samples from 45 women during normal pregnancy. Day of ovulation was estimated as 14 days after the start of the last menstrual period. To accommodate individual variation, lines were drawn from 9-day (±4 days) averages.

continuously from this time until term. At term, the serum free alpha level averages 54% of the hCG concentration (nmol/nmol). Other laboratories have also observed higher proportions of free beta to hCG in early and of free alpha to hCG in late gestation. Hay[32] examined serum from women with early in vitro fertilization pregnancies, and found free beta levels close to those of hCG. Elegbe et al.,[29] Vaitukaitis,[30] and Ashitaka et al.[54] have all reported a rise in free alpha levels in the 2nd and 3rd trimesters of pregnancy, similar to that shown in Figure 2. Considering the levels of free beta in early and of free alpha in late pregnancy, codetermination of hCG and free subunits in sera may aid the general evaluation of placental function, and in the diagnosis threatened abortion and ectopic pregnancy.

The metabolic clearance rate of uncombined alpha subunit is 23 times greater and of uncombined beta subunit is 9 times greater than that of hCG.[55] Figure 3 shows hCG and free-subunit levels secreted by the placenta each day, calculated from serum concentrations (nmol/ℓ) and metabolic clearance rates (l/day). As shown, free beta secretion averages 1/3 that of hCG in the 1st trimester of pregnancy and free alpha level is 12 times greater than hCG at term.[56] The presence of hCG and 10 to 50% free beta (% of free + combined) in the extracts or explant culture fluids of early 1st trimester placentae,[28,35,55] and of 5 to 10 times more free than combined alpha-subunit in explants of term tissues,[28,55,56] confirms the finding of high proportions of free subunit in early pregnancy and at term. These studies illustrate the magnitude of free subunit production in normal gestation and, considering the differing timings of free alpha and beta production, suggest the independent regulation of subunit synthesis. The existence of free or superfluous subunit production illustrates the intricacy, and possibly the inefficiency, of a multigene heterodimeric hormone-generating system.

hCG free subunits are also found, sometimes in abnormal proportions, in the circulation of subjects with gestational disorders. Several investigators have reported higher proportions of free to combined beta subunit in the sera of trophoblast disease patients compared with those with normal pregnancy.[35,57-60,98-100] Ozturk et al.,[59,98] using a free beta assay with zero hCG cross-reactivity, found hCG and an average of 0.62% free beta in pregnancy sera (n = 62), and an average of 1.8% free beta in hydatidiform mole sera (n = 9), and an average of 9.2% free beta in patients with choriocarcinoma (n = 7). Fan et al.[99] measured free beta using an assay with 4% hCG cross-reactivity, and detected higher levels of free beta in trophoblast disease patients than Ozturk and colleagues. While not detecting any free beta in pregnancy sera, they measured hCG and 8.8% free beta in hydatidiform mole (n = 15) and hCG and 19% free beta in patients with choriocarcinoma (n = 28). In our laboratory we used the radioimmunoassay with monoclonal antibody 1E5 (gift from Hybritech) to measure free beta (hCG cross-reactivity <0.2%) in serum samples. As shown in Figures 1, 2, and Table 1, hCG and an average of 1.1% free beta was found in pregnancy serum (n = 196), hCG and an average of 0.27% free beta was detected in hydatidiform mole patients preevacuation (n = 12, all underwent spontaneous remission), and hCG and an average of 5.0% free beta in patients pretherapy for choriocarcinoma (n = 5). While a significant difference (p <0.0001 in t test) was found between pregnancy or hydatidiform mole and the choriocarcinoma group, no statistical difference (p <0.05) was apparent between the former two groups (pregnancy or benign disease). We examined serum samples from subjects (all underwent spontaneous remission) 2 weeks after the evacuation of hydatidiform mole (Table 2), and found that that the percent free beta in the circulation had increased more than 20-fold (average 0.14% free beta preevacuation, 3.4% 2 weeks post evacuation). We speculate that after all hCG-producing tissue has been evacuated, slowly clearing hormone can become dissociated, leading to an increased percent free beta. Khazaeli et al.[60] have shown a relationship between elevated percent free beta, at the time of diagnosis of hydatidiform mole, and persistent disease. Of 12 woman with levels of >4 ng per IU hCG/mℓ, 10 developed persistent or metastatic disease. The studies of Ozturk and Khazaeli and colleagues[60,98] and our laboratory together suggest that minimal proportions free beta (<4%) are found in normal pregnancy and

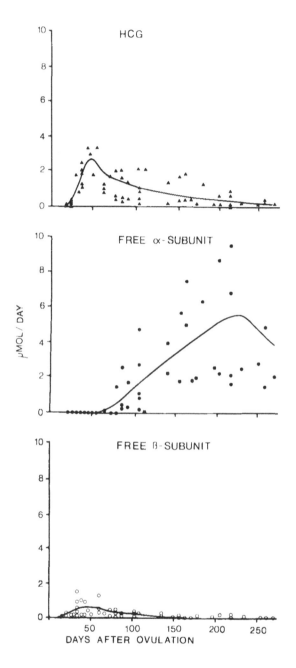

FIGURE 3. Daily hCG, free alpha, and free beta subunit production by the trophoblast in pregnancy. Levels were calculated from serum concentrations (Figure 2) using metabolic clearance rates.[49] To accommodate individual variation, lines were drawn from 9-day (±4 days) averages.

hydatidiform mole patients who undergo spontaneous remission. Higher proportions of free beta (>4%) may accompany the persistent hydatidiform mole or malignant disease (chorio-carcinoma).

Trophoblast disease is a rare disorder in the U.S. (1 in 1500 to 2000 pregnancies), however, the occurrence can exceed 1 in 400 pregnancies among Alaska natives and in Equatorial Africa and the Far East. Elegby et al.[29] measured hCG and free alpha in the sera of 283 pregnant black

Table 1
SERUM HCG AND FREE BETA LEVELS IN PATIENTS WITH TROPHOBLAST DISEASE, PRETHERAPY

Patient	hCG (nmol/ℓ)	Free beta	
		hCG (nmol/ℓ)	(% of hCG)
Complete or partial hydatidiform mole			
1	530	3.6	0.67
2	810	1.1	0.13
3	240	0.38	0.15
4	1650	0.86	0.05
5	3400	1.9	0.06
6	1380	3.6	0.26
7	1320	1.9	0.14
8	1320	2.1	0.15
9	81	0.95	1.2
10	260	0.64	0.25
11	530	1.0	0.19
12	83	<0.03	<0.04
		mean	0.27
Choriocarcinoma or persistant trophoblast disease			
13	2.1	0.09	4.3
14	20	2.0	10
15	48	1.5	3.1
16	143	9.1	6.4
17	24	0.29	1.2
		mean	5.0[a]

Note: Samples from women with mole are preevacuation, those from patients with choriocarcinoma or persistent disease were collected at the time of diagnosis.

[a] $p < 0.0001$ when compared to mole in double-sided Student's t test.

subjects, 135 from the U.S. and 148 from Nigeria where there is a high risk of trophoblast disease. Similar hCG levels were found throughout gestation in both populations, however, slightly higher (5 to 20%) free alpha levels were consistently detected among the Nigerian group in each month of pregnancy. It was proposed that the Nigerian group, having higher free alpha levels, may be at increased risk for gestation trophoblast disease.

Glycoprotein hormone common free alpha subunit and hCG free beta have also been detected in the sera or urines of subjects with a wide variety of other or nontrophoblastic neoplasms.[42-46] Measurements of free alpha and to a lesser extent hCG have been used to differentiate malignant and benign islet-cell tumors (17 of 27 with malignancy were positive for free alpha or hCG, none of 43 subjects with benign disease was positive for either).[44] hCG free beta has been detected in the sera of some subjects with gastrointestinal and gynecologic cancers, particularly those of squamous cell origin.[43,45,46] In our own studies,[39,43] a modified form of free beta was identified in the fluids of two lines of cultured squamous cervical cancer cells

Table 2
SERUM hCG AND FREE BETA LEVELS PRIOR TO
AND FOLLOWING THE EVACUATION OF
HYDATIDIFORM MOLE

Patient	hCG Prior to evacuation (nmol/ℓ)	Free beta	
		Prior to evacuation (% of hCG)	2 weeks post evacuation (% of hCG)
5	3400	0.06	0.27
4	1650	0.05	8.9
6	1380	0.26	3.9
7	1320	0.14	3.9
2	810	0.13	2.5
1	530	0.19	1.0
mean		0.14	3.4[a]

Note: Study limited to subjects with complete or partial mole having hCG and free beta measurements prior to and 2 weeks following evacuation.

[a] $p < 0.0001$ when compared to preevacuation percent free beta (double-sided Student's t test).

(DoT and CaSki). This free beta-like molecule, although reactive in the hCG beta subunit radioimmunoassay, was not detected in assays using antisera directed to the COOH-terminal segment of the subunit.

Glycoprotein hormone free alpha can also be detected in the pituitary tissue, serum, and urine of nonpregnant adults. Normal human adult free alpha secretion parallels the release of LH dimer. It is secreted in an LH-like pulsatile pattern,[3,61] and its levels are increased concordant with those of LH under gonadotropin releasing hormone (GnRH) influence.[61,64] Similarly, free alpha secretion parallels that of TSH, and thyrotropin releasing hormone (TRH) administration elevates circulating levels of both TSH and free alpha.[64] Although free alpha production in response to TRH is less than that in response to GnRH,[64] production of the glycoprotein hormone free alpha by both gonadotropes and thyrotropes is suggested.

Serum free alpha levels are elevated in postmenopausal women and in men with primary testicular failure.[33,34,62,63] Elevated levels of pituitary free alpha have also been reported in uremic subjects.[67] Free alpha and less commonly LH free beta subunit are detected in the sera of subjects with pituitary tumors.[37,65,66] Elevated free alpha and, to a lesser extent, TSH free beta subunit levels have been associated with thyroid disorders,[33] and are secreted by cultured thyrotropic tumor cells.[9,12,36] Free beta subunits of both FSH and LH are secreted by rat anterior pituitary cells in primary culture.[4,5] Pituitary LH, FSH, and TSH, like placental hCG, are all produced together with glycoprotein hormone free alpha and/or their corresponding free beta subunit. Just as abnormally high hCG free subunit serum levels may be associated with trophoblastic or nontrophoblastic tumors, abnormal elevations of circulating pituitary glycoprotein free subunits could be associated with pituitary tumor, or possibly a primary gonadal dysfunction. The low and varying levels of free circulating pituitary glycoprotein hormone subunits in each of the reproductive stages (prepuberty, puberty, normal adult cycles, menopause) and in normal thyroid function has not clearly been established. When sensitive and specific free subunit assays are developed, parallel measurements of pituitary hormones and free subunits may aid the assessment of normal pituitary-gonad and pituitary-thyroid function.

III. SEPARATION AND ASSAY OF FREE SUBUNITS

For qualitative and preparative studies, gel filtration on Sephadex® G100, Sephacryl® S200, or Bio-Gel® P100 can be used to separate free subunits and dimer hCG,[6,16,21,25,30,35] LH,[10] or FSH.[37] Eluates can be detected according to their radioactivity in biosynthetically labeled preparations, or analyzed by radioimmunoassay (RIA) using ^{125}I-hCG/LH/FSH and the appropriate beta subunit-directed antisera to detect hormone and free beta, and similar tracer with alpha subunit-directed antisera to quantitate hormone and free alpha. Figure 4 shows the elution volumes from Bio-Gel® P100 of hCG and its alpha and beta subunits. Immobilized antisera can also be used in the separation of hormone and free subunits. In studies by the author, immobilized beta subunit antisera has been used in free alpha purification schemes to remove contaminating hCG and free beta.[16] Conversely, immobilized alpha subunit antisera has been used as part of free beta isolation protocols.[35] Glycoprotein hormones subunits may also be separated by polyacrylamide gel electrophoresis.[10,11,23,24,35,41] Figure 5 (lane 1) illustrates the electrophoretic migration of labeled hCG subunits on sodium dodecyl sulfate 5 to 20% gradient polyacrylamide gels.

The general problem in quantitating the free alpha and free beta subunits of the gonadotropins in sera, urines, and tissues by immunoassays is that antisera developed to one subunit (using dissociated component of hormone as immunogen) similarly recognizes combined and free components, preventing their independent quantitation in the presence of dimer.[68] Several laboratories have used the data from alpha subunit- and beta subunit-directed assays with corresponding tracers and standards to measure total alpha (hormone + free alpha) and total beta subunit (hormone + free beta) immunoreactivities.[4-6,31,41,45,46,68] The proportions of hormone and free subunit have then been deduced from the total immunoreactivities, or from the similarity of log-logit dose-response lines to those of dimer, alpha or beta component standards. If, for instance, there is twice the alpha than beta subunit immunoreactivity, it is assumed that a mixture of somewhat similar amounts of hormone and free alpha are present. There are major limitations to this method, and to the deductions of hormone and free subunit proportions. First, hormone and both free subunits can coexist.[24,27,32,35,56] Second, affinities of hormone, free subunits, and mixtures thereof for antisera in competitive assays can vary. Hay[32] and Shapiro et al.[58] have, in part, surmounted these difficulties in quantitating mixtures of hCG and free subunits by using a dimer-specific anti alpha/^{125}I-anti beta immunoradiometric assay ("Tandem" assay kit, available from Hybritech) to determine the absolute amount of hormone, and beta and alpha subunit-directed RIAs to determine the level hCG plus the corresponding free subunit. The amount of free subunit is then deduced by subtracting the absolute hormone level. This clearly is not the most accurate method. Optimally, absolute hCG, absolute free beta, and absolute free alpha should be measured in separate assays. Aiming to achieve absolute specificity for hCG, free alpha, and free beta, monoclonal antibodies have now been generated.[32,69,71]

Several laboratories have been successful in generating monoclonal free beta subunit-specific antibodies.[32,69] Hormone has minimal cross-reactivity in free beta immunoassays using these monoclonal antibodies. For instance, hCG has <0.2% of the potency of free beta in the 1E5 monoclonal anti hCG beta assay.[69] There has been mixed success in generating alpha subunit-specific monoclonal antibodies. Gupta et al.[70] found the products of hybrid cell clones from fusion of myeloma cells with splenocytes of mouse immunized with hCG alpha subunit recognize all four glycoprotein hormone dimers (hCG, LH, FSH, and TSH) in addition to free alpha subunit; Thotakura and Bahl,[71] however, successfully isolated antibodies with limited dimer crossreactivity (0.17 to 1.75%). To generate alpha subunit antisera with minimal hCG affinity, efforts have also been directed at using the naturally free subunit, rather than the dissociation product of dimer, as immunogen. As will be discussed in the following sections of this review, the free gonadotropin alpha subunit is slightly larger than that combined; it contains additional sugar residues which block dimer formation. In the homologous RIA using ^{125}I-alpha

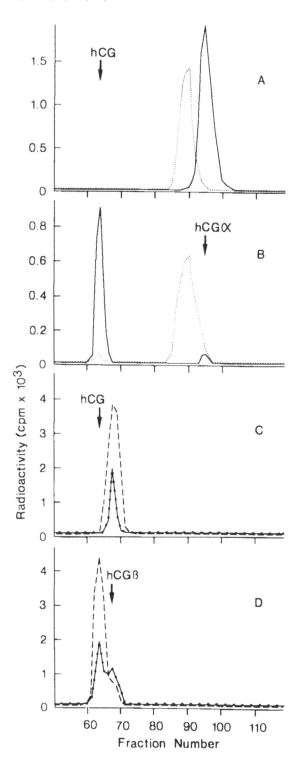

FIGURE 4. Combination studies of [³H]-Pro/Leu/Man biosynthetically labeled hCG components and free subunits from placental explant cultures with unlabeled hCG alpha and beta preparations (batch CR123). Detailed procedures have been published elsewhere.[35] Briefly, labeled subunits (20 pmol) were incubated 60 hr at 37°C with excess of corresponding component (20 nmol). Products were analyzed by gel filtration on Bio-Gel® P100 columns. Panel A, gel filtration of placental explant hCG alpha component (——) and free alpha (•••). Panel B, hCG alpha component and free alpha following incubation with unlabeled beta subunit. Panel C, gel filtration of placental explant hCG beta component (—) and free beta (-•-). Panel D, hCG beta component and free beta following incubation with unlabeled alpha subunit.

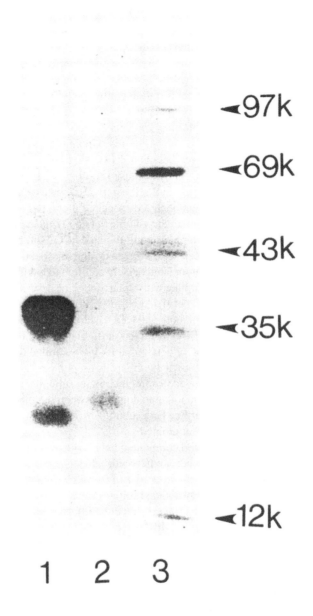

◄97k
◄69k
◄43k
◄35k
◄12k

1 2 3

FIGURE 5. Electrophoresis in sodium dodecyl sulfate on 5 to 20% polyacrylamide gels. Methods have been previously described in detail.[35] Arrowheads indicate molecular weights of standard proteins (lane 3).

tracer and antisera generated to the BeWo choriocarcinoma cell line free alpha (antisera H7, generated by R. Hussa of The Medical College of Wisconsin), hCG has less than 3% the potency of free alpha.[27,29] In studies in our laboratory with antisera H7, we were able to reduce the hCG cross-reactivity further, to <0.5%, by using ^{125}I-dissociated hCG (>90% of ^{125}I is on alpha subunit) instead of ^{125}I-alpha tracer. Apparently, a radioiodine is present on Tyr 37, part of the free subunit-specific epitope on ^{125}I-alpha, but not on the alpha in ^{125}I-dissociated hCG. The presence of radioiodine at Tyr 37 may lower the affinity of the tracer for free subunit-specific clones, diminishing the overall sensitivity and specificity of the assay.

IV. STRUCTURE AND PROPERTIES OF FREE ALPHA AND FREE BETA

As previously published,[35] the dissociated beta component of hCG and the placental free beta subunit, whether from sera or culture fluids, pregnancy or choriocarcinoma, comigrate on sodium dodecyl sulfate polyacrylamide gels ($M_r = 35,000$), suggesting size similarity. Size similarity is also suggested by their like gel filtration elution patterns. We purified radioactive free beta and hCG from the medium of placental explant cultures, maintained in fluids containing [^{3}H] mannose, leucine, and proline. The beta component of dimer was isolated by dissociation of hCG.[35] Figure 4, panel C shows the overlapping elution patterns of the two radioactive beta subunit preparations (free beta and dissociated beta) on Bio-Gel[®] P100 gel filtration columns. The abilities of the two preparations to combine with alpha subunit was also addressed. As shown in panel D, when 20 pmol radioactive free beta or dissociated beta was incubated 60 hr at 37°C with 20 nmol unlabeled dissociated alpha (NIH batch CR123), the product eluted from a gel filtration column in the volume of hCG. This indicated that free beta, like dissociated beta, can combine with alpha subunit to make hCG. Similar results have been observed with JAr choriocarcinoma cell line free beta and dissociated beta component preparations.[35] Placental free beta and dissociated beta, in addition to having similar molecular weights, Stoke's radii, and abilities to combine with alpha (under the described conditions), also have similar potencies in monoclonal anti beta RIAs (using ^{125}I-hCG beta and antibody 1E5) and in assays using antisera to the beta subunit core and COOH-terminal segments (antisera R126, R529, and R525, kindly supplied by S. Birken and R. Canfield of Columbia University). This suggests that the free and combined hCG beta subunits have similar or possibly identical structures. To the best of the author's knowledge, there is no published data on the size and combination properties of pituitary gonadotropin free beta subunits. However, with the numerous established structural and functional analogies between the placental and pituitary glycoprotein hormones,[1,2] it is likely that free beta subunits of TSH, FSH, and LH, if secreted, are of similar size and structure to those combined.

A very different story to that of free and component beta is uncovered when comparing the free and combined alpha subunits. In other studies with labeled preparations from placental explant cultures,[16,35] the size and combining properties of free alpha and dissociated alpha were examined. As shown in Figure 5, on sodium dodecyl sulfate polyacrylamide gels free alpha (lane 2), migrates slower than dissociated alpha (lane 1, lower band), suggesting a larger molecular size ($M_r = 24,000$ vs. 22,000). A similar size difference has been observed between the alpha components of the other glycoprotein hormones and pituitary free alpha.[10,12,14] On Bio-Gel[®] P100 columns, placental free alpha elutes prior to dissociated alpha component (Figure 4, panel A), suggesting a difference in Stoke's radius. Free alpha and dissociated alpha (20 pmol) were separately incubated 60 hr at 37°C with 20 nmol unlabeled dissocated beta (NIH batch CR123), then applied to the gel filtration column. After incubation, the dissociated alpha preparation eluted from the column in the volume of hCG, however, the free alpha preparation eluted as free alpha (Figure 4, panel B). This suggested that the large free alpha was unable to combine with dissociated beta to form hCG. Similar findings have been reported with the free alpha from cultured choriocarcinoma cells,[35] and with the free alpha of pituitary origin.[10] Comparisons of free and combined alpha subunits of glycoprotein hormones, their molecular size, and ability to combine with beta subunit suggest structural differences. In the following paragraphs, studies from the author's and other laboratories will be presented, which compare the structures of the two forms of alpha subunit and try to explain the reasons for the size difference and the inability of the free molecule to associate with beta subunit.

Several investigators have examined the pituitary free alpha and placental free alpha (from serum or urine), and investigated the effects of neuraminidase and other glycosidase treatments on the size difference with dissociated alpha from glycoprotein hormones. Neuraminidase and other glycosidases reduced the size difference, as shown by gel filtration or gel electrophoresis,[14,17,18,20,22] suggesting that additional *N*-acetylneuraminic acid (sialic acid) and possibly other sugar residues, on the free molecule, are in part responsible for its size difference with that dissociated from dimer. That the size difference resides in the sugar rather than the peptide moieties was also indicated by sequencing studies, which showed that the N-terminal 16 amino acids of choriocarcinoma cell line dissociated alpha and free alpha were identical.[72] While not excluding the possibility of RNA splicing or posttranslational processing, the presence of a single human alpha subunit gene[73] supports the existence of a single alpha subunit polypeptide.

The sequence -Pro-Thr-Pro- is present at residues 38 to 40 on the alpha subunit.[74] This is a potential site for O-glycosylation (Thr or Ser, commonly with at least one adjacent Pro[75]). To look for an O-linked sugar unit at this site on free alpha, we purified radioactive free alpha from placental organ cultures maintained in fluids containing [³H] glucosamine.[16,95] The free alpha was digested with trypsin and peptide 36 to 42, containing this site, isolated. Using serial glycosidase digestion and Bio-Gel® P4 gel filtration (to monitor changes in size), we identified a radioactive sugar unit on this peptide (radioactive glucosamine is incorporated into *N*-acetylgalactosamine and *N*-acetylneuraminic acid residues on O-linked sugar units).[16,95] Beta-elimination methods were used to release the O-linked sugar unit on the free alpha tryptic peptide. The size of the released sugar unit was compared with that similarly released from the beta subunit of hCG. A heterogeneous mixture of sugar units, varying in size, was found in the free alpha preparation. A similar mixture was found in the preparation from beta subunit, suggesting that the same oligosaccharide structures are found at the single site on free alpha as at the four sites on hCG beta subunit.[16,95] We concluded that an O-linked oligosaccharide can be attached at Thr 39 on the free alpha subunit, and that this may account for the size difference with dissociated alpha. Parsons et al. have identified a similar O-linked sugar unit on bovine pituitary free alpha,[10] and have shown that cleavage of this O-linked sugar unit by treatment with mixed glycosidases eliminates the size difference and permits combination with dissociated beta.[15] This indicates that an O-linked sugar unit on glycoprotein hormone free alpha can contribute to its large size relative to alpha component of dimer, and can also be responsible for its inability to combine with dissociated beta.

Recent studies also suggest differences in the N-linked sugar structures on the free and combined alpha subunit of glycoprotein hormones. As shown by Peters et al.[24] in chorio-carcinoma cell line preparations, the N-linked oligosaccharides on hCG alpha component are a mixture of endo-*N*-acetylglucosaminidase H-sensitive and -resistant structures, whereas those on free alpha are strictly endo-*N*-acetylglucosaminidase-resistant. Blithe et al.[21] have shown that pregnancy urine free alpha molecules contain a mixture of di- and trisialyl N-linked oligosac-charides, of which 25 to 40% bind ConA. This is different from the oligosaccharides from dissociated alpha, which are primarily monosialyl structures, and predominantly bind ConA. Both groups suggest that larger, more negatively charged N-linked oligosaccharide structures than those on dissociated alpha may be found on the free alpha. Not all placental free alpha molecules contain the O-linked sugar (5 to 50% of total), yet none combine,[77] suggesting that other alpha subunit structural variations may also block alpha-beta association. Seemingly, a combination of larger N-linked sugar units, a single O-linked oligosaccharide, and possibly other molecular variations differentiate the free and combined alpha subunits.

V. ORIGINS OF FREE GLYCOPROTEIN HORMONE SUBUNITS

Are free subunits native synthetic products, or the outcome of circulating or intracellular hormone dissociation? This question has been addressed using the JAr line of trophoblast cancer cells.

Peters et al.[24] used cultures pulsed 10 min with fluids containing [³H]mannose and chased 1 to 480 min with nonradioactive medium, to examine the kinetics of hCG and free subunit biosynthesis and secretion. Following specific chase periods, hCG and free subunits were

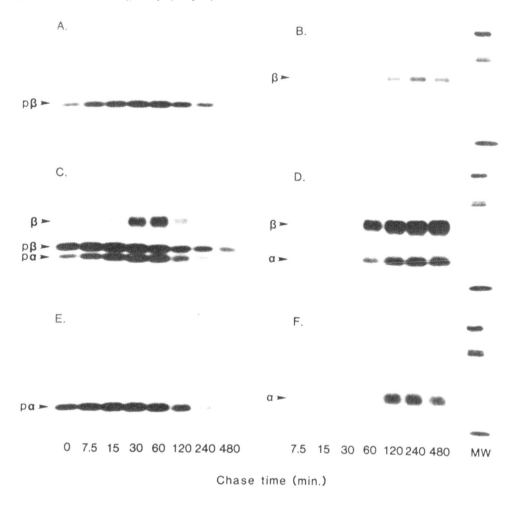

FIGURE 6. The incorporation of [³H]Man into free beta, alpha-beta dimer, and free alpha forms of hCG synthesized and secreted by JAr cells. Cultures containing 2×10^7 cells were pulsed for 10 min with [³H]Man and chased for intervals ranging from 0 to 480 min. The hCG forms were immunoprecipitated from cell lysates and chase media. One third of each immunoprecipitate was used for sodium dodecyl sulfate polyacrylamide gel electrophoresis and the radioactive bands were visualized by fluorography (5-day exposure). The cellular precursors (pα, pβ) and mature forms (α, β) of hCG are identified by the arrows. Panel A, cellular free beta; Panel B, secreted free beta; Panel C, cellular alpha-beta dimer; Panel D, secreted alpha-beta dimer; Panel E, cellular free alpha; Panel F, secreted free alpha. Molecular weight markers (MW) are (from top) bovine serum albumin (M_r = 69,000), ovalbumin (M_r = 46,000), carbonic anhydrase (M_r = 30,000), and cytochrome c (M_r = 12,000). (From Peters, B. P., Krzesicki, R. F., Hartle, R. J., Perini, F., and Ruddon, R.W., *J. Biol. Chem.*, 259, 15123, 1984. With permission.)

purified from cell lysates and spent medium using specific mono- and polyclonal antibodies. Purified products were visualized and analyzed by sodium dodecyl sulfate polyacrylamide gel electrophoresis (Figure 6). As shown, free beta (panel A), hCG (panel C), and free alpha (panel E) appear and accumulate in parallel in lysates and medium (panels B, D, and F), suggesting their cobiosynthesis and secretion.

We have studied the secretion of hCG and free subunits by 70% confluent flasks of JAr cells.[35] Over a 20-hr period (aliquots of medium assayed at 0, 2.5, 5, 10, and 20 hr) hCG, free alpha, and free beta were secreted in parallel. This suggested that free alpha and free beta are primary secretory products, and absence of postsecretion subunit dissociation or combination. In a further experiment, spent fluids containing hCG and free beta were incubated 20 hr in 70% confluent flasks of cells, with or without cycloheximide (blocks further translation). A change in the proportion of hCG and free subunit was not observed, confirming the absence of postsecretion dissociation. Although these studies do not exclude dissociation of hormone circulating in vivo, they clearly indicate the native production of free subunits. That dissociation in the circulation is not a major source of free subunits is also suggested by the studies of Wehmann and Nisula,[78] which show negligible excretion of free alpha and free beta by subjects infused with highly purified hCG.

VI. BIOLOGICAL FUNCTIONS OF FREE SUBUNITS

Early studies with glycoprotein hormone subunits suggested minimal biological activities. It was suggested that hCG and human LH subunits competed with ^{125}I-hCG in binding receptors on cell membranes of bovine or human corpus luteum, and rat testis, but only in 100- to 1000-fold greater concentrations than that required for hormone.[88-90] Similarly, it was shown that FSH and TSH subunits have minimal activities in hormone radioreceptor assays.[90,91] Although these minimal receptor affinities have been ascribed to contaminant dimer in subunit preparations,[90,92] there is still some debate concerning low level free alpha and free beta biological activities. To address the issue of intrinsic subunit activities, Williams et al.[92] used immobilized antisera to superpurify and remove the last traces of contaminant hormone from standard preparations of bovine LH and TSH free subunits. While LH free subunits were shown to have no demonstrable affinity for the rat testis LH receptor, TSH alpha subunit was found to have a minimal binding activity in the bovine thyroid TSH system (potency of 0.02% of that of TSH), suggesting a very small, but intrinsic subunit activity. Recent studies by Moudgal and Li,[93] also with superpurified preparations, contradict the findings of Williams et al. and show that hCG and ovine LH beta subunits give dose-dependent responses in the Leydig cell testosterone assay and in the rat granulosa cell progesterone system. The activity of the beta subunit preparations was found to be 0.01 to 0.06% of that of hormone. As to whether glycoprotein hormone subunits have intrinsic receptor-binding or steroidogenic activities is still open to debate, however, other activities not normally ascribed to glycoprotein hormones need to be considered. Recent studies by Begeot et al.[94] suggest a role for LH alpha subunit in inducing lactotrope differentiation. In adenohypophysial tissues explanted from rat fetuses, GnRH was shown to induce lactotrope differentiation (demonstrated by increased numbers of immunoreactive lactotropes). The action of GnRH was blocked by added rat LH antiserum, but not by rat LH beta antiserum which binds hormone and free beta subunit. This suggested, by deduction, a role of free alpha subunit in lactotrope differentiation. This role was confirmed by showing a direct action of differing concentrations of porcine LH alpha on the number of immunoreactive lactotropes.

VII. METABOLISM AND CLEARANCE OF FREE SUBUNITS

The NeuAc content of glycoprotein hormones has a major effect on their blood half-lives. hCG (approximately 10% NeuAc by weight) injected into normal subjects has a plasma half-life of 6 hr (rapid component), human FSH (approximately 5% NeuAc) has a half-life of 4 hr, and human LH, with only 2% NeuAc, has a half-life of just 1 hr.[1,78] Removal of NeuAc from the glycoprotein hormones by neuraminidase treatment ablates their plasma half-lives. Asialo-hCG, for instance, infused in normal subjects, has a plasma half-life of minutes, compared with several hours for the intact hormone.[79] Rosa et al.[96] determined the plasma half-life of asialo hCG to be 3.6 min, 100 times quicker than the rapid component of intact hormone. The rapid clearance of asialo-hCG is, however, the result of uptake by hepatic receptors for galactose-terminated glycoproteins, a pathway that has little, if any, role in the clearance of the intact hormones.[80] What pathway causes the differing clearance rates of the glycoprotein hormones? To address this question, we examined uptake of glycoprotein hormones by rat kidneys continuously perfused in vitro. Consistent with their differing plasma half-lives, uptake of human LH by the

Table 3
UPTAKE OF GLYCOPROTEIN HORMONES AND SUBUNITS BY THE PERFUSED RAT KIDNEY

Perfused sample	FR (mℓ/min)	FE (% ± SEM)
hCG (batch CR121), 1, 10, and 100 μg	0.058	9.7 ± 0.51
hCG-beta (CR123), 10 and 100 μg	0.29	47 ± 3.7[a]
hCG-alpha (CR123), 10 and 82 μg	0.74	104 ± 2.9[a]
hLH, 3 and 6 μg	0.20	38 ± 4.2[a]

Note: Male rats weighing 250 to 350 g were perfused in vitro according to the method of Terao and Tannen.[82] A Krebs-Hanseleit perfusate (140 mℓ) containing 6.7% albumin was recirculated through the renal vascular system for 3 hr. [^{14}C]inulin was added for determining glomerular filtration rate (GFR), as was differing concentrations of glycoprotein hormone or subunit, ^{3}H-labeled by reductive methylation. After 1, 2, and 3 hr perfusion, accumulated urines and aliquots of perfusates were removed and radioactivity determined. FR is the filtration rate of hCG or subunit radioactivity, determined as cpm/min in urine divided by remaining cpm/m ℓ in perfusate. FE is the filtration efficiency, and FR is expressed as a percentage of the GFR.

a p >0.001 when compared to hCG in double-sided Student's t test.

perfused kidney was shown to be four times faster (filtration rate) than uptake of hCG (Table 3). This suggested a role for the kidney, possibly repulsion of more negatively charged (or more sialylated) hormones by the anionic glomerular charge barrier,[81] in the maintenance of glycoprotein hormone half-lives (L. Cole, unpublished observations).

The plasma disappearance rate (metabolic clearance rate, MCR) of the glycoprotein hormone subunits also appears to be NeuAc content-related. hCG alpha subunit, for instance, with approximately 7% NeuAc (MCR = 91 m ℓ/min), is cleared two to three times faster than hCG beta subunit (MCR = 35 ml/min) which has a two to three times greater NeuAc component.[55] That the kidney, and possibly the glomerular charge barrier, is also responsible for the difference in MCRs of free subunits is suggested by our studies with the continuously perfused rat kidney, showing a two- to threefold faster uptake (filtration) of hCG alpha subunit than beta subunit (Table 3). It should be noted that our studies used ^{3}H-labeled hormones and subunits, and involved the measurement of radioactivity in urines and perfusates. Although this does not limit interpretation with regard to disappearance rates or subunit uptake, it does with regard to the nature of the excreted product. Radioactivity in urine could represent degraded rather than intact hormone or subunit. In studies of clearance of infused dissociated hCG alpha and beta in humans, less than 1% of total disposal was accounted for by urinary excretion of "authentic" alpha and beta (that of the same molecular size, potency in a variety of immunoassays, and percent ConA Sepharose®-binding as administered preparations).[83] Several recent studies show that COOH-terminal and core fragments of hCG beta subunit can be detected in urines.[38,49,53,84-86,102] Furthermore, when beta subunit, rather than hCG, is injected into normal humans, beta subunit core fragment can be excreted.[50,85] Studies in this laboratory and others show that core fragment can account for as much as 90% of the total beta subunit (hCG + free beta + core fragment) in urine.[53,84,97] It appears that renal uptake has a major role in the clearance of glycoprotein hormone subunits and may be responsible for the their differing plasma half-lives.

VIII. BETA CORE FRAGMENT

In addition to free alpha and free beta, degraded core fragments of glycoprotein hormone beta subunits have been detected in urines. Krichevsky and colleagues at Columbia University have identified a core fragment of LH beta subunit in postmenopausal female urines (personal communication). Numerous investigators have detected a small form of hCG beta subunit (beta core fragment) in pregnancy and cancer patient urines.[48-54,85,86,97,101,102] Beta core fragment has been purified from urines and characterized as a molecule of approximately 10,000 mol wt, and found to consist of two polypeptide chains, beta 6 to 40 disulfide-linked to beta 55 to 92, and to be missing the sialic acid and most of the galactose found on the beta subunit of hCG.[48] As discussed under "Metabolism and Clearance of Free Subunits", beta core fragment may be originate as a renal degradation product of free beta.[50,85] We investigated the possibility that beta core fragment is also secreted directly by the placenta.[97]

We examined 24 hr organ cultures of trophoblast tissue from 1st, 2nd, and 3rd trimester pregnancies. Each secreted into medium hCG, free beta and, as was observed in the urine of pregnant women, higher concentrations of the beta core fragment. The beta core fragment immunoreactive material secreted by trophoblast tissue was compared with that purified from a pregnancy urine preparation. The secreted material was identical to the purified urinary beta core fragment in its elution pattern from a HPLC gel filtration column, a reverse-phase HPLC column, an ion-exchange column, by *Ricinus communis* agglutinin-agarose affinity chromatography, and by electrophoresis and immunoblotting with beta fragment-reactive monoclonal antibodies. It was concluded that beta core fragment can also originate in trophoblast tissue, and may be the principal hCG beta-immunoreactive molecule secreted.[97]

Several laboratories have generated beta core fragment-specific immunoassays, with minimal free beta and hCG cross-reactivities.[52,53,102] Using an immunoradiometric assay with the fragment-specific monoclonal antibody B204, generated by Drs. A. Krichevsky and G. Armstrong of Columbia University,[102] and the beta subunit monoclonal antibody HCO-514 (a kind gift from Hybritech), we measured levels of beta core fragment immunoreactivity in urines.[51,97] We examined 64 morning urines from 17 woman with normal pregnancy. The mean hCG values (nmol/ℓ ± standard error) in the periods 3 to 4, 5 to 6, 7 to 9, 10 to 13, 13 to 17, 18 to 21, and 22 to 40 weeks after ovulation were 9.1 ± 1.9, 30 ± 8.0, 36 ± 6.2, 26 ± 4.0, 12 ± 1.9, 8.2 ± 1.3, and 6.9 ± 0.95, respectively. Free beta levels averaged 1/4 to 1/5 those of hCG until week 18 to 21, after which they became undetectable (<0.1 nmol/ℓ). The mean beta core fragment levels were 6.9 ± 3.2, 35 ± 6.5, 74 ± 5.5, 87 ± 4.9, 81 ± 4.3, 61 ± 5.7 and 21 ± 8.7, respectively. In samples from women with early pregnancy (3 to 7 weeks after ovulation) beta core fragment levels were less than or equivalent to those of hCG. In samples obtained after this period, however, beta fragment exceeded hormone levels. hCG and free beta levels reached a peak at 7 to 9 weeks after ovulation, as they do in serum (Figure 2). The highest beta core fragment levels, however, were detected 4 to 8 weeks after the hCG peak, when they as much as sevenfold exceeded hormone concentration. Although concerned by sample to sample variations in urine density, deviation was small enough to demonstrate a significant drop in hCG and free beta values from 5 to 9 to 10 to 17 weeks of pregnancy ($p < 0.05$ and $p < 0.005$ in Student's t test) and a corresponding increase in beta core fragment levels during the same period ($p < 0.005$). A significant reduction ($p < 0.05$) in serum hCG and free beta levels (55 and 54%, respectively) was also observed in the same periods (Figure 2).

Several investigators have found beta core fragment-like molecules in the urines of cancer patients.[50,51,53,101] Using an assay that measures both free beta and beta core fragment,[51] urines from 67 nonpregnant cancer-free women were examined. An average level of 0.13 ng/mℓ free beta plus beta core fragment was detected. Of the 67, 6 (8.9%) had levels exceeding a selected cut-off value, 0.2 ng/mℓ Of 112 woman with active gynecologic cancer, pretherapy, 72 (64%) had urine levels exceeding this cut-off value. When urines were limited to those with creatinine

>0.5 mg/mℓ, or to first morning samples (mean creatinine 1.0 ng/mℓ), the overall sensitivity for gynecologic cancer was raised to 76% (73% for cervical, 65% for endometrial, and 83% for ovarian maligancies).[51,104] When separate immunoradiometric assays were used for free beta and beta core fragment, 15% of the cancer group had levels of free beta and 65% had levels of beta core fragment exceeding 0.2 ng/mℓ. We conclude that beta core fragment measurements may be useful in the detection and management of nontrophoblastic malignancies.

ACKNOWLEDGMENTS

Many thanks to Drs. Ernest I. Kohorn, Peter E. Schwartz, and Alan DeCherney of Yale University, to L.C. Wong and Ho K. Ma of the University of Hong Kong, to Roland Pattillo and Robert Hussa of the Medical College of Wisconsin, and Katsumi Yazaki of Gumna University in Japan, for serum and urine samples. Without these samples none of the clinical studies presented would have been possible. Many thanks also to Drs. S. Birken, A. Krichevsky, G. Armstrong, and R. Canfield of Columbia University, and to Hybritech Inc. (San Diego, CA) for the monoclonal antibodies used in the clinical assays. The effort of Dr. R. Tannen and colleagues of the University of Michigan, Ann Arbor, who performed the rat kidney perfusion studies, is greatly appreciated.

Research presented was supported by grant PDT-299 from the American Cancer Society and CA44131 from NIH.

REFERENCES

1. **Catt, K. J. and Pierce, J. G.,** Gonadotropic hormones of the adenohypophysis, in *Reproductive Endocrinology,* 2nd ed., Yen, S. C. and Jaffe, R. B., Eds., W. B. Saunders, Philadelphia, 1986, chap. 3.
2. **Hussa, R. O.,** Human chorionic gonadotropin, a clinical marker: review of its biosynthesis, *Ligand Rev.,* 3, 6, 1981.
3. **Winters, S. J. and Troen, P.,** Pulsatile secretion of immunoreactive alpha subunit in man, *J. Clin. Endocrinol. Metab.,* 60, 344, 1985.
4. **Grotjan, H. E., Berkowitz, A. S., and Keel, B. A.,** Minimal quantities of FSH beta subunit are released by rat anterior pituitary cells in primary culture, *Mol. Cell. Endocrinol.,* 41, 205, 1985.
5. **Grotjan, H. E., Leveque, N. W., Berkowitz, A. S., and Keel, B. A.,** Quantitation of LH subunits released by rat anterior pituitary cells in primary culture, *Mol. Cell. Endocrinol.,* 35, 121, 1984.
6. **Taliadouros, S., Louvet, J. P., Birken, S., Canfield, R. E., and Nisula, B. C.,** Biological and immunological characterization of crude commercial human choriogonadotropin, *J. Clin. Endocrinol. Metab.,* 54, 1002, 1982.
7. **Pierce, J. G.,** Eli Lilly lecture: The subunits of pituitary thyrotropin — their relationships to other glycoprotein hormones, *Endocrinology,* 89, 1331, 1971.
8. **Swaminathan, N. and Bahl, O. P.,** Dissociation and recombination of the subunits of human chorionic gonadotropin, *Biochem. Biophys. Res. Comm.,* 40, 422, 1970.
9. **Ronin, C., Stannard, B. S., Rosenbloom, I. L., Magner, J. A., and Weintraub, B. D.,** Glycosylation and processing of high-mannose oligosaccharides of thyroid-stimulating hormone subunits: comparison to nonsecretory cell glycoproteins, *Biochemistry,* 23, 4503, 1984.
10. **Parsons, T. F., Bloomfield, G. A., and Pierce, J. G.,** Purification of an alternate form of the alpha subunit of the glycoprotein hormones from bovine pituitaries and identification of its O-linked oligosaccharide, *J. Biol. Chem.,* 258, 240, 1983.
11. **Chin, W. W., Maloof, F., and Habener, J. F.,** Thyroid-stimulating hormone biosynthesis, *J. Biol. Chem.,* 256, 3059, 1981.
12. **Magner, J. A., Ronin, C., and Weintraub, B. D.,** Carbohydrate processing of thyrotropin differs from that of free alpha subunit and total glycoproteins in microsomal subfractions of mouse pituitary tumor, *Endocrinology,* 115, 1019, 1984.
13. **Chin, W. W., Habener, J. F., Martorana, M. A., Keutmann, H. T., Kieffer, J. G., and Maloof, F.,** Thyroid-stimulating hormone: isolation and partial characterization of hormone and subunits from a mouse thyrotrope tumor, *Endocrinology,* 107, 1384, 1980.

14. **Kourides, I. A., Hoffman, B. J., and Landon, M. B.,** Difference in glycosylation between secreted and pituitary free alpha subunit of glycoprotein hormones, *J. Clin. Endocrinol. Metab.,* 51, 1372, 1980.

15. **Parsons, T. F. and Pierce, J. G.,** Free alpha like material from bovine pituitaries, *J. Biol. Chem.,* 259, 1, 1984.

16. **Cole, L. A., Perini, F., Birken, S., and Ruddon, R. W.,** An oligosaccharide of the O-linked type distinguishes the free from the combined form of hCG alpha subunit, *Biochem. Biophys. Res. Comm.,* 122, 1260, 1984.

17. **Posillico, E. G., Handwerger, S., and Tyrey, L.,** Demonstration of intracellular and secreted forms of large human chorionic gonadotropin alpha subunit in cultures of normal placental tissue, *Placenta,* 4, 439, 1983.

18. **Nishimura, R., Hamamoto, T., Utsonomiya, T., and Mochizuki, M.,** Heterogeneity of free alpha subunit in term placenta, *Endocrinol. Jpn.,* 30, 663, 1983.

19. **Cole, L.A.,** O-glycosylation of proteins in the normal and neoplastic trophoblast, *Troph. Res.,* 2, 139, 1987.

20. **Cox, S. G.,** Nature of the difference in apparent molecular weights between the alpha subunit of urinary human chorionic gonadotropin and the alpha protein secreted by HeLa cells, *Biochem. Biophys. Res. Comm.,* 98, 942, 1981.

21. **Blithe, D. L. and Nisula, B. C.,** Variations in the oligosaccharides on free and combined alpha subunits of human choriogonadotropin in pregnancy, *Endocrinology,* 117, 2218, 1985.

22. **Jones-Brown, Y. R., Wu, C. Y., Weintraub, B. D., and Rosen, S. W,** Synthesis of chorionic gonadotropin subunits in human choriocarcinoma clonal cell line JEG-3: carbohydrate differences in glycopeptides from free and combined alpha subunits, *Endocrinology,* 155, 1439, 1984.

23. **Weintraub, B. D., Stannard, B. S., Linnekin, D., and Marshall, M.,** Relationship of glycosylation to *de-novo* thyroid-stimulating hormone biosynthesis and secretion by mouse pituitary tumor cells, *J. Biol. Chem.,* 255, 5715, 1980.

24. **Peters, B. P., Krzesicki, R. F., Hartle, R. J., Perini, F., and Ruddon, R. W.,** A kinetic comparison of the processing and secretion of the ab dimer and the uncombined a and b subunits of chorionic gonadotropin synthesized by human choriocarcinoma cells, *J. Biol. Chem.,* 259, 15123, 1984.

25. **Cole, L. A. and Hussa, R. O.,** The carbohydrate on human chorionic gonadotropin produced by cancer cells, *Adv. Exper. Med. Biol.,* 176, 245, 1984.

26. **Styne, D. M., Kaplan, S. L., and Grumbach, M. M.,** Plasma glycoprotein alpha subunit in the neonate and in the prepubertal and pubertal children: effects of luteinizing hormone-releasing hormone, *J. Clin. Endocrinol. Metab.,* 50, 450, 1980.

27. **Cole, L. A., Kroll, T. G., Ruddon, R. W., and Hussa, R. O.,** Differential occurrence of free beta and free alpha subunits of human chorionic gonadotropin (hCG) in pregnancy sera, *J. Clin. Endocrinol. Metab.,* 58, 1200, 1984.

28. **Rozmus, V. M. and Skalba, P.,** Die Konzentrationsbestimmung des Choriogonadotropin (HCG) und seiner freien Untereinheiten im Plazentagewebe in verschiedenen Perioden der normalen Schwangerschaft, *Zbl. Gynakol.,* 106, 834, 1984.

29. **Elegbe, R. A., Pattillo, R. A., Hussa, R. O., Hoffmann, R. G., Damole, I. O., and Finlayson, W. E.,** Alpha subunit and human chorionic gonadotropin in normal pregnancy and gestational trophoblastic disease, *Obstet. Gynecol.,* 63, 335, 1984.

30. **Vaitukaitis, J. L,** Changing placental concentrations of human chorionic gonadotropin and its subunits during gestation, *J. Clin. Endocrinol. Metab.,* 38, 755, 1974.

31. **Hussa, R. O. and Pattillo, R. A.,** Predominance of gonadotropin alpha subunit resulting from preferential loss of hCG-beta production in an established cell line, *In Vitro,* 16, 585, 1980.

32. **Hay, D. L.,** Discordant and variable production of human chorionic gonadotropin and its free alpha and beta subunits in early pregnancy, *J. Clin. Endocrinol. Metab.,* 61, 1195, 1985.

33. **Kourides, I. A., Weintraub, B. D., Ridgway, E. C., and Maloof, F.,** Pituitary secretion of free alpha and beta subunit of human thyrotropin in patients with thyroid disorders, *J. Clin. Endocrinol. Metab.,* 40, 872, 1975.

34. **Styne, D. M., Conte, F. A., Grumbach, M. M., and Kaplan, S. L.,** Plasma glycoprotein hormone alpha subunit in the syndrome of gonadal dysgenesis: the effect of estrogen replacement in hypergonadotropin hypogonadism, *J. Clin. Endocrinol. Metab.,* 50, 1049, 1980.

35. **Cole, L. A., Hartle, R. J., Laferla, J. J., and Ruddon, R. W.,** Detection of the free beta subunit of human chorionic gonadotropin (hCG) in cultures of normal and malignant trophoblast cells, pregnancy sera, and sera of patients with choriocarcinoma, *Endocrinology,* 113, 1176, 1983.

36. **Ridgway, E. C., Kieffer, J. D., Ross, D. S., Downing, M., Mover, H., and Chin, W. W.,** Mouse pituitary tumor line secreting only the alpha subunit of the glycoprotein hormones: development from a thyrotropic tumor, *Endocrinology,* 113, 1587, 1983.

37. **Snyder, P. J., Bashey, H. M., Kim, S. U., and Chappel, S. C.,** Secretion of uncombined subunits of luteinizing hormone by gonadotroph cell adenomas, *J. Clin. Endocrinol. Metab.,* 59, 1169, 1984.

38. **Amr, S., Rosa, C., Birken, S., Canfield, R., and Nisula, B.,** Carboxyterminal peptide fragments of the beta subunit are urinary product of the metabolism of desialylated human choriogonadotropin, *J. Clin. Invest.,* 76, 350, 1985.

39. **Cole, L. A., Birken, S., Sutphen, S., Hussa, R. O., and Pattillo, R. A.,** Absence of the COOH-terminal peptide on ectopic human chorionic gonadotropin beta subunit (hCGbeta), *Endocrinology,* 110, 2198, 1982.

40. **Cole, L. A. and Hussa, R. O.,** Use of glycosidase digested human chorionic gonadotropin beta subunit to explain the partial binding of ectopic glycoprotein hormones to Con A, *Endocrinology,* 109, 2276, 1981.

41. **Ruddon, R. W., Hanson, C. A., Bryan, A. H., and Anderson, C.,** Synthesis, processing and secretion of human chorionic gonadotropin subunits by cultured human cells, in *Chorionic Gonadotropin,* Segal, S. J., Ed., Plenum Press, New York, 1980, 295.

42. **Nishimura, R., Hamamoto, T., Morimoto, N., Ozawa, M., Ashitaka, Y., and Tojo, S.,** The characterization of alpha subunit of glycoprotein hormone produced by undifferentiated carcinoma, *Endocrinol. Jpn.,* 29, 11, 1982.

43. **Hussa, R. O., Fein, H. G., Pattillo, R. A., Nagelberg, S. B., Rosen, S. W., Weintraub, B. D., Perini, F., Ruddon, R., and Cole, L. A.,** A distinctive form of human chorionic gonadotropin beta subunit-like material produced by cervical carcinoma cells, *Cancer Res.,* 46, 1948, 1986.

44. **Kahn, C. R., Rosen, S. W., Weintraub, B. D., Fajans, S. S., and Gorden, P.,** Ectopic production of chorionic gonadotropin and its subunits by islet-cell tumors, *N. Engl. J. Med.,* 297, 565, 1977.

45. **Weintraub, B.D. and Rosen, S. W.,** Ectopic production of the isolated beta subunit of human chorionic gonadotropin, *J. Clin. Invest.,* 52, 3135, 1973.

46. **Nagelberg, S. B., Cole, L. A., and Rosen, S. W.,** A novel form of ectopic human chorionic gonadotrophin beta subunit in the serum of a woman with epidermoid cancer, *J. Endocrinol.,* 107, 403, 1985.

47. **Cole, L. A., Restrepo-Candelo, H., Lavy, G., and DeCherney, A.,** hCG free beta-subunit an early marker of outcome of in vitro fertilization clinical pregnancies, *J. Clin. Endocrinol. Metab.,* 64, 1328, 1987.

48. **Birken, S., Armstrong, E. G., Kolks, M. A. G., Cole, L. A., Agosto, G. M., Krichevsky, A., and Canfield, R. E.,** The structure of the human chorionic gonadotropin beta core fragment, *Endocrinology,* in press, 1988.

49. **Schroeder, H. R. and Halter, C. M.,** Specificity of human beta-choriogonadotropin assays for the hormone and for an immunoreactive fragment present in urine during normal pregnancy, *Clin. Chem.,* 29, 667, 1983.

50. **Papapetrou, P. D. and Nicopoulou, S. C.,** The origin of a human chorionic gonadotropin beta-fragment in the urine of patients with cancer, *Acta Endocrinol.,* 112, 415, 1986.

51. **Cole, L. A., Wang, Y., Elliott, M., Latif, M., Chambers, J. T., Chambers, S. K., and Schwartz, P. E.,** Urinary human chorionic gonadotropin free beta-subunit and core fragment: a new marker of gynecologic cancers, *Cancer Res.,* 48, 1356, 1988.

52. **O'Connor, J., Krichevsky, A., Birken, S., Armstrong, E., Schlatterer, J., and Canfield, R.,** Development of an immunoradiometric assay for hCG beta-fragment, *Cancer Res.,* 48, 1361, 1988.

53. **Akar, A. H., Wehmann, R. E., Blithe, D. L., Blacker, C., and Nisula, B.,** A radioimmunoassay for the core fragment of the human chorionic gonadotropin beta-subunit, *J. Clin. Endocrinol. Metab.,* 66, 538, 1988.

54. **Ashitaka, Y., Nishimura, R., Takamori, M., and Tojo, S.,** Production and secretion of HCG and HCG subunits by trophoblastic tissue, in *Chorionic Gonadotropin,* Segal, S. J., Ed., Plenum Press, New York, 1980, 147.

55. **Wehmann, R. E. and Nisula, R. E.,** Metabolic clearance rates of the sub-units of human chorionic gonadotropin in man, *J. Clin. Endocrinol. Metab,* 48, 753, 1979.

57. **Khazaeli, M. B., Haghighi-Rad, F., Laferla, J. J., Beierwaltes, W. H., and Beer, A. E.,** Comparison of serum levels of human chorionic gonadotropin (hCG) and its free beta subunit in normal and gestational trophoblastic disease using monoclonal antibodies, presented at 2nd Int.Congr. Reproductive Immunology, Kyoto, Japan, 1983.

58. **Shapiro, A. I., Tsai-Feng, W., Ballon, S. C., and Lamb, E. J.,** Use of an immunoradiometric assay and a radioimmunoassay for the detection of free beta human chorionic gonadotropin, *Obstet. Gynecol.,* 65, 545, 1985.

59. **Ozturk, M., Bellet, D., Manil, L., Hennen, G., Frydman, R., and Wands, J.,** Physiologic studies of human chorionic gonadotropin (hCG), alpha hCG, and beta hCG as measured by specific monoclonal immunoradiometric assays, *Endocrinology,* 120, 549, 1987.

60. **Khazaeli, M. B., Hedayat, M. M., Hatch, K. D., To, A. C. W., Soong, S. J., Shingleton, H. M., Boots, L. R., and LoBuglio, A. F.,** Radioimmunoassay of free beta subunit of human chorionic gonadotropin as a prognostic test for persistent trophoblastic disease in molar pregnancy, *Am. J. Obstet. Gynecol.,* 155, 320, 1986.

61. **Spratt, D. I., Chin, W. W., Ridgway, E. C., and Crowley, W. F.,** Administration of low dose pulsatile gonadotropin releasing hormone (GnRH) to GnRH-deficient men regulates free alpha subunit secretion, *J. Clin. Endocrinol. Metab.,* 62, 102, 1986.

62. **Hagen, C. and McNeilly, A. S.,** The specificity and application of a radioimmunoassay for the alpha subunit of luteinizing hormone in man, *Acta Endocrinol. (Copenhagen),* 78, 664, 1975.

63. **Edmonds, M., Molitch, M., Pierce, J. G., and Odell, W. D.,** Secretion of alpha subunits of luteinizing hormone (LH) by the anterior pituitary, *J. Clin. Endocrinol. Metab.,* 41, 551, 1975.

64. **Hagen, C. and McNeilly, A. S.,** Changes in circulating levels of LH, FSH, LH beta and alpha subunit after gonadotropin-releasing hormone, and TSH, LH beta and alpha subunit after thyrotropin-releasing hormone, *J. Clin. Endocrinol. Metab.,* 41, 466, 1975.

65. **Kourides, I. A., Weintraub, B. D., Rosen, S. W., Ridgway, E. C., Kliman, B., and Maloof, F.,** Secretion of alpha subunit of glycoprotein hormones by pituitary adenomas, *J. Clin. Endocrinol. Metab.,* 43, 97, 1976.

66. **Kourides, I. A., Ridgway, E. C., Weintraub, B. D., Bigos, S. T., Gershengorn, M. C., and Maloof, F.,** Thyrotropin-induced hyperthyroidism: use of alpha and beta subunit levels to identify patients with pituitary tumors, *J. Clin. Endocrinol. Metab.,* 45, 534, 1977.

67. **Hagen, C., Gilby, E. D., McNeilly, A. S., Olgaard, K., Bondy, P. K., and Rees, L. H.,** Comparison of circulating glycoprotein hormones and their subunits in patients with oat cell carcinoma of the lung and uraemic patients on chronic dialysis, *Acta Endocrinol. (Copenhagen),* 83, 26, 1976.

68. **Cole, L. A., Hussa, R. O., and Rao, C. V.,** Discordant synthesis and secretion of human chorionic gonadotropin and subunits by cervical carcinoma cells, *Cancer Res.,* 41, 1615, 1981.

69. **Khazaeli, M. B., England, B. G., Dieterle, R. C., Nordblom, G. D., Kabza, G. A., and Beierwaltes, W. H.,** Development and characterization of a mono-clonal antibody which distinguishes the beta subunit of human chorionic gonadotropin (beta hCG) in the presence of hCG, *Endocrinology,* 109, 1290, 1981.

70. **Gupta, S. K., Singh, O., Kauer, I., and Talwar, G. P.,** Characterization of monoclonal anti-alpha-human chorionic gonadotropin antibody, *Indian J. Med. Res.,* 81, 281, 1985.

71. **Thotakura, N. R. and Bahl, O .P.,** Highly specific and sensitive hybridoma antibodies against the alpha subunit of human glycoprotein hormones, *Endocrinology,* 117, 1300, 1985.

72. **Ruddon, R. W., Bryan, A. H., Hanson, C. A., Perini, F., Ceccorulli, L. M., and Peters, B .P.,** Characterization of the intracellular and secreted forms of the glycoprotein human chorionic gonadotropin produced by human malignant cells, *J. Biol. Chem.,* 256, 5189, 1981.

73. **Boothby, M., Ruddon, R. W., Anderson, C., McWilliams, D., and Boime, I.,** A single gonadotropin a-subunit gene in normal tissue and tumor-derived cell lines, *J. Biol. Chem.,* 256, 5121, 1981.

74. **Morgan, F. J., Birken, S., and Canfield, B.,** The amino acid sequence of human chorionic gonadotropin, *J. Biol. Chem.,* 250, 5247, 1975.

75. **Sadler, J. E.,** Biosynthesis of glycoproteins: formation of O-linked oligosaccharides, in *Biology of Carbohydrates,* Vol. 2, Ginsburg, V. and Robbins, P. W., Eds., John Wiley & Sons, New York, 1984, 199.

76. **Cole, L. A., Birken, S., and Perini, F.,** The structures of the serine linked sugar chains on human chorionic gonadotropin, *Biochem. Biophys. Res. Comm.,* 126, 333, 1985.

77. **Coorless, C. and Boime, I.,** personal communication, 1986.

78. **Wehmann, R. E. and Nisula, B. C.,** Metabolic and renal clearance rates of purified human chorionic gonadotropin, *J. Clin. Invest.,* 68, 104, 1981.

79. **Van Hall, E. V., Vaitukaitis, J. L., Ross, G. T., Hickman, J. W., and Ashwell, G.,** Effects of progressive desialylation on the rate of disappearance of immunoreactive HCG from plasma in rats, *Endocrinology,* 89, 11, 1971.

80. **Lefort, G., Stolk, J., and Nisula, B. C.,** Evidence that desialylation and uptake by hepatic receptors for galactose-terminated glycoproteins are immaterial to the metabolism of human chorionic gonadotropin in the rat, *Endocrinology,* 115, 1151, 1984.

81. **Kanwar, Y. S. and Farquar, M. G.,** Role of glycosaminoglycans in the permeability of glomerular basement membrane, *Fed. Proc. Fed. Am. Soc. Exp. Biol.,* 30, 334, 1980.

82. **Terao, N. and Tannen, R. L.,** Characterization and acidification by the distal nephron using the isolated perfused rat kidney, *Kidney Int.,* 20, 36, 1981.

83. **Wehmann, R. E. and Nisula, B. C.,** Renal clearance rates of the subunits of human chorionic gonadotropin in man, *J. Clin. Endocrinol. Metab.,* 50, 674, 1980.

84. **Lefort, G. P., Stolk, J. M., and Nisula, B. C.,** Renal metabolism of the beta subunit of human choriogonadotropin in the rat, *Endocrinology,* 119, 924, 1986.

85. **Wehmann, R. E. and Nisula, B. C.,** Characterization of a discrete degradation product of human chorionic gonadotropin beta subunit in humans, *J. Clin. Endocrinol. Metab.,* 51, 101, 1980.

86. **Masure, H. R., Jaffe, W. L., Sickel, M. A., Birken, S., Canfield, R. E., and Vaitukaitis, J. L.,** Characterization of a small molecular size urinary immunoreactive human chorionic gonadotropin like substance produced by the normal placenta and by hCG secreting neoplasms, *J. Clin. Endocrinol. Metab.,* 53, 1014, 1981.

87. **Amr, S., Wehmann, R. E., Birken, S., Canfield, R. E., and Nisula, B.,** Characterization of a carboxyterminal peptide fragment of the human choriogonadotropin beta subunit excreted in the urine of a woman with choriocarcinoma, *J. Clin. Invest.,* 71, 329, 1983.

88. **Rao, C. V.,** Properties of gonadotropin receptors in the cell membranes of bovine corpus luteum, *J. Biol. Chem.,* 249, 2864, 1974.

89. **Rao, C. V.,** Gonadotropin receptors in human corpora lutea of the menstrual cycle and pregnancy, *Am. J. Obstet. Gynecol.,* 128, 146, 1977.

90. **Nisula, B. C., Taliadouros, G. S., and Carayon, P.,** Primary and secondary biologic activities intrinsic to the human chorionic gonadotropin molecule, in *Chorionic Gonadotropin,* Segal, S. J., Ed., Plenum Press, New York, 1980, 17.

91. **Wolff, J. R., Winand, R. J., and Kohn, L. D.,** The contribution of the subunits of thyroid stimulating hormone to the binding and biological activity of thyrotropin, *Proc. Natl. Acad. Sci. U.S.A,* 71, 3460, 1974.

92. **Williams, J. F., Davies, T. F., Catt, K. J., and Pierce, J. G.,** Receptor- binding activity of highly purified bovine luteinizing hormone and thyrotropin, and their subunits, *Endocrinology,* 106, 1353, 1980.

93. **Moudgal, N. R. and Li, C. H.,** Beta subunits of human choriogonadotropin and ovine lutropin are biologically active, *Proc. Natl. Acad. Sci. U.S.A.,* 79, 2500, 1982.

94. **Begeot, M., Hemming, F. J., Dubois, P. M., Combarnous, Y., Dubois, M. P., and Aubert, M. L.,** Induction of pituitary lactotrope differentiation by luteinizing hormone alpha subunit, *Science,* 226, 566, 1984.

95. **Cole, L. A.,** Distribution of O-linked sugar units on hCG and its free α-subunit, *Mol. Cell. Endocrinol.,* 50, 45, 1987.

96. **Rosa, C., Amr, S., Birken, S., Wehmann, R., and Nisula, B.,** Effect of desialylation of human chorionic gonadotropin on its metabolic clearance rate in human, *J. Clin. Endocrinol. Metab.,* 59, 1215, 1984

97. **Cole, L. A. and Birken, S.,** Beta-subunit core fragment may originate in the placenta and be the predominant hCG-form produced (Abstr. 1123), presented at 70th Annual Meeting of The Endocrine Society, New Orleans, 1988.

98. **Ozturk, M., Berkowitz, R., Goldstein, D., Bellet, D., and Wands, J,** Differential production of human chorionic gonadotropin and free subunits in gestational trophoblast disease, *Am. J. Obstet. Gynecol.,* 158, 193, 1988.

99. **Fan, C., Goto, S., Furuhashi, Y., and Tomoda, Y.,** Radioimmunoassay of the serum free beta-subunit of human chorionic gonadotropin in trophoblastic disease, *J. Clin. Endocrinol. Metab.,* 64, 313, 1987

100. **Vaitukaitis, J. L. and Ebersole, E. R.,** Evidence for altered synthesis of human chorionic gonadotropin in gestational trophoblastic tumors, *J. Clin. Endocrinol. Metab.,* 42, 1048, 1976.

101. **Vaitukaitis, J. L.,** Immunologic and physical characterization of human chorionic gonadotropin (hCG) secreted by tumors, *J. Clin. Endocrinol. Metab.,* 37, 505, 1973.

102. **Krichevsky, A., Armstrong, E. G., Schlatter, J., Birken, S., Lustbader, J., Bikel, K., Silverberg, S., and Canfield, R. E.,** Preparation and characterization of antibodies to the urinary fragment of the human chorionic gonadotropin beta-subunit, *Endocrinology,* 122, in press, 1988.

103. **Cole, L. A., Wang, Y., Chambers, J., Chambers, S., and Schwartz, P.,** Urinary gonadotropin fragments (UGF): a new gynecologic cancer marker (Abstr. 18), presented at the Society of Gynecologic Oncologists Annual Meeting, Miami, 1988.

Chapter 4

CHARGE HOMOGENEITY OF HUMAN LUTEINIZING HORMONE DOES NOT IMPLICATE CHARGE HOMOGENEITY OF ITS SUBUNITS

L. A. van Ginkel and J. G. Loeber

I. INTRODUCTION

During the last 10 years there has been increasing awareness of molecular heterogeneity of a number of hormones, including luteinizing hormone (LH). During the 1950s and 1960s many investigators were primarily concerned with isolation and purification of hormones using, for the most part, rather crude fractionation systems. Human pituitary glands obtained at autopsy were preserved frozen or acetone-dried and subsequently extracted. Purification often comprised ammonium sulfate or ethanol fractionation followed by various ion exchange chromatography steps.[1-8] Many investigators were not aware of the need to include protease inhibitors from the very beginning of the procedure. Consequently, when characterization of the highly purified preparations by electrophoresis or isoelectric focusing revealed a multitude of protein bands, this was later interpreted as being purification artifacts.[9-10]

However, during the 1970s, immunoassay systems with ever-increasing specificity and sensitivity became available. Hence it was possible to assess the immunoreactive profile after isoelectric focusing of gently extracted single pituitary glands or even plasma samples. The observations in such cases of a variety of molecular forms gradually led to the conclusion that there exists a "natural" heterogeneity, at least for LH and the other pituitary glycoprotein hormones, follicle-stimulating hormone (FSH), and thyroid-stimulating hormone (TSH).

From studies in the rat and monkey it was concluded for FSH that the pituitary gland modulates its own secretion, i.e., it secretes different molecular forms of the hormone in relation to the endocrine status of the body.[11,12] Since then, a large number of studies have reported that in various species there are structural differences in LH and FSH secreted by the pituitary gland, depending on age, sex, and reproductive cycle.[13-21] Although some investigators have found differences in molecular size, the most prominent distinction between the molecular forms appears to be charge, with variability being attributed to the degree of desialylation of the carbohydrate moieties.

Commonly, investigations of LH heterogeneity have been carried out by electrophoresis,[22] isoelectric focusing,[23-24] or chromatofocusing,[25] and in recent years followed by fast protein liquid chromatography[26] as well as high performance liquid chromatography.[27] Resolution has been monitored by radioimmunoassay, receptor assay, or in vitro bioassay.

In an attempt to reduce the data accumulating from the various assay systems, the concept of the ratio of biological activity over immunochemical activity (B/I ratio) has been introduced. Unfortunately, this ratio depends largely on the reference preparations being used in each particular study. These preparations are very heterogeneous themselves,[9,28] thus making comparisons between investigations rather cumbersome. However, within any one study this ratio may give information about the relative purity of the various molecular forms.

Most of the established (international) reference preparations for human LH have been shown to have isoelectric points in the pH region between 7 and 9.[9,28-30] This is probably due to the classical schemes used for isolation and purification of the glycoproteins from the pituitary gland which were designed to obtain maximal yields of, in particular, FSH. This hormone is much less abundant and less stable than LH (and TSH). Its pI values vary between 3 and 6. Most schemes contain an ion exchange chromatography step at about pH = 6 to separate FSH from LH. However, the FSH fraction still contains a substantial amount of LH. Usually, in the process of FSH purification this LH activity has been destroyed with chymotrypsin,[31] urea,[32] or removed by partitioning[33] or electrophoresis.[34] In contrast, we have purified this "acidic" form (or forms) of LH and coined it "Type I" LH in distinction from "Type II" consisting of the more common "basic" form (or forms).[35]

LH and the other glycoprotein hormones consist of two noncovalently bound subunits of about equal molecular size, the α-subunit being common and the β-subunit carrying the hormonal specificity. In human LH, the α-subunit has two carbohydrate chains attached to the protein core, whereas the β-subunit has one.[36] So far, in studying heterogeneity of human

LH, little attention has been paid to its consisting parts, the subunits. This is the more striking since it has been shown that LH spontaneously dissociates into its subunits at only slightly elevated temperatures (37°C) at which, for example, enzymatic desialylation experiments have been carried out.[37,38] In this chapter the contribution of the subunits to LH heterogeneity will be emphasized.

Experiments designed to shed some light on the phenomenon of molecular heterogeneity are centered around the use of isoelectric focusing in sucrose density gradients combined with radioimmunoassay and in vitro bioassay.[39,40] Recently, high performance hydrophobic interaction chromatography has come into use.[41,42] Thus, two more or less complementary techniques, one concerning the charged groups and the other regarding the hydrophobic groups of the molecules, were available.

Much thought has been given to the question of which LH preparation(s) to take as a model. Studies have been described in which a single preparation derived from a large number of pituitary glands was used.[23,24,30,39] The advantage then is that a series of experiments can be performed with the same material. The drawback is that in the process of isolation and purification, in fact, a number of molecular forms have been selected. In other studies, therefore, the emphasis has been on the use of a large number of individual pituitary glands, with the concomitant opposite arguments pro and contra.[43] In this chapter results of both lines of approach will be presented — highly purified LH preparations as well as material from individual glands.

Apart from making an inventory of the heterogeneity of the model preparations using the techniques mentioned above, experiments have been carried out to explore possible relationships between the observed molecular forms by making use of incubation at elevated temperatures (37 and 56°C) and digestion by neuraminidase.[44]

II. MATERIALS

A. Hormone Preparations

Three highly purified human LH preparations were included in these studies. The first two, NM14 Type I and NM14 Type II, had been prepared previously by ion exchange chromatography on CM- and DEAE-Sephadex® followed by Sephadex®-G100 gel filtration starting from crude glycoprotein material, kindly donated by the Dutch Growth Foundation. These preparations have been characterized extensively.[35] Their biological activities as assessed by ovarian ascorbic acid depletion test were 4690 (3320 to 6930) and 8400 (6600 to 11,100) IU/mg, respectively, in terms of the first International Reference Preparation for human luteinizing hormone for immunoassay (MRC 68/40). This latter preparation itself, a gift from Dr. P. L. Storring (National Institute for Biological Standards and Control, London, U.K.) was the third preparation included. It can be classified as a "Type II LH" since its purification scheme is very similar to the one used for NM14 Type II, the main difference being that at the time, Hartree et al.[6] included an additional cation exchange step using Amberlite® IRC-50.

With regard to LH subunits the following preparations were included. Subunits from NM14 Type II and a previous batch (NM04 Type II) had been prepared by chemical dissociation (8 mol/ℓ urea) of the intact hormone[8] and kindly donated by Drs. G. Hennen and J. Closset, Liège, Belgium. The first International Standards for α- and β-subunits of human pituitary luteinizing hormone (NIBSC 78/554 and 78/556, respectively) were a gift from Dr. P. L. Storring (NIBSC, London).

B. Pituitary Tumor Material

Material from patients was obtained from Prof. Dr. S. W. J. Lamberts, Academic Hospital, Rotterdam, The Netherlands, in a collaborative project on the characterization of tumor-LH.

Table 1
SPECIFICITIES (EXPRESSED AS PERCENTAGE CROSS REACTION) OF THE ANTISERA USED

	Calibrating preparation		
RIA system	NM14	NM14 α-subunit	NM04 β-subunit
LH$_i$	1	0.27[a]	0.04[a]
		(0.16—0.50)	(0.03—0.05)
LH$_\alpha$	0.20	1	0.002
	(0.18—0.21)		(0.002—0.003)
LH$_\beta$	0.24[a]	0.006	1
	(0.22—0.26)	(0.005—0.006)	

Note: 95% confidence limits are given between parentheses.

[a] Nonparallelism between the two calibration preparations in this system.

Pituitary (tumor) material, obtained after surgery or autopsy, was brought into culture. After incubation the medium was separated from the cells. The cells were homogenized and the cell content was prepared by centrifugation. The incubation medium was concentrated and freed from salt. An aliquot was focused as described below. No attempt was made to separate intact hormone and free subunits, since all tumors had been shown to produce only subunits.[45]

C. Other Reagents

Bovine serum albumin (BSA) and hemoglobin were purchased from Sigma, Sephadex® G100 from Pharmacia, Sac-cel® from Wellcome, cellulose from Whatman, and Na-^{125}I (IMS-30) from Amersham International. The 110 mℓ IEF column and Ampholines® were from LKB. All other chemicals were reagent grade from Merck.

III. METHODS

A. Radioimmunoassay (RIA)

Three different RIA systems were used. The first was designed to measure "intact" LH, i.e., undissociated LH, employing an antiserum raised against hCG (code 8383). The other two were specific for LH$_\alpha$ (antiserum code 8082) and LH$_\beta$ (antiserum code 8095), respectively. For each system the preparation used for calibration was also the one employed as tracer, i.e., NM14 Type II, NM14 α-subunit and NM04 β-subunit, respectively. The procedure has been described before[46] except that for separation of antibody-bound and free hormone now Sac-cel® was used. The tracer was prepared by means of the Chloramine T method.[47] This method yielded for the intact hormone and the α-subunit a specific activity of 100 μCi/μg and for the β-subunit a specific activity of 50 μCi/μg. Prior to use the tracer was purified by chromatography on cellulose.[48]

Table 1 shows the specificities of the different antisera with respect to the intact hormone, the α-subunit and the β-subunit. In some cases there is nonparallelism between the calibrating preparation used in that system and the preparation being tested. A severe case of nonlinearity is observed when testing NM14 α-subunit in the RIA-system LH$_i$. From Figure 1 it can be concluded that NM14 α-subunit cannot compete fully with iodinated intact LH for all the binding sites in the antiserum used (code 8383).

By testing samples in all three RIA systems response ratios can be calculated. A ratio α/I is defined here as the ratio of the response in the RIA system LH$_\alpha$ to the response in the system LH$_i$. Likewise there is a ratio β/I. Based on the average crossreactivities an intact

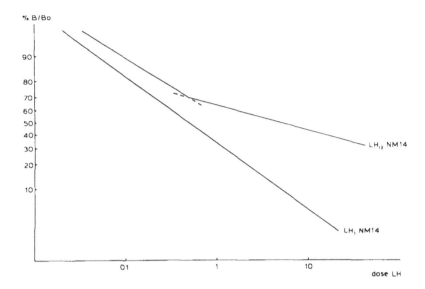

FIGURE 1. Dilution curves of the calibration preparations LH$_i$ (NM14) and LH$_\alpha$ (NM14) obtained with the RIA system LH$_i$. (From van Ginkel, L. A. and Loeber, J. G., *Acta Endocrinol. (Copenhagen)*, 110, 182, 1985. With permission.)

LH component is expected to have a ratio α/I of 0.2 and a ratio β/I of 0.24. On the other hand an α-subunit component is expected to have a ratio α/I of 3.7, ranging from 2 to 6.3 due to nonparallelism (Figure 1). A β-subunit component should have a ratio >5. With samples having an unknown composition, the situation is even more complicated. Intermediate values indicate the presence of several different components. These criteria are only given as guidelines and must be used with great care. For instance, the presence of a relatively large amount of α-subunit will not affect the response β itself, but the ratio β/I will be affected because of an increased value of I. Fortunately, in many cases a more specific in vitro biological system (see below) could give more decisive information.

For data reduction and evaluation a computer program developed by Dr. D. Rodbard was used, which is based on the logit-log transformation of the dose-response curve.[49] For all systems the within-assay variation was <11% and the between assay variation <14%. The working range was 0.08 to 5 ng of the calibrating preparation (LH$_i$ system: NM14; LH$_\alpha$ system: NM14 α-subunit; LH$_\beta$ system: NM04 β-subunit).

B. In Vitro Bioassay (TPA)

An in vitro bioassay based on the production of testosterone by mouse Leydig cell preparation[50] was employed. The testosterone produced by graded doses of LH calibrator or samples was quantified by radioimmunoassay. Samples (0.1 mℓ) were incubated with 0.4 mℓ cell suspension which contained the equivalent of one fifth of a testis. Results obtained with this assay system were expressed in terms of nanogram NM14.

For calculation of the LH bioactivity of a sample, two methods were used. If the samples were measured at a single dose level, the dose-response curve for LH was fitted by a computer program using the mathematical expression $y = A + B \ln (x + C)$ in which x is the dose of LH, y the amount of testosterone produced and A and B are constants. The value of C is calculated by an iterative process such as to minimize the residual sum of squares[51] and is a constant for each assay. By using this type of regression analysis the correlation between x and y improved as compared to standard logarithmic regression. The LH potencies of the samples were then found by interpolation. Each result was the mean of three estimates. Each top fraction, containing a component in its "purest" form, was assayed again at two

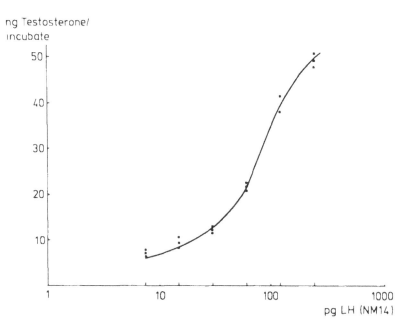

FIGURE 2.   Testosterone production as a function of the amount of bioactive LH present in a mouse Leydig cell incubation mixture.

dose levels in triplicate and standard parallel-line procedures were applied.[52] Using the latter method the assay had a within-assay variation of 7% and a between-assay variation of 12%. The working range was 16 to 500 pg LH NM14. A typical dose-response curve is shown in Figure 2.

C. Isoelectric Focusing (IEF)

Isoelectric focusing (IEF) is a technique by means of which molecules are separated on a charge basis. Prior to the experiment, carrier ampholites are added which form a pH gradient under the influence of an electric field. Due to the buffering capacity of these carrier ampholites, sample molecules adapt a charge corresponding to the pH at that particular place in the gradient. They move through the gradient as long as they carry a charge. When the pH value in the gradient corresponds with the isoelectric point of the sample molecules, they focus at that spot and remain there as long as the gradient is stable.

LH samples, together with the internal marker hemoglobin, were mixed with the gradient solutions. High-speed isoelectric focusing using constant power was used. The power was limited to 15 W and the voltage to 1600 V. The temperature of the cooling water was 10 to 15°C. After 18 hr the current was switched off and the column emptied using a peristaltic pump. Fractions of 1.1 mℓ were collected. The UV absorbance of the effluent was measured on-line (Uvicord® 2, LKB) to determine the elution volume of the hemoglobin. The pH of each fifth fraction was measured immediately (Radiometer). In 12 consecutive experiments the hemoglobin maximum was located at pH = 7.49 ± 0.04 (mean ± SD). After pH measurement all fractions were neutralized with 2.5 mol/ℓ phosphate buffer pH 7.4. BSA was added to a final concentration of 1% (W/V). Since the sucrose in the column fractions in concentrations above 1% (W/V) shows significant inhibition of the testosterone production in the in vitro bioassay, samples were diluted accordingly or dialyzed-ultrafiltered in an Amicon® MMc cell. The Ampholines® did not show interference in any of the assay systems used.

D. Hydrophobic Interaction Chromatography (HIC)

Hydrophobic interaction chromatography (HIC) separates molecules on the basis of differences in strengths of hydrophobic interactions with a matrix bearing hydrophobic groups.[41] Due to the relatively mild conditions used in this technique the molecules do not denature. This means that the separation is, in contrast to the separation mechanism in reversed-phase chromatography, based on the number and strength of the hydrophobic groups only on the surface of the molecule. Under conditions of high salt concentration, the sample molecules are absorbed on the chromatographic material, after which they are eluted by a decreasing salt gradient.

Here the technique in its "high performance" form (TSK-5PW phenyl column, LKB) (HPHIC) was used. All runs were performed with a debiet of 0.5 mℓ/min in 0.02 mol/ℓ phosphate buffer saline (pH = 7.0) with an ammonium sulfate gradient, decreasing from 0.5 to 0 mol/ℓ. Fractions of 0.65 mℓ were collected into tubes containing 0.5 mℓ PBS plus 0.25% (W/V) BSA.

E. Procedure

All our experiments to acquire information about the nature of LH heterogeneity were performed according to the scheme below.

Solution of preparation
under investigation
$\downarrow$
G100-gel filtration as
further purification step
$\downarrow$
Incubation:
at 37°C,
at 56°C, or
in the presence of neuraminidase
$\downarrow$
Fractionation
according to
- size
- charge
- hydrophobic groups
$\downarrow$
Investigation of
different fractions
by: RIA
system LH$_i$
system LH$_\alpha$
system LH$_\beta$
in vitro bioassay

As has been shown before,[35] LH preparations usually contain some "free" subunits. In those experiments in which it was essential to remove these, a gel filtration step was carried out prior to incubation.

Table 2
COMPONENTS IDENTIFIED IN THE pH
GRADIENT pH 3.5 TO 10 USING
IMMUNOCHEMICAL SYSTEMS AND AN
IN VITRO BIOASSAY (TPA)

pI	R-α/I	R-β/I	R-B/I
4.75 ± 0.02	2.2 ± 0.1	0	0
5.04 ± 0.02	1.6 ± 0.2	0.26 ± 0.07	0.4 ± 0.1
5.60 ± 0.01	—[a]	0	0.9 ± 0.2
6.06 ± 0.04	0.87 ± 0.03	0.35 ± 0.05	0.7 ± 0.2
6.57 ± 0.05	—[a]	0.39 ± 0.05	0.7 ± 0.2
6.87 ± 0.03	—[a]	0.47 ± 0.03	—[b]

Note: pI values are the mean of two experiments; activities are expressed as response ratios (see text).

[a] No response maximum detected with the RIA system LHα.
[b] No response maximum detected with the TPA.

IV. RESULTS

A. IEF of LH Type I

1. Preparation NM14

In the pH range 3.5 to 10 approximately 1 μg LH (NM14) Type I was focused. Column fractions were collected and assayed with the three RIA systems and the TPA. The results of this experiment and one in which 10 μg was focused are summarized in Table 2. Components are identified by their pI value and characterized by the immunochemical ratios α/I and β/I as described in Section III, Methods.

On the basis of these criteria it is concluded that a total of four biologically active components, pI values being 5.04, 5.60, 6.06, and 6.57, three α-subunits with pI values 4.75, 5.04, and 6.06, and four β-subunits with pI values 5.04, 6.06, 6.57, and 6.87 are present. The coappearence of biologically active forms with α- and/or β-subunits is, as will be discussed below, a common phenomenon. The ratios B/I, being the ratio of in vitro bioactivity to immunochemical activity (RIA system LH$_i$), are all less than 1. For the components with pI = 5.04 and 6.06 this value is affected by the presence of cross-reacting α-subunits, which enhance the response I but do not affect B.

2. NM14 After Gel Filtration Chromatography

Because of evidence of the presence of free α-subunits in the NM14 Type I preparation, 60 μg was gel filtered on Sephadex® G100. In order to obtain significant amounts of both subunits the preparation was incubated for 20 hr at 37°C, making the simultaneous study of the intact and subunit molecules possible. The chromatogram as detected with the RIA systems is represented in Figure 3. On the basis of this chromatogram three pools were prepared. Pool LH$_i$, containing the intact molecules (fractions 38 to 45); pool LH$_s$, containing the subunits (fractions 50 to 60), and a pool containing the fractions in between (pool LH$_{i/s}$, fraction 46 to 49).

The IEF results for the pools LH$_i$ and LH$_s$ are summarized in Table 3. Pool LH$_i$ contains five different components (Figure 4) with pI values of 5.15, 5.51, 6.08, 6.42, and 6.88. The components with pI = 6.42 and 6.88 had not been observed in the experiment without gel filtration, probably due to the presence of interfering subunits in that experiment. The pool LH$_s$ (Figure 5) contains four α-subunits with pI values 4.45, 4.77, 5.09, and 6.05. The component with pI = 4.45 was not observed in the experiment without gel filtration.

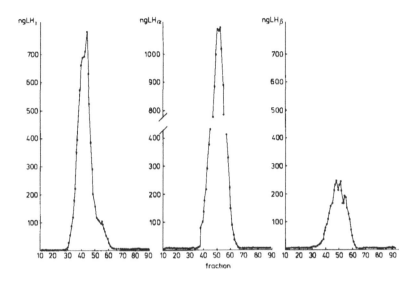

FIGURE 3. Gel filtration (G100) of LH Type I (NM14) after incubation for 20 hr at 37°C as detected with the three different RIA systems.

Table 3

COMPONENTS DETECTED IN THE POOLS LH$_i$ AND LH$_s$ OBTAINED BY GEL FILTRATION OF NM14 TYPE I PRIOR TO ISOELECTRIC FOCUSING

Pool-LH$_i$			Pool-LH$_s$		
pI	R-α/I	R-β/I	pI	R-α/I	R-β/I
			4.45	6.4	0
			4.77	5.5	0
5.15	0.66	0.20	5.09	5.3	0.63
5.51	0	0			
6.08	0.38	0.24	6.05	2.7	0.98
6.42	0	0.20			
			6.64	0	2.1
6.88	0	0.40			
			6.96	0	2.4
			7.21		—ᵃ
			7.43	0	3.4

ᵃ Only detected with the RIA system LH$_\beta$.

β-Subunits with pI values 5.09, 6.05, 6.64, 6.96, 7.21, and 7.43 are detected. Of these six components the two most basic were not observed in the experiment without gel filtration. On the basis of the G100-chromatogram obtained with the RIA system LH$_\beta$ the assumption that the pools LH$_s$ and LH$_{i/s}$ do not contain the same charge population of β-subunits was tested. However, the conclusion was that the subunits populations of these pools are identical with respect to charge.[53]

3. NM14 After Incubation at 56°C

A different approach to obtain information about the subunit content of a preparation is to dissociate it as far as possible. For this purpose 5 μg of the preparation LH (NM14) Type

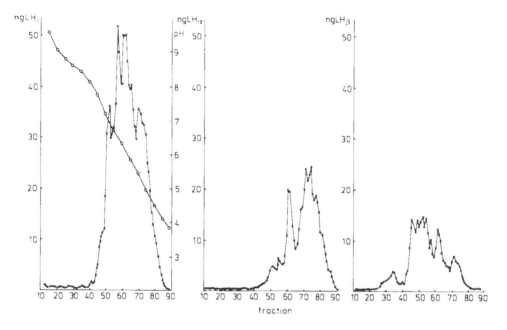

FIGURE 4.   IEF of LH (NM14) Type I after gel filtration (G100) (Pool LH₁) as detected with the three RIA systems. The pH gradient is shown in the left panel.

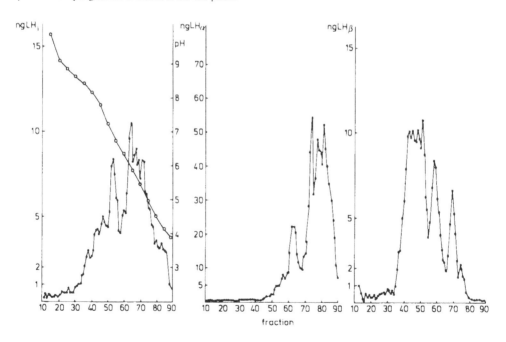

FIGURE 5.   IEF of LH (NM14) Type I after gel filtration (G100) (Pool LHₛ) as detected with the three RIA systems. The pH gradient is shown in the left panel.

I was incubated for 20 hr at 56°C. The potency of this preparation in terms of LH (NM14) Type II decreased from 0.32 ± 0.04 to 0.10 ± 0.02. This decrease, as well as severe nonparallelism and incomplete competition in the system LH_i, proves the almost complete absence of LH_i. The focusing patterns assessed with the three RIA systems are shown in Figure 6.

The response obtained with the RIA system LH_i is indeed very low and clearly a result

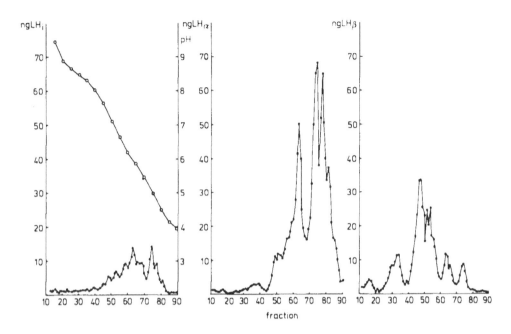

FIGURE 6. IEF of LH Type I (NM14) after incubation for 20 hr at 56°C. The pH gradient is shown in the left panel.

Table 4
OBSERVED β-SUBUNITS AFTER INCUBATION OF NM14 TYPE I FOR 20 HR AT 56°C COMPARED WITH VALUES OBTAINED AFTER INCUBATION AT 37°C AND GEL FILTRATION

pI	56°C	37°C (mean ± SD [N = 2])
5.10	6	14 ± 3
6.00	8	23 ± 1
6.64	16	17 ± 3
7.0—7.5	47[a]	49 ± 4
8.32—8.49	21[b]	
9.23	3	

Note: The quantitative distribution is expressed as the percentage of the total immunoreactivity detected (% t.d.).

[a] Mainly a component with pI = 7.49.
[b] Mainly a component with pI = 8.32.

of cross-reactive α-subunits. With the RIA system LH_α a pattern very much the same as observed for the G100 pool LH_s is obtained (Table 3). These results with regard to LH_i and α-subunits are as expected on the basis of the experiments described above.

The pattern obtained with the RIA system LH_β is different, though. Compared with Table 3 the resolution between pH = 7.0 and 7.5 seems less in favor of the component with pI = 7.43. In addition, at the basic end of the gradient, components with pI = 8.32, 8.49, and 9.23 are observed. The results are summarized in Table 4. Comparing the % t.d.

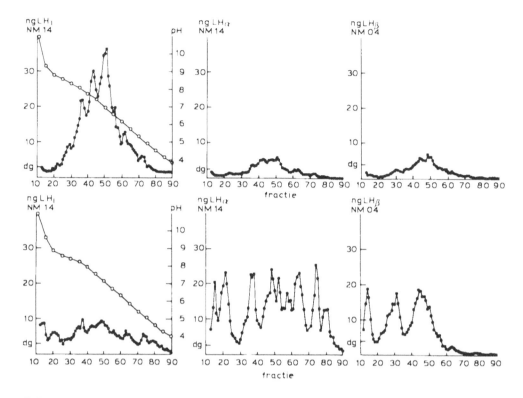

FIGURE 7. IEF of LH Type II (NM14) after gel filtration as detected with the three RIA systems. Top half: Pool LH$_i$, bottom half: Pool LH$_x$. The respective pH gradients are shown in the left panels. (From van Ginkel, L. A. and Loeber, J. G., *Acta Endocrinol. (Copenhagen)*, 110, 182, 1985. With permission.)

(percentage fraction of the total immunochemical response detected) after incubation (see below) at 37°C followed by gel filtration with these results, it is concluded that the basic β-subunits seem to develop from the two most acidic forms with pI values 5.10 and 6.02. The component with pI = 6.66 and the group of components between pI = 7.0 and 7.5 does not change quantitatively.

B. IEF of LH Type II

1. Preparation NM14

The results obtained for this preparation have been published in detail previously.[39] The procedure used was similar to the one described above for LH NM14 Type I. Combining the results obtained with and without gel filtration chromatography, four biologically active components were identified with pI values of 7.09 ± 0.05, 7.71 ± 0.05, 8.13 ± 0.04, and 8.46 ± 0.05 (mean ± SD, n >5). A total of seven α-subunits, pI values being 4.60 ± 0.09, 5.15 ± 0.05, 5.99 ± 0.04, 7.08 ± 0.05, 8.09 ± 0.04, 8.83 ± 0.01, and 9.7 ± 0.3 were identified. Additional proof of the subunit nature of these components was obtained by incubation of solutions of the preparation at 37°C, after which a clear increase of the immunoreactivity as a function of time was assessed. Similar results were found for the detected β-subunits, with pI values of 7.65 ± 0.04, 8.37 ± 0.02, 8.54 ± 0.02, and 10.4 ± 1.6. The IEF pattern of the G100 pools LH$_i$ and LH$_x$ have been reproduced in Figure 7.

2. Preparation MRC 68/40

Figure 8 shows the IEF patterns for LH MRC 68/40 as obtained with the three RIA systems.[40] Over 90% of the total LH$_i$ immunoreactivity is observed to be located between

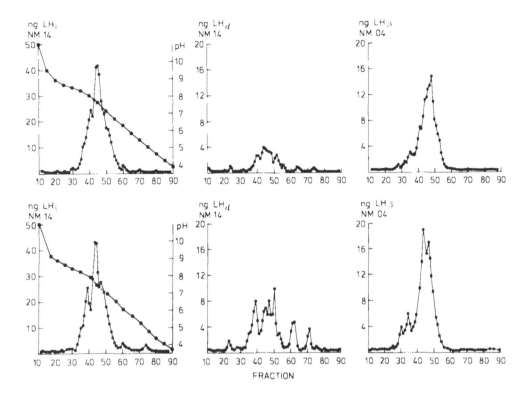

FIGURE 8.   IEF of LH MRC 68/40 (Type II) as detected with the three RIA systems. Top half: control, bottom half: after incubation for 20 hr at 37°C. The respective pH gradients are shown in the left panels. (From van Ginkel, L. A. and Loeber, J. G., *Acta Endocrinol. (Copenhagen)*, 114, 572, 1987. With permission.)

pH = 6.8 and pH = 8.2, comprising two maxima. The major proportion (85%) of the LH_α immunoreactivity is located again between pH = 6.8 and pH = 8.2, but, contrary to LH_i, there were also minor response maxima at either side of this pH interval. The major proportion (84%) of the LH_β immunoreactivity is located at pH = 7.35 with minor maxima around pH = 8.4 and pH = 8.5.

In Figure 8 the corresponding IEF patterns obtained after incubation of the sample during 20 hr at 37°C are shown as well. The LH_i pattern is only slightly different from the one without incubation. There is increased resolution at the basic end of the pH gradient. Furthermore, there appear to be some maxima at the acidic end of the gradient. Changes due to incubation are more pronounced when either subunit RIA system is used. Six different α-subunits were detected, ranging in pI value from pI = 5.17 to pI = 8.80. β-Subunits were only detected above pH = 7.5. The total number of β-subunits was three, ranging in pI value from pI = 7.63 to pI = 8.43.

Combining the results for the two LH preparations shows that the pI values for the detected intact components as well as for the individual α- and β-subunits are very similar (Table 5). Actually, it seems that differences are only in a quantitative sense. Nonetheless, some differences can be shown. For example, NM14 contains material with a lower pI value (7.09) than present in MRC 68/40. A similar observation can be made for the α-subunits: a component with pI = 4.60 is only present in NM14.

3. Individual LH_i Components Before and After Incubation at 37°C.

The most interesting question was which subunit components originated from each of the four LH_i components during incubation at 37°C. For this purpose a relatively large amount of 100 μg was focused and top fractions of each of these components were pooled, con-

Table 5
pI VALUES (MEAN ± SD,
N≥2) OF LH COMPONENTS
AS DETECTED IN THE
PREPARATIONS MRC 68/40
AND NM14 TYPE II

Preparation (MRC 68/40)	Preparation (NM14 Type II)
	7.09 ± 0.05
7.69 ± 0.02	7.71 ± 0.05
8.05 ± 0.03	8.13 ± 0.04
	8.46 ± 0.05

α-Subunits

	4.60 ± 0.09
5.17 ± 0.09	5.15 ± 0.05
6.07 ± 0.07	5.99 ± 0.04
7.04 ± 0.05	7.08 ± 0.05
7.33 ± 0.05	
8.02 ± 0.02	8.09 ± 0.04
8.80 ± 0.04	8.83 ± 0.01
	9.7 ± 0.3

β-Subunits

7.63 ± 0.05	7.65 ± 0.04
8.33 ± 0.03	8.37 ± 0.02
8.43 ± 0.02	8.54 ± 0.02
	10.4 ± 1.6

Note: Values for subunits were obtained after incubation at 37°C.

centrated, and freed from sucrose and Ampholines® by ultrafiltration. An aliquot of these pools was incubated at 37°C for 20 hr and focused while a nonincubated aliquot served as a control. These experiments show that a charge-homogeneous intact component does itself contain a heterogeneous population of subunits. In fact, each LH_i component contains virtually all the α- and β-subunits detected in this preparation. However, quantitatively there is a clear shift to the more basic subunits when the "starting" component becomes more basic. Thus the α-subunit with pI = 4.60 is not observed when incubating the intact components with pI values 7.71, 8.13, and 8.46 and the α-subunit with pI = 9.7 is only observed on incubating the component with pI = 8.46.

C. Composition of Subunit Reference Preparations

Information about subunit heterogeneity described so far was obtained indirectly by temperature-induced dissociation of intact LH molecules. It was interesting to compare these results with those from "established" subunit preparations, obtained by chemical dissociation (8 mol/ℓ urea),[8] namely the International Standards for LH_α and LH_β. The differences were negligible (Table 6) although two additional α-subunits were present with pI values 6.70 and 7.35, the first one only in trace amounts.

D. Effect of Neuraminidase Treatment

In the previous sections it was shown that intact LH as well as the α- and β-subunit consists of strongly heterogeneous populations of components with different isoelectric

Table 6
α-SUBUNIT AND β-SUBUNIT
COMPONENTS DETECTED IN
THE PREPARATION LH$_α$ NIBSC
78/554 or LH$_β$ NIBSC 78/556

LH$_α$		LH$_β$	
pI	% t.d.	pI	% t.d.
4.76	5.3	7.60	16.3
5.04	9.9	8.40	21.1
5.94	17.2	8.55	21.5
6.70	—[a]	9.61	20.3
6.96	21.0		
7.35	13.3		
8.02	14.7		
8.72	4.4		
9.32	0.4		

Note: Responses are expressed as a percentage
of the total immunoreactivity detected (%
t.d.).

[a] Component present in trace amount only.

points. The next step was to determine whether this charge heterogeneity is located in the
carbohydrate chain, caused by differences in sialic acid content, or is located in the protein
core of the molecule.[44]

1. Neuraminidase Treatment of LH NM14 Type I

Of this preparation 5 μg was incubated with 75 mU neuraminidase. With the RIA system
LH$_i$ four major components could be detected with pI values of 7.67, 8.35, 8.98, and 9.43.
Additional components with pI values 8.08 and 8.53 were present in trace quantities. This
set of four components is different from the set detected without neuraminidase treatment
(Tables 2 and 3), using the criteria described in Section III, Methods.

For the α-subunit, too, a clear shift to more basic components was observed: pI values
of 5.22, 6.10, 7.57, 8.35, 8.89, and 9.43. Only the two most acidic components were
detected in non-neuraminidase incubated samples. Additional components with pI values of
6.75, 6.91, and 7.29 were detected.

Five LH$_β$ components were detected after neuraminidase treatment, with pI values of
7.67, 8.37, 8.53, 9.32, and 9.86. Of the β-subunit components present in control experiments
only trace amounts were present.

2. Neuraminidase Treatment of LH NM14 Type II

From this preparation the four earlier-mentioned components (paragraph B1) were isolated
and treated separately with neuraminidase. It was concluded that the relatively acidic components can be transferred into the more basic ones. Two additional components with pI
values 8.9 and 9.3 were detected. Furthermore, the component with pI = 9.26 apparently
does not contain removable sialic acid.

It appears that neuraminidase treatment yields a collection of in total five different biologically active components, with the same pI values irrespective from which of the four
original components had been started. Therefore, including the most "acidic" material (pI
= 7.07), the total number of distinct biologically active components is at least six. The

Table 7
α-SUBUNIT COMPONENTS
DETECTED IN INCUBATION MEDIA
OF CELLS FROM THREE
"NONFUNCTIONING" PITUITARY
TUMORS

pI (mean ± SD)	P50 (% t.d.)	P122 (% t.d.)	P144 (% t.d.)
<3.5		13.5	3.8
3.92 ± 0.01		6.6	2.9
4.22 ± 0.02		4.0	5.3
4.46 ± 0.02	a	9.8	7.7
4.80 ± 0.08	4.9	10.0	9.4
5.17 ± 0.05	10.4	8.7	6.8
6.04 ± 0.06	13.9	8.4	7.2
7.25 ± 0.05	20.5	10.5	18.3
8.10			16.6
8.34 ± 0.02	10.2	4.7	
8.92		1.7	
9.37	10.0		

Note: Responses are expressed as a percentage of the total immunochemical response (% t.d.).

a   Trace component.

ratio of in vitro biological activity to immunoreactivity (B/I) of these components was 1.43 ± 0.05 (mean ± SD, n = 2) and 1.24 (n = 1), respectively.

With regard to the subunits it was concluded that the population in a neuraminidase-treated sample is qualitatively the same as in a corresponding control sample.[44] A significant quantitative shift to the more basic components is clearly observed, though. In one experiment after incubation at 56°C two "new" α-subunits with pI values 7.62 and 8.44 were discovered, probably developed from the components with pI values 7.08 and 8.09.

E. Pituitary Tumor Material

In Table 7 the results obtained for three different incubation media of cells from so-called "nonfunctioning" pituitary tumors are summarized. In these media relatively large amounts of α-subunits can be detected.[45] The pI values were statistically homogeneous and were combined accordingly. Comparing the results with those for α-subunit components in LH and subunit preparations shows that the media contain additional acidic components with pI values below 4.5. An extra component with pI = 8.34 is observed in two experiments. Previously, a component with this pI value was only observed after neuraminidase treatment of LH Type I.

Another interesting phenomenon is illustrated in Figure 9. The top half shows the IEF patterns for the incubation medium of one of the pituitary tumors (P50), whereas the bottom half shows the patterns for the corresponding cell contents. Qualitatively the α-subunit components are the same in both profiles, but the cells contain a relatively larger amount of the more acidic components.

F. Hydrophobic Interaction Chromatography (HIC)

Hydrophobic interaction chromatography, performed in its "high performance" mode (HPHIC), was used to perform separations which were not dictated by the presence of

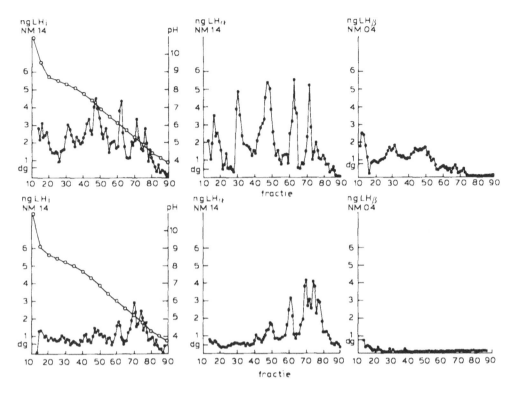

FIGURE 9. IEF of culture media from a pituitary tumor (P50). Top half: medium, bottom half: cell content. The respective pH gradients are shown in the left panels.

differing amounts of sialic acid.[53] As described above for IEF studies, prior to chromatography, LH preparations were incubated for 20 hr at 37 or 56°C.

After incubation of LH Type II at 37°C for the α-subunit (α1 and α2) as well as the β-subunit (β1 and β2) two components could be detected. The α-subunits are clearly stronger retained on the column, indicating a more hydrophobic character. After incubation at 56°C the pattern for the α-subunit remains qualitatively the same. For the β-subunit a third (β3) component develops which has a larger retention time on the column. (Figure 10). To determine whether β3 originates directly from the intact LH molecule by dissociation or from the "free" β-subunit (e.g., the first-eluting (β1), and quantitatively most important β-subunit), β1 and β3 were isolated in a preparative experiment. Component β1 was incubated for 20 hr at 56°C. An aliquot of this sample, as well as a corresponding nonincubated control, were reanalyzed by HPHIC. The two obtained chromatograms were identical to each other and were a perfect rechromatography of β1. Component β3, which was now incubated at 4°C (control) or 37°C, gave a different result. This component appeared to be unstable and was transformed into β1. This process was clearly observed in the control sample, but was even more profound in the sample incubated at 37°C.

This phenomenon was the same for LH NM14 Type I and indicates the presence of at least two pools of intact LH molecules, one of which is highly unstable and dissociates rapidly at 37°C whereas the other dissociates only at 56°C, yielding the unstable β-subunit (β3). Other experiments[53] showed that even prolonged incubation (several weeks) at 37°C is not sufficient to completely dissociate this "stable" material.

V. DISCUSSION

In recent years analytical techniques for proteins have advanced rapidly. Electrophoretic techniques such as isoelectric focusing in sucrose density gradients as well as high perform-

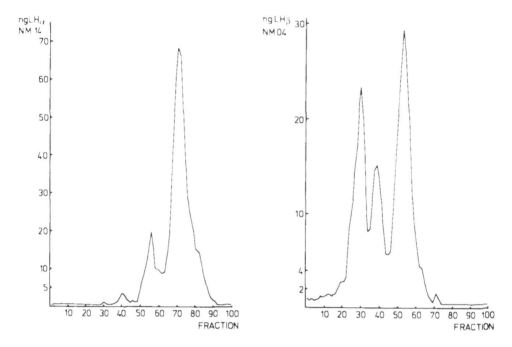

FIGURE 10.   HPHIC of LH Type II (NM14) after incubation for 20 hr at 56°C as detected with the subunit RIA systems.

ance liquid chromatography have shown to be potent techniques. Combining these "separation techniques" with highly sensitive and specific "detection techniques" such as immunoassays and in vitro bioassays has shown to be very useful in the research of heterogeneity of LH preparations. In our present studies, too, these techniques appeared to be highly informative.

A. IEF of Intact LH

Three LH preparations were used: MRC 68/40, NM14 Type I, and NM14 Type II. When combining the results obtained for these preparations it must be borne in mind that NM14 Type I contains components with pI values <6.2 and so cannot be compared with the other two preparations directly. From the experiments with NM14 Type II and MRC 68/40 it is concluded that although there is a quantitative difference in the detected components the pI values are very much alike (Table 5). A total of four biologically active components, with pI values of 7.1, 7.7, 8.1, and 8.5 are observed. These results for MRC 68/40 are in agreement with those obtained by Robertson et al.[28] It is therefore unlikely that the observed heterogeneity is just an artifact, even less in view of almost identical pI values of LH in plasma as reported by Strollo et al.[54]

As an explanation for LH heterogeneity usually the varying sialic acid contents of the compounds are mentioned. Indeed, it was observed that after neuraminidase treatment of isolated individual LH components all of the more basic components detected previously were obtained. Thus, after neuraminidase treatment of the component with pI = 7.1, the components with pI = 7.7, 8.1, and 8.5 are observed. In addition, two more basic components developed with pI values of 8.9 and 9.3. When starting with any one of the three other components, the same observations were made. Experiments in which LH was treated with a relatively large amount of neuraminidase followed by focusing in a pH gradient 9 to 11 did not give evidence for components of an even more basic character. It is therefore concluded that the component with pI = 9.3 does not contain removable sialic acid residues and can be regarded as asialo LH.

The other preparation, NM14 Type I, yielded as expected, a different focusing pattern. The components detected have pI values of 5.0, 5.6, 6.1, and 6.6 (Table 2). The latter component consists probably of two components closely adjacent with regard to pI value. Neuraminidase treatment of this preparation yielded components with pI values considerably shifted to the basic end of the gradient. The most basic component had pI values of 9.0 and 9.4, not significantly different from the components observed after neuraminidase treatment of the LH Type II preparations.

According to Pierce and Parsons[36] the protein core of LH is synthesized in the classical manner on the ribosomes. Subsequently, the carbohydrate moieties are attached to the asparagine residues. As such, disregarding protein heterogeneity, asialo LH can be regarded as the "origin" of all LH molecules, whether Type I or Type II. It is interesting to see that these "original" components are not detected in any of the preparations investigated. In fact, at this moment three regions of LH bioactivity can be discriminated: Type I with pI values <7.0, Type II with pI ≤7.0 to ≤8.5, and "original LH" with pI >8.5. When looking at the transition of LH Type I to "original LH" due to neuraminidase digestion it is observed that the intermediate components show resemblance to LH Type II: a component with pI = 7.7 is present while the components with pI = 8.1 and 8.5 are detected in trace amounts. However, there also exists a difference: a component pI = 8.35 was not observed in LH Type II. LH Type I must contain additional sialic acid residues apart from the ones occupied in LH Type II. When such an extra sialic acid residue is not removed in the beginning of the desialylation process, the development of additional components can be explained.

When comparing the B/I values for the various components[39] it was concluded that the differences in B/I values within Type II and "original" LH are small and not significant. On the other hand, Type I components seem to have smaller B/I values. However, after correcting the values in Table 2 for the presence of α-subunits, which enhance I, but do not affect B and hence decrease B/I, the differences with those in Type II are only small.

B. IEF of LH Subunits

So far the results discussed pertained to biologically active (intact) LH. The spontaneous dissociation of LH into its subunits at elevated temperatures is a known phenomenon. For this reason the study of subunit heterogeneity was undertaken. From the results obtained by isoelectric focusing of either LH Type II after thermal incubation (Table 5) or the international reference preparation for LH$_\alpha$ (NIBSC 78/554) and LH$_\beta$ (NIBSC 78/556) (Table 6) it was concluded that between different preparations a close agreement exists in the subunit population.[39,40] This is interesting since the subunit reference preparations were not obtained by incubation at an elevated temperature but by chemical dissociation.[8] α-Subunits appear to have pI values ranging from 4.5 to 9.7 whereas only "basic" (pI >7.5) β-subunits are observed.

On the other hand, LH Type I gives a more complicated picture. No basic α-subunits are present in this preparation. In addition, the β-subunits are very different from the ones observed for Type II and the international reference preparations. Below pH = 7.0, three components with pI values 5.1, 6.0, and 6.7 are observed whereas between pH 7.0 and 7.5 three additional components are present (Table 3). None of these β-subunits was observed in any of the above-mentioned LH Type II and international subunit reference preparations.

Neuraminidase treatment of LH or subunit preparations showed for the α-subunits a steady shift to the more basic components. The most basic α-subunit, with pI = 9.4, was not observed in any of the untreated preparations. After treatment of LH Type I, two additional α-subunits were detected at intermediate pH values, pI = 7.6 and 8.4. These two components were not observed in NM14 Type II, MRC 68/40 (Table 5), or the international α-subunit reference preparation. As discussed above for intact LH Type I, these two components might have sialic acid residues in additional positions.

With regard to the β-subunit, neuraminidase treatment of NM14 Type II and MRC 68/40 shows a shift to the more basic components. The two most basic β-subunits, with pI values 9.3 and 9.8, were present in the untreated material, the component with pI = 9.8 in trace amount. By neuraminidase treatment of NM14 Type I precisely all β-subunit components present in Type II were formed.

From the observations mentioned here for neuraminidase treatment of intact LH, α-, and β-subunits it is concluded that the difference in charge between LH Type I and LH Type II is caused by additional sialic acid residues which are located mainly at the β-subunit.

Other experiments indicate that there are more aspects. During the study of subunit heterogeneity two techniques to obtain subunits were used. The first was to incubate a preparation at 37°C and subsequently to isolate the subunit fraction by Sephadex® G100 gel filtration chromatography. The second approach was to incubate the preparation for 20 hr at 56°C after which no bioassayable LH was left. For the α-subunit this resulted in a similar population of components irrespective of the incubation temperature used in the dissociation step. However, in the case of 56°C two additional components with pI = 7.6 and 8.4 were observed after neuraminidase treatment. These α-subunits happen to be the same as those which appeared after treatment of intact LH Type I. Since LH NM14 Type I contains free α-subunits it is concluded that some sialic acid residues on the α-subunit within the intact molecule are poor substrates for neuraminidase.

With regard to the β-subunit a different observation was made. After incubation of LH Type I at 56°C but not at 37°C, basic components were seen (Table 4) which had pI values identical to those originated after neuraminidase treatment of LH Type I. In principle two explanations are possible: LH Type I contains heat-labile sialic acid residues which are liberated at 56°C or there exist multiple conformations of the β-subunit, either within the intact molecule or the free subunit.

In summary, it was shown that each investigated LH preparation does contain heterogeneous populations of LH_i, LH_α, and LH_β. Moreover, each individual charge homogeneous LH_i component can be dissociated into a set of subunits with discrete pI values, the relative amounts being the only difference. This means that an individual LH_i component contains a heterogeneous population of α- and β-subunits, a conclusion also reached by Hartree et al.[36] after fractionation of LH by fast protein liquid chromatography.

C. IEF of Pituitary Tumors

The results discussed so far were always obtained for preparations isolated from large pools of pituitary glands. The α-subunit heterogeneity was also investigated by focusing culture media of cells from "nonfunctioning" pituitary tumors. The results (Table 7) show that in the culture media of a single pituitary tumor almost the same collection of pI values exists as was shown for highly purified LH preparations. However, the culture media contain forms of an even more acidic nature with pI values <4.5. It is interesting to note that the cell content contains relatively more of the more acidic components than the medium of the same tumor (Figure 9). A possible explanation might be a difference in excretion rate of the various components.

D. HPHIC After Incubation at 37 or 56°C

With HPHIC the heterogeneity observed for both the α- and β-subunits was not as extensive as seen with IEF. After incubation at 37°C for both subunits two components are found. In each instance there is a major component comprising 80 to 90% of the LH_α or LH_β activity. For the β-subunit an additional third component was seen only after incubation at 56°C. This component is very unstable and was rapidly transformed to the major component.

Hence, there seems to be a molecular form of intact LH containing a β-subunit in such a conformation that it is stable with regard to incubation at 37°C, but less so at 56°C. As far as the HPHIC results are regarded there appears to be no difference between LH Type I and Type II. The β-subunit concerned was not observed in the reference preparations.

E. Conclusions

Reviewing all available data for LH_i, LH_α, and LH_β the following number of components have been detected: LH_i, 11 components ranging from pI = 5 to 9.4; LH_α, 15 components ranging from pI <3.5 to 9.4; and LH_β, 11 components ranging from pI 5.1 to 9.8. According to Pierce and Parsons[36] LH has three carbohydrate moieties attached to the protein core, two at the α-subunit chain and one at the β-subunit chain. Each carbohydrate chain is terminated by a biantennary structure, each antenna ending with a sialic acid residue. Therefore, according to this model, LH can carry a maximum of six sialic acid residues. How can the large number of components detected be fit into this model? For the α-subunit there are 16, 9, or 5 theoretical ways to distribute zero to four sialic acid residues depending on whether or not a discrimination can be made between the two antennae of each carbohydrate chain, the two chains, or none at all. Likewise for the β-subunit there are four, three, or two possibilities to distribute zero to two sialic acid residues. The total number of α-subunits does not exceed the maximum number possible. However, the total number of detected β-subunits exceeds four, indicating that the situation here is more complicated.

For LH_i it was previously concluded[39] that the component with pI = 7.1 contains five sialic acid residues, leaving only one to explain the presence of at least four more acidic forms. The presence of various β-subunits in LH Type I but not in LH Type II indicates that the source of these more acidic LH_i components is located in these β-subunits. Several authors have shown that the carbohydrate chains of human luteinizing hormone, in addition to sialylated oligosaccharides, partly are terminated by sulfated hexoses.[36,55] The precise oligosaccharide structure in our LH preparations is unknown, but the presence of sulfated hexoses, especially in the preparation NM14 Type I cannot be ruled out. The presence of such groups in the preparation NM14 Type II is less likely since there is a systematic shift to very high pI values after neuraminidase which appear to be a step-by-step process from pI = 7.1 upward.[44]

Charge differences appear to be the major cause for human LH heterogeneity as was reported by many investigators.[22-24,26,28-30] By isoelectric focusing it is possible to select charge homogeneous LH components, which, in addition, do not differ in size as assessed by gel filtration. However, such components still contain a variety of α- and β-subunits.[39]

Moreover, a new technique, i.e., high performance hydrophobic interaction chromatography, revealed heterogeneity in the hydrophobic parts of the molecule. The most prominent results were at least two forms for the α-subunit and at least three forms for the β-subunit. There are probably two different populations of LH_i molecules with different thermostability at 37°C.

The study of human LH and LH-subunit heterogeneity is far from being completed where the nature as well as the physiological meaning is concerned. The great resemblance among different preparations and culture media in the qualitative distributions of components is a promising observation encouraging us to continue our work on LH heterogeneity.

REFERENCES

1. **Koenig, V. L. and King, E.**, Extraction studies of sheep pituitary gonadotropic and lactogenic hormones in alcoholic acetate buffers, *Arch. Biochem. Biophys.*, 26, 219, 1950.
2. **Roos, P. and Gemzell, C. A.**, The isolation of human pituitary follicle-stimulating hormone, *Biochim. Biophys. Acta (Amsterdam)*, 82, 218, 1964.
3. **Squire, P. G. and Li, C. H.**, Purification and properties of interstitial cell-stimulating hormone from sheep pituitary glands, *J. Biol. Chem.*, 234, 520, 1959.
4. **Ward, D. N., Adams-Mayne, M., and Wade, J.**, Association of luteinizing hormone activity with an acidic protein from sheep pituitary glands, *Acta Endocrinol. (Copenhagen)*, 36, 73, 1961.
5. **Reichert, L. E. and Parlow, A. F.**, Partial purification and separation of human pituitary gonadotropins, *Endocrinology*, 74, 236, 1964.
6. **Hartree, A. S., Butt, W. R, and Kirkham, K. E.**, The separation and purification of human luteinizing and thyrotropic hormones, *J. Endocrinol.*, 29, 61, 1964.
7. **Rathnam, P. and Saxena, B. B.**, Isolation and physicochemical characterization of luteinizing hormone from human pituitary glands, *J. Biol. Chem.*, 245, 3725, 1970.
8. **Closset, J., Vandalem, J. L., Hennen, G., and Lequin, R. M.**, Human luteinizing hormone; isolation and characterization of the native hormone and its α- and β-subunits, *Eur. J. Biochem.*, 57, 325, 1975.
9. **Storring, P. L., Bangham, D. R., Cotes, P. M., Gaines Das, R. E., and Jeffcoate, S. L.**, The international reference preparation of human pituitary luteinizing hormone for immunoassay, *Acta Endocrinol. (Copenhagen)*, 88, 250, 1978.
10. **Bangham, D. R.**, Assays and standards, in *Hormones and Blood*, Gray, C. H. and James, V. H. T., Eds., Academic Press, London, 1983, 255.
11. **Peckham, W. D., Yamaji, T., Dierschke, D. J., and Knobil, E.**, Gonadal function and the biological and physico-chemical properties of follicle-stimulating hormone, *Endocrinology*, 92, 1660, 1973.
12. **Bogdanove, E. M., Nolin, J. M., and Campbell, G. T.**, Qualitative and quantitative gonad-pituitary feedback, *Rec. Prog. Horm. Res.*, 31, 567, 1975.
13. **Robertson, D. M., Foulds, L. M., Ellis, S.**, Heterogeneity of rat pituitary gonadotropins on electrofocusing; differences between sexes and after castration, *Endocrinology*, 111, 385, 1982.
14. **Burstein, S., Schaff-Blass, E., Blass, J., and Rosenfield, R. L.**, The changing ratio of bioactive to immunoreactive luteinizing hormone (LH) through puberty principally reflects changing LH radioimmunoassay dose-response characteristics, *J. Clin. Endocrinol. Metab.*, 61, 508, 1985.
15. **Buckingham, J. C. and Wilson, C. A.**, Peripubertal changes in the nature of LH, *J. Endocrinol.*, 104, 173, 1985.
16. **Reader, S. C. J., Robertson, W. R., and Diczfalusy, E.**, Microheterogeneity of luteinizing hormone in pituitary glands from women of pre- and postmenopausal age, *Clin. Endocrinol.*, 19, 355, 1983.
17. **Marrama, P., Zaidi, A. A., Montanini, V., Celani, M. F., Cioni, K., Carani, C., Morabito, F., Resentini, M., Bonati, B., and Baraghini, G. F.**, Age and sex related variations in biologically active and immunoreactive serum luteinizing hormone, *J. Endocrinol. Invest.*, 6, 427, 1983.
18. **Chappel, S. C., Ulloa-Aguirre, A., and Coutifaris, C.**, Biosynthesis and secretion of follicle-stimulating hormone, *Endocr. Rev.*, 4, 179, 1983.
19. **Lucky, A. W., Rich, B. W., Rosenfield, R. L., Fang, V. S., and Roche-Bender, B. A.**, LH bioactivity increases more than immunoreactivity during puberty, *J. Paediatr.*, 97, 205, 1980.
20. **Lucky, A. W., Rich, B. W., Rosenfield, R. L., Fang, V. S., and Roche-Bender, B. A.**, Bioactive LH: a test to discriminate true precocious puberty from premature thelarche and adrenarche, *J. Paediatr.*, 97, 214, 1980.
21. **Warner, B. A., Dufau, M. L., and Santen, R. J.**, Effects of aging and illness on the pituitary testicular axis in men: qualitative as well as quantitative changes in luteinizing hormone, *J. Clin. Endocrinol. Metab.*, 60, 263, 1985.
22. **Roos, P., Nyberg, L., Wide, L., and Gemzell, C. A.**, Human pituitary luteinizing hormone; isolation and characterization of four glycoproteins with luteinizing activity, *Biochim. Biophys. Acta (Amst.)*, 405, 363, 1975.
23. **Reichert, L. E.**, Electrophoretic properties of pituitary gonadotropins as studied by electrofocusing, *Endocrinology*, 88, 1029, 1971.
24. **Robertson, D. M. and Diczfalusy, E.**, Biological and immunological characterization of human luteinizing hormone. II. A comparison of the immunological and biological activities of pituitary extracts after electrofocusing using different standard preparations, *Mol. Cell. Endocrinol.*, 9, 57, 1977.
25. **Keel, B. A. and Grotjan, H. E., Jr.**, Characterization of rat lutropin charge microheterogeneity using chromatofocusing, *Anal. Biochem.*, 142, 267, 1984.
26. **Hartree, A. S., Lester, J. B., and Shownkeen, R. C.**, Studies of the heterogeneity of human pituitary LH by fast protein liquid chromatography, *J. Endocrinol.*, 105, 405, 1985.

27. Hallin, P., Madej, A., and Edqvist, L. E., Subunits of bovine lutropin; hormonal and immunological activity after separation with reverse-phase high pressure liquid chromatography, *J. Chromatogr.*, 319, 195, 1985.

28. Robertson, D. M., Fröysa, B., and Diczfaiusy, E., Biological and immunological characterization of human luteinizing hormone. IV. Biological and immunological profile of two international reference preparations after electrofocusing, *Mol. Cell. Endocrinol.*, 11, 91, 1978.

29. Storring, P. L., Zaidi, A. A., Mistry, Y. G., Lindberg, M., Stenning, B. E., and Diczfalusy, E., A comparison of preparations of highly purified human pituitary luteinizing hormone: differences in the luteinizing hormone potencies as determined by in vivo bioassays, in vitro bioassay and immunoassay, *Acta Endocrinol. (Copenhagen)*, 101, 339, 1982.

30. Zaidi, A. A., Qazi, M. H., and Diczfalusy, E., Molecular composition of human luteinizing hormone: biological and immunological profiles of highly purified preparations after electrofocusing, *J. Endocrinol.*, 94, 29, 1982.

31. Reichert, L. E., Selective inactivation of the luteinizing hormone contamination in human pituitary follicle-stimulating hormone preparations by digestion with α-chymotrypsin, *J. Clin. Endocrinol. Metab.*, 27, 1065, 1967.

32. Saxena, B. B. and Rathnam, P., Purification and properties of human pituitary FSH, in *Gonadotropins*, Rosenberg, E., Ed., Geron-X, Los Altos, Calif., 1968, 3.

33. Butt, W. R., Crooke, A. C., and Wolf, A., Some problems related to the investigation of the immunological properties of human pituitary follicle-stimulating hormone, in *Gonadotropins*, Wolstenholme, G. E. W. and Knight, J., Eds., J. & A. Churchill, London, 1965, 85.

34. Graesslin, D., Weise, H. Chr., and Bettendorf, G., Isolation of human FSH and human LH controlled by isoelectic focusing, in *Gonadotropins*, Saxena, B. B., Beling, C. G., and Gandy, H. M., Eds., Interscience, New York, 1972, 159.

35. Loeber, J. G., Human luteinizing hormone; structure and function of some preparations, *Acta Endocrinol. (Copenhagen)*, Suppl. 210, 1, 1977.

36. Pierce, J. G. and Parsons, T. F., Glycoprotein hormones: structure and function, *Annu. Rev. Biochem.*, 50, 465, 1981.

37. Loeber, J. G., Nabben-Fleuren, J. W. G. M., Elvers, L. H., Segers, M. F. G., and Lequin, R. M., Spontaneous dissociation of human pituitary luteinizing hormone in solution, *Endocrinology*, 103, 2240, 1978.

38. Strickland, T. W. and Puett, D., The kinetic and equilibrium parameters of subunit association and gonadotropic dissociation, *J. Biol. Chem.*, 257, 2954, 1982.

39. van Ginkel, L. A. and Loeber, J. G., Heterogeneity of human lutropin; detection and identification of α- and β-subunits, *Acta Endocrinol. (Copenhagen)*, 110, 182, 1985.

40. van Ginkel, L. A. and Loeber, J. G., Heterogeneity of human luteinizing hormone; detection and identification of α- and β-subunits in international reference preparations, *Acta Endocrinol. (Copenhagen)*, 114, 572, 1987.

41. Fausnaugh, J. L., Pfannkoch, E., Gupta, S., and Regnier, F. E., High-performance hydrophobic interaction chromatography of proteins, *Anal. Biochem.*, 137, 464, 1984.

42. Kato, Y., Kitamura, T., and Hashimoto, T., Operational variables in high-performance hydrophobic interaction chromatography of proteins on TSK gel Phenyl-5PW, *J. Chromatogr.*, 298, 407, 1984.

43. Wide, L., Median charge and charge heterogeneity of human pituitary FSH, LH and TSH. I. Zone electrophoresis in agarose suspension, *Acta Endocrinol. (Copenhagen)*, 109, 181, 1985.

44. van Ginkel, L. A. and Loeber, J. G., Heterogeneity of human luteinizing hormone; effect of neuraminidase treatment on biologically active hormone and α- and β-subunits, *Acta Endocrinol. (Copenhagen)*, 114, 577, 1987.

45. Lamberts, S. W. J., Verleun, T., Oosterom, R., Hofland, L., van Ginkel, L. A., Loeber, J. G., van Vroonhoven, C. C. J., Stefanko, S. Z., and de Jong, F. H., The effects of bromocriptine, TRH and GnRH on hormone secretion by gonadotropin-secreting pituitary adenomas in vivo and in vitro *J. Clin. Endocrinol. Metab.*, 64, 524, 1987.

46. Lequin, R. M., Segers, M. F. G., Closset, J., and Hennen, G., Immunochemical characterization of antisera to human luteinizing hormone and its subunits, *Horm. Metab. Res.*, Suppl. 5, 49, 1974.

47. Greenwood, F. C., Hunter, W. M., and Glover, J. S., The preparation of ^{131}I-labelled human growth hormone of high specific radioactivity, *Biochem. J.*, 89, 114, 1963.

48. Hunter, W. M., Assessment of radioiodinated hormone preparations, *Acta Endocrinol. (Copenhagen)*, Suppl. 142, 134, 1969.

49. Rodbard, D. and Lewald, J. E., Computer analysis of radioligand assay and radioimmunoassay data, *Acta Endocrinol. (Copenhagen)*, Suppl. 147, 79, 1970.

50. van Damme, M. P., Robertson, D. M., and Diczfalusy, E., An improved in vitro bioassay method for measuring luteinizing hormone (LH) activity using Leydig cell preparations, *Acta Endocrinol. (Copenhagen)*, 77, 655, 1974.

51. **Bates, W. K. and MacAllister, D. F.**, Some methods of non-linear regression using desk-top calculators with application to the Lowry protein method, *Anal. Biochem.*, 59, 190, 1973.
52. **Finney, D. J.**, *Statistical Methods in Biological Assay*, 2nd ed., C. Griffin, London, 1964.
53. **van Ginkel, L. A. and Loeber, J. G.**, unpublished data, 1987.
54. **Green, E. D., Baenziger, J. U., and Irving, B.**, Cell-free sulfation of human and bovine pituitary hormones, *J. Biol. Chem.*, 260, 15631, 1985.
55. **Strollo, F., Harlin, J., Hernandez-Montes, H., Robertson, D. M., Zaidi, A. A., and Diczfalusy, E.**, Qualitative and quantitative differences in the isoelectrofocusing profile of biologically active lutropin in the blood of normally menstruating and post-menopausal women, *Acta Endocrinol. (Copenhagen)*, 97, 166, 1981.

Chapter 5

PROLACTIN MICROHETEROGENEITY

E. Markoff, Y. N. Sinha, and U. J. Lewis

TABLE OF CONTENTS

I. INTRODUCTION

The presence of multiple forms of immunoreactive prolactin (PRL) variants is well known. Gel filtration and electrophoretic patterns of PRL from serum, amniotic fluid, pituitary extracts, and media from in vitro pituitary incubations have all demonstrated the existence of various molecular size forms of PRL. However, little is known about the physiological significance of these multiple forms. In this report, we will briefly summarize some of the characteristics of the major PRL variants.

II. GLYCOSYLATED PROLACTIN

A. Identification

Isolation and characterization of a glycosylated form of PRL was first reported for the ovine hormone (G-oPRL).[1] The carbohydrate unit was found to be attached to the asparagine at position 31. In collaboration with Drs. J. G. Pierce and T. W. Strickland, the composition of the carbohydrate unit was shown to be N-acetylglucosamine$_2$, mannose$_2$, and fucose$_1$. A glycosylated form of porcine PRL has recently been reported by Pankov and Butner[2] and here too the carbohydrate was linked to asparagine at position 31. In this case, however, the composition of the carbohydrate was more complex than that found for G-oPRL. Both N-acetylglucosamine and N-acetylgalactosamine were detected as was N-acetylneuraminic acid. The latter carbohydrate imparts a more negative charge to the glycosylated porcine PRL as compared to unmodified porcine PRL. This was not the case for either G-oPRL[1] or glycosylated human PRL (G-hPRL),[3] where there was no evidence for an acidic carbohydrate residue.

Other species have now been found to produce a glycosylated PRL. Pankov and Butner[2] detected this form in the whale and Papkoff[4] found various avian, reptilian, and amphibian PRLs to be glycosylated.

A 25,000-dalton protein from mouse pituitary glands, having structural similarities to murine PRL, has been described.[5] In binding experiments with ^{125}I-labeled concanavalin A (ConA), the mPRL-like material bound to the labeled ConA, indicating that the material is a glycosylated protein. In addition, two bands on either side of the 25,000 band also bound ConA. Tyrosine peptide mapping of these areas produced PRL-like fingerprints. Thus, several glycosylated forms of murine PRL might exist. Using a recently developed lectin-binding radioimmunoassay,[6] G-PRL has been measured in mouse plasma. Preliminary analysis of plasma from two strains of mice showed higher levels of G-PRL in the C3H/St strain than in the C57BL/6J, and in females than in males.[6]

Synthesis of G-oPRL in a cell-free system was demonstrated by Strickland and Pierce.[7] Of special interest was the observation of two glycosylated forms. To be determined is whether the second larger form is a result of glycosylation of oPRL at more than one site or whether there is a single glycosylation site but the carbohydrate units differ in size.

The amount of glycosylated PRL in the pituitary gland is considerable. Sinha and Gilligan[5] estimated that 10 to 15% of the PRL in human and ovine glands was glycosylated. Pankov and Butner[3] reported a value of 30 to 40% for glycosylated porcine PRL.

Although not identified as glycosylated PRL, the 25,000 dalton form of hPRL was first detected in pituitary extracts by Meuris et al.[8] Immunoblotting of sodium dodecyl sulfate electrophoresis gels was used as the detection method. The unidentified hPRL migrated as a slightly larger form of the hormone and was later characterized by Meuris[9] as a glycosylated modification on the basis of its binding behavior to ConA. Sinha et al.[10] and Markoff et al.[11] extended these observations by showing that the slightly larger form of hPRL was also in serum. As yet, the composition and point of attachment of the carbohydrate unit are unknown.

B. Bioactivity

That G-hPRL is a circulating form of PRL has been suggested by the work of several investigators. Shoupe et al.[12] demonstrated that the sera of pregnant women in all 3 trimesters of pregnancy contained a 60,000-dalton, immunologically active PRL that bound to ConA. This glycosylated form ranged from 10 to 30% of the total PRL detected by radioimmunoassay. This 60,000-dalton G-hPRL was isolated by chromatography over Sephadex® G-100 under nondissociating, nondenaturing conditions. It is unknown whether the 60,000-dalton form may represent some type of aggregate that could be dissociated into a 25,000-dalton form. Sinha et al.,[10] using polyacrylamide gel electrophoresis under reducing and denaturing conditions, showed that the serum of men and nonpregnant women contained a 25,000-dalton protein that crossreacted with anti-hPRL serum in Western blot analysis. This 25,000-dalton hPRL was subsequently shown to bind to ConA.[11] Using polyacrylamide gel electrophoresis and immunoblotting, the 25,000-dalton G-hPRL has also been identified in the serum of term pregnant women.[13,49] In addition, primary monolayer cultures of human anterior pituitary cells will synthesize and secrete G-PRL as well as PRL.[11]

Glycosylated PRL has also been identified in human amniotic fluid from 32 to 40 weeks of pregnancy.[13] As the probable source of amniotic fluid PRL, human decidual tissue was tested for the ability to synthesize G-hPRL. Human term decidua, in vitro, synthesizes and secretes about 50% as much G-hPRL as hPRL.[13] In contrast to term decidua however, luteal phase endometrium, which has also been shown to synthesize G-hPRL, seems to secrete the majority of its PRL in the glycosylated form.[14]

Both G-oPRL and G-hPRL showed reduced immunological activity as compared to the nonglycosylated forms when tested in radioimmunoassays. The glycosylated forms showed only 20 to 30% of the immunoreactivity of the nonglycosylated forms.[1,3]

In all of the tests of bioactivity in which G-oPRL has been used, the glycosylated form consistently showed reduced activity. In the original report on the isolation of G-oPRL, the glycosylated form had only one third the lactogenic activity of oPRL as measured by the pigeon crop sac assay.[1] G-oPRL has also been tested in the in vitro mouse mammary gland assay and in the Nb_2 assay. In these assays G-oPRL was only 20% as effective as oPRL in stimulating casein synthesis[15] and Nb_2 cell proliferation.[16]

In contrast to G-oPRL, glycosylated porcine PRL has shown increased activity compared to porcine PRL in the pigeon crop sac assay.[2] A difference in the composition of the carbohydrate unit may be responsible for the discrepancies in the lactogenic activity (pigeon crop sac) reported for glycosylated ovine and porcine PRL. The glycosylated oPRL had reduced lactogenic activity whereas the porcine hormone with a more complex carbohydrate unit had enhanced biological activity.

III. AGGREGATES AND DISULFIDE OLIGOMERS

Both the chemistry and biological activity of the polymeric forms of PRL have been reviewed recently.[17-19] Briefly, at least three immunoreactive molecular weight variants have been described for serum from normal individuals and patients with pituitary tumors or functional hyperprolactinemia.[20-25] Although the existance of this molecular weight heterogeneity has been well documented, very little is understood about the differences in bioactivity between these forms. Subramanian and Gala[18] in their recent review article point out the general conclusion that the higher molecular weight species are less biologically active but have similar radioimmunoassay activity when compared to PRL. In a growing number of clinical reports, there are indications that the radioimmunoassay estimate of serum PRL does not correlate well with the clinical findings. In one example, there is a large number of women with galactorrhea who have normal serum PRL levels as measured by radioimmunoassay.[26] On the other hand, there are some cases where hyperprolactinemia was present, as measured by radioimmunoassay, but no obvious clinical correlations were present. In some of these cases where the serum was analyzed by Sephadex® chromatography, it appears that the majority of the PRL was a large molecular weight species.[27-29]

There is still disagreement about the relative bioactivities of the molecular weight PRL variants. Using the Nb$_2$ bioassay, Whitaker et al.[30] showed good correlation between radioimmunoassay and bioassay levels of PRL for each of the three molecular weight variants in sera from both normal patients and patients with PRL-secreting tumors. In contrast, Jackson et al.,[31] also using the Nb$_2$ assay, showed diminished bioactivity of the large molecular size PRL.

Sephadex® chromatography has revealed dimers, polymers, and aggregates of PRL in murine pituitary gland and plasma.[32] The proportions of the different forms are markedly influenced by different physiological, pharmacological, and pathological states. Interestingly, under resting states, mice of the high mammary tumor C3H/St strain contain predominantly larger forms of PRL, whereas mice of the low mammary tumor C57BL/St strain have monomeric PRL.

IV. CHARGE ISOMERS

Preparations of PRL always show electrophoretic heterogeneity unless special efforts are made to separate the charge isomers.[33-35] There can be no doubt that PRL readily deamidates to form more acidic forms but evidence that these acidic components are a result of deamidation is indirect. The evidence for deamidation under alkaline conditions is based on (1) a correlation of formation of the more acidic forms with evolution of ammonia;[33,35] (2) no alteration in amino acid composition other than for asparagine and glutamine residues; and (3) the comigration of the acidic forms when analyzed by gel electrophoresis at a pH near 4 where ionization of the side chain COOH groups is suppressed.

Although there is this evidence for deamidation of isolated PRL, as yet there have been no studies done that demonstrate that the acidic components of PRL seen in extracts of fresh pituitary glands are actually desamido forms. All that is known is that the acidic components comigrate with known desamido forms. There is the possibility that there are acidic forms which are not a result of deamidation. This conclusion was made after comparing the bioactivity and immunoreactivity of the acidic forms as they are found in pituitary extracts with the values found for intentionally deamidated PRL.[36] With both types of acidic forms, the immunological reactivity was decreased. However, there was a difference in bioactivity of the two types. Intentionally deamidated forms showed a progressive decrease in bioactivity as deamidation progressed whereas the most acidic form isolated from pituitary extracts showed enhanced bioactivity. One can postulate, therefore, that the two types of acidic forms are different. The identity of the acidic form with enhanced activity is unknown and only recently has it been possible even to make a guess as to its chemical nature. The new information is that PRL is phosphorylated in vivo in the pituitary gland[37] and that the phosphorylated PRL migrates electrophoretically similarly to the PRL with enhanced bioactivity. Finding that acidic forms of PRL are phosphorylated does not minimize the potential importance of deamidation in modulating physiologic actions of the hormone. Both deamidation and phosphorylation may be involved in this process. Proposed mechanisms whereby deamidation could influence bioactivity are (1) altering receptor binding,[36] (2) directing proteolytic cleavage during production of metabolically active fragments,[38] or (3) serving as a timer of development and aging.[39] Compared to the main, nondeamidated PRL, the pigeon crop sac activity of the deamidated PRL was found to be lower both in the mouse[40] and the rat.[41]

The report of Hashimoto et al.[42] of the sexual dimorphism in the amino acid composition of mouse PRL may be another example of variant charge isomers of PRL in the pituitary gland. In this case the implication would be that the resulting heterogeneity is not a post-translational phenomenon but rather stems from different levels of gene expression.

V. CLEAVED FORMS

Cellular processing of PRL to cleaved forms with biological activity is receiving increased attention. Mittra[43] has reviewed reports that would support the view that proteolysis is a step in the bioactivation of PRL. Nolin,[44] using immunohistochemical evidence, has postulated that tissue proteinases process PRL to specific metabolically active peptides. Compton and Witorsch[45] have studied in detail the proteolytic modification of PRL by the rat ventral prostate gland. They showed that the degradation of PRL by various subcellular fractions was qualitatively different and that stable products were generated. This suggests specificity and not a general nonspecific proteolysis of the hormone. Oetting and Walker[46] have found that PRL is reproducibly processed to smaller forms in the pituitary gland and that these substances are secreted into the medium during culturing of pituitary cells.

All these studies point to the possibility that cellular processing of PRL is of physiologic importance. The reports also indicate that investigators are aware that the cleaved forms of PRL must be isolated in quantities sufficient for chemical and physiologic studies and for production of specific radioimmunoassays. Only then can the biologic significance of the proteolytic processing of PRL be better assessed.

Following Mittra's[47] original observations, cleaved PRL has been detected in murine,[48] human,[49] bovine, and ovine[50] pituitary extracts. In addition, analysis of human plasma by immunoblotting methods has shown a 16,000- and an 8000-dalton PRL-immunoreactive component in the plasma of pregnant women.[49] The highest concentration of these components was observed in plasma sampled closest to the time of delivery. Cleaved PRL has not yet been detected in mouse or rat plasma, however, this might be due to the sensitivity of the techniques employed rather than an absence of the variant.

In in vitro studies, the ratio of cleaved to intact PRL in the medium following a 6 hr incubation of mouse pituitaries was slightly but consistently greater for mice of the C3H/St strain, which have a high natural incidence to mammary tumors, than for mice of the C57BL/6J strain, in which mammary tumor incidence is nil.[48]

VI. THE "21K" PRL

Evidence is now emerging for the existence of a smaller molecular weight form of PRL,[51] similar to the 20,000 M_r form of hGH (Lewis et al.[52]) in which a chain of amino acids is missing. Immunostaining of pituitary extracts, analyzed by polyacrylamide gel electrophoresis in the presence of sodium dodecyl sulfate, shows a band just ahead of the main PRL band that crossreacts with PRL antibodies.[5,48] This band appears in both mouse and rat pituitary extracts[51] as well as in bovine and ovine pituitary tissues.[53] One experiment revealed a similar band in human plasma.[49] This band becomes labeled when murine pituitary glands are incubated with radioactive amino acids. Partial peptide maps, based on tyrosine-containing peptides,[54] resembled that of the major 23,000-dalton form of PRL, with one spot missing, suggesting that if a chain of amino acids is missing, the deleted region includes a tyrosine residue. The structural modifications and biological properties of this variant have not been characterized. However, the smaller form of hGH exhibits significant alterations in biological, immunological, and receptor-binding properties.[55]

VII. THE "31K" PRL

When mouse pituitary extracts are analyzed by disc electrophoresis in nondenaturing gels under basic conditions (pH 8.5), which separates the proteins according to charge as well as size, three protein bands appear in the area where PRL usually migrates.[40] The most prominent of these is the major form of pituitary PRL. The band just ahead of this major band is a deamidated form of PRL. The third band, the fastest migrating and most acidic of the three, has a molecular weight of approximately 31,000, as determined by a Ferguson plot of its R_f in gels of varying concentrations. This material, isolated from both mouse[40] and rat[41] pituitary glands, had significant pigeon crop sac-stimulating activity, but very little crossreactivity in the radioimmunoassay. From in vitro labeling experiments, the protein appeared not to be a precursor of PRL.[56] Partial peptide map showing only tyrosine-containing peptides revealed no apparent resemblance to the fingerprint of the main mouse PRL. Thus, it appears to be a separate gene product that could represent a ''proliferin''-like protein of the prolactin-growth hormone family.[57] Asawaroengchai et al.[58] also reported an acidic, fast-migrating protein in culture medium after incubation with rat pituitary glands; this protein exhibited a high bioactivity to immunoactivity ratio, but it is not clear whether this component and the 31,000 M_r form are the same substance.

REFERENCES

1. **Lewis, U. J., Singh, R. N. P., Sinha, Y. N., and VanderLaan, W. P.,** Glycosylated ovine prolactin, *Proc. Natl. Acad. Sci. U.S.A.*, 81, 385, 1984.
2. **Pankov, Yu. A. and Butnev, V. Yu.,** Multiple forms of pituitary prolactin. Glycosylated form of prolactin with enhanced biological activity, *Int. J. Peptide Protein Res.*, 28, 113, 1986.
3. **Lewis, U. J., Singh, R. N. P., Sinha, Y. N., and VanderLaan, W. P.,** Glycosylated human prolactin, *Endocrinology*, 116, 359, 1985.
4. **Papkoff, H.,** personal communication, 1986.
5. **Sinha, Y. N. and Gilligan, T. A.,** Identification and partial characterization of a 25K protein structurally similar to prolactin, *Proc. Soc. Exp. Biol. Med.*, 178, 505, 1985.
6. **Sinha, Y. N. and Lewis, U. J.,** Lectin-binding radioimmunoassay: a new method for measuring glycosylated variants of prolactin (PRL) and other proteins in plasma, *Endocrinology*, 118 (Suppl.), 918, 1986.
7. **Strickland, T. W. and Pierce, J. G.,** Glycosylation of ovine prolactin during cell-free biosynthesis, *Endocrinology*, 116, 1295, 1985.
8. **Meuris, S., Svoboda, M., Vilamala, M., Christophe, J., and Robyn, C.,** Monomeric pituitary growth hormone and prolactin variants in man characterized by immunoperoxidase electrophoresis, *FEBS Letters*, 154, 111, 1983.
9. **Meuris, S.,** Contribution a l'Etude, de l'Ubiquite cellulaire et de l'Heterogeneite moleculaire des Hormones Lactogeniques Thesis, Univ. Libre de Bruxelles, Faculte de Medecine, Brussels, 1984.
10. **Sinha, Y. N., Gilligan, T. A., and Lee, D. W.,** Detection of a high molecular weight variant of prolactin in human plasma by a combination of electrophoretic and immunological techniques, *J. Clin. Endocrinol. Metab.*, 58, 752, 1984.
11. **Markoff, E., Lee, D. W., and Roper, L.,** Synthesis and release of glycosylated prolactin *in vitro* and *in vivo*, presented at 66th Annu. Meet. Endocrine Society, Baltimore, Md., 1985, 1168A.
12. **Shoupe, D., Montz, F. J., Kletzky, O. A., and deZerega, G. S.,** Response to thyrotropin-releasing hormone stimulation of concanavalin A-bound and -unbound immunoassayable prolactin during human pregnancy, *Am. J. Obstet. Gynecol.*, 147, 482, 1983.
13. **Lee, D. W. and Markoff, E.,** Synthesis and release of glycosylated prolactin by human decidua *in vitro*, *J. Clin. Endocrinol. Metab.*, 62, 990, 1986.
14. **Heffner, L. J., Iddenden, D. A., and Lyttle, C. R.,** Electrophoretic analysis of secreted human endometrial proteins: identification and characterization of luteal phase prolactin, *J. Clin. Endocrinol. Metab.*, 62, 1288, 1986.
15. **Markoff, E.,** unpublished data, 1985.
16. **Worsley, I. G. and Friesen, H. G.,** personal communication, 1986.
17. **Lewis, U. J., Sinha, Y. N., Markoff, E., and VanderLaan, W. P.,** Multiple forms of prolactin: properties and measurement, in *Neuroendocrine Perspectives*, Vol. 4, Muller, E. E., MacLeod, R. M., and Frohman, L. A., Eds., Elsevier Science, Amsterdam, 1985, 43.
18. **Subramanian, M. G. and Gala, R. R.,** Do prolactin levels measured by RIA reflect biologically active prolactin, *J. Clin. Immun.*, 9, 42, 1986.

19. **Sinha, Y. N.**, Structural variants of prolactin, in *Proc. of the Int. Symp. on the Pituitary Gland*, Yoshimura, F. and Gorbman, A., Eds., Elsevier Science, Amsterdam, 399, 1986.

20. **Suh, H. K. and Frantz, A. G.**, Size heterogeneity of human prolactin in plasma and pituitary extracts, *J. Clin. Endocrinol. Metab.*, 39, 928, 1974.

21. **Guyda, H. J.**, Heterogeneity of human growth hormone and prolactin secreted *in vitro:* immunoassay and radioreceptor assay correlations, *J. Clin. Endocrinol. Metab.*, 41, 953, 1975.

22. **Garnier, P. E., Aubert, M. L., Kaplan, S. L., and Grumback, M. M.**, Heterogeneity of pituitary and plasma prolactin in man: decreased affinity of "big" prolactin in a radioreceptor assay and evidence for its secretion, *J. Clin. Endocrinol. Metab.*, 47, 1273, 1978.

23. **Farkouh, N. H., Packer, M. G., and Frantz, A. G.**, Large molecular size prolactin with reduced receptor activity in human serum: high proportion in basal state and reduction after thyrotropin-releasing hormone, *J. Clin. Endocrinol. Metab.*, 48, 1026, 1979.

24. **Soong, Y. K., Ferguson, K. M., McGarrick, G., and Jeffcoate, S. L.**, Size heterogeneity of immunoreactive prolactin in hyperprolactinemic serum, *Clin. Endocrinol.*, 16, 259, 1982.

25. **Suh, H. K. and Frantz, A. G.**, Big and little prolactin in human plasma, *Clin. Res.*, 21, 960, 1973.

26. **Kleinberg, D. L., Noel, G. L., and Frantz, A. G.**, Galactorrhea: a study of 235 cases including 48 with pituitary tumor, *New Eng. J. Med.*, 296, 579, 1977.

27. **Whittaker, P. G., Wilcox, T., and Lind, T.**, Maintained fertility in a patient with hyperprolactinemia due to big, big prolactin, *J. Clin. Endocrinol. Metab.*, 53, 863, 1981.

28. **Anderson, A. N., Pedersen, H., Djursing, H., Andersen, B. N., and Friesen, H. G.**, Bioactivity of prolactin in a woman with an excess of large molecular size prolactin, persistent hyperprolactinemia and spontaneous conception, *Fertil. Steril.*, 38, 625, 1982.

29. **Andino, N. A., Bidot, C., Valdes, M., and Machado, A. J.**, Chromatographic pattern of circulating prolactin in ovulatory hyperprolactinemia, *Fertil. Steril.*, 44, 600, 1985.

30. **Whitaker, M. D., Klee, G. G., Kao, P. C., Randall, R. V., and Heser, D. W.**, Demonstration of biological activity of prolactin molecular weight variants in human sera, *J. Clin. Endocrinol. Metab.*, 58, 826, 1983.

31. **Jackson, R. D., McDonald, M. A., Wortsman, J., and Malarkey, W. B.**, Characterization of big, big prolactin in two patients with hyperprolactinemia, galactorrhea and normal menses, *Endocrinology*, 110 (Suppl.), 358, 1982.

32. **Sinha, Y. N.**, Molecular size variants of prolactin and growth hormone in mouse serum: strain differences and alteration of concentrations by physiological and pharmacological stimuli, *Endocrinology*, 107, 1959, 1980.

33. **Lewis, U. J., Cheever, E. V., and Hopkins, W. C.**, Kinetic study of the deamidation of growth hormone and prolactin, *Biochim. Biophys. Acta*, 214, 498, 1970.

34. **Nyberg, F., Roos, P. and and Wide, L.**, Human pituitary prolactin. Isolation and characterization of three isohormones with different bioassay and radioimmunoassay activities, *Biochim. Biophys. Acta*, 625, 255, 1980.

35. **Haro, L. S. and Talamantes, F. J.**, Secreted mouse prolactin (PRL) and stored ovine PRL. I. Biochemical characterization, isolation and purification of their electrophoretic isoforms, *Endocrinology*, 116, 346, 1985.

36. **Haro, L. S. and Talamantes, F. J.**, Secreted mouse prolactin (PRL) and stored ovine PRL. II. Role of amides in receptor binding and immunoreactivity, *Endocrinology*, 116, 353, 1985.

37. **Oetting, W. S., Tuazon, P. T., Traugh, J. A., and Walker, A. M.**, Phosphorylation of prolactin, *J. Biol. Chem.*, 261, 1649, 1986.

38. **Lewis, U. J., Singh, R. N. P., Bonewald, L. F., and Seavey, B. K.**, Altered proteolytic cleavage of human growth hormone as a result of deamidation, *J. Biol. Chem.*, 256, 11645, 1981.

39. **Robinson, A. B., McKerrow, J. H., and Cary, P.**, Controlled deamidation of peptides and proteins: an experimental hazard and a possible biological timer, *Proc. Natl. Acad. Sci. U.S.A.*, 66, 753, 1970.

40. **Sinha, Y. N. and Baxter, S. R.**, Identification of a nonimmunoreactive but highly bioactive form of prolactin in the mouse pituitary by gel electrophoresis, *Biochem. Biophys. Res. Commun.*, 86, 325, 1979.

41. **Sinha, Y. N. and Gilligan, T. A.**, Identification of a less immunoreactive form of prolactin in the rat pituitary, *Endocrinology*, 108, 1091, 1981.

42. **Hashimoto, H., Yasuhara, T., Nakajima, T., Harigaya, T., and Hoshino, K.**, Sexual dimorphism in amino acid compositions of mouse prolactin, *Biochem. Biophys. Res. Commun.*, 130, 1209, 1985.

43. **Mittra, J.**, Somatomedins and proteolytic bioactivation of prolactin and growth hormone, *Cell*, 38, 347, 1984.

44. **Nolin, J. M.**, Molecular homology between prolactin and ovarian peptides: evidence for physiologic modification of the parent molecule by the target, *Peptides*, 3, 823, 1982.

45. **Compton, M. M. and Witorsch, R. J.**, Proteolytic degradation and modification of rat prolactin by subcellular fractions of the rat ventral prostate gland, *Endocrinology*, 115, 476, 1984.

46. **Oetting, W. S. and Walker, A. M.**, Intracellular processing of prolactin, *Endocrinology*, 117, 1565, 1985.

47. **Mittra, I.**, A novel "cleaved prolactin" in the rat pituitary. I. Biosynthesis, characterization and regulatory control, *Biochem. Biophys. Res. Commun.*, 95, 1750, 1980.
48. **Sinha, Y. N. and Gilligan, T. A.**, A cleaved form of prolactin in the mouse pituitary gland: Identification and comparison of *in vitro* synthesis and release in high and low incidences of mammary tumors, *Endocrinology*, 114, 2046, 1984.
49. **Sinha, Y. N., Gilligan, T. A., Lee, D. W., Hollingsworth, D., and Markoff, E.**, Cleaved prolactin: evidence for its occurrence in human pituitary gland and plasma, *J. Clin. Endocrinol. Metab.*, 60, 239, 1985.
50. **Sinha, Y. N.**, unpublished data, 1985.
51. **Sinha, Y. N. and VanderLaan, W. P.**, A "21K" variant of prolactin: detection by Western blot analysis, *Clin. Res.*, 34, 94A, 1986.
52. **Lewis, U. J., Bonewald, L. F., and Lewis, L. J.**, The 20,000-dalton variant of human growth hormone: location of the amino acid deletion, *Biochem. Biophys. Res. Commun.*, 92, 511, 1980.
53. **Sinha, Y. N.**, unpublished data, 1985.
54. **Elder, J. H., Pickett, R. A., II., Hampton, J., and Lerner, R. A.**, Radioiodination of proteins in single polyacrylamide gel slices, *J. Biol. Chem.*, 252, 6510, 1977.
55. **Lewis, U. J., Dunn, J. T., Bonewald, L. F., Seavey, B. K., and VanderLaan, W. P.**, A naturally occurring structural variant of human growth hormone, *J. Biol. Chem.*, 253, 2679, 1978.
56. **Sinha, Y. N.**, unpublished data, 1986.
57. **Linzer, D. I. H. and Nathans, D.**, Nucleotide sequence of a growth-related mRNA encoding a member of the prolactin-growth hormone family, *Proc. Natl. Acad. Sci. U.S.A.*, 81, 4255, 1984.
58. **Asawaroengchai, H., Russell, S. M., and Nicoll, C. S.**, Electrophoretically separable forms of rat prolactin with different bioassay and radioimmunoassay activities, *Endocrinology*, 102, 407, 1978.

Chapter 6

FOLLICLE-STIMULATING HORMONE MICROHETEROGENEITY

David Mark Robertson

TABLE OF CONTENTS

I. INTRODUCTION

It is now recognized that LH, FSH, TSH, and hCG obtained from pituitary, plasma, and urinary sources are heterogeneous in terms of both structure and a number of functional aspects such as their in vivo, in vitro, and immunological activities. Furthermore, the profile of these hormone isoforms is influenced by the endocrine status of the animal.

Using a variety of fractionation techniques, a number of isoforms of FSH from a wide range of species have been identified. As the isoform distribution is dramatically affected by neuraminidase treatment which specifically removes terminal sialic acid residues it would appear that hormone heterogeneity is largely due to its sialic acid content. Since the desialylated hormone is cleared more rapidly from the circulation than the native hormone, the various isoforms may have different biological activities in vivo. Examination of the immunological, in vitro bioactivity, and receptor-binding activity of these isoforms also shows differences indicating that the various isoforms may also differ in their receptor binding and subsequent biological activity.

These findings suggest that hormone action in vivo is regulated both by the amount and type of hormone secreted. These two facets may be regulated by different control mechanisms.

The structure of FSH, and in particular its carbohydrate composition, will be considered in detail elsewhere in this volume (Chapter 2). However, several aspects will be emphasized for consideration in the present chapter. The α and β chains of FSH contain up to two oligosaccharide chains attached through N-linked glycosylation sites to asparagine in both subunits.[1,2] The oligosaccharide consists of the sugars *N*-acetylglucosamine, fucose, mannose, galactose, or *N*-acetylgalactosamine and sialic acid. Removal of sialic acid residues from FSH by neuraminidase treatment, which results in a shorter circulating half-life, has variable effects on its biological activity (hFSH,[3] eFSH[4]) as assessed by in vitro bioassay systems and little effect on its immunological activity.[5] Removal of sugars from gonadotrophins either sequentially with appropriate glycosidases or by using hydrogen fluoride treatment (which removes all terminal sugars except *N*-acetylglucosamine) has little effect on their immunological activity or receptor binding but has dramatic inhibitory effects on its biological activity as assessed in both in vivo and in vitro assay systems.[6] However, removal of the remaining *N*-acetylglucosamine molecule does lead to a loss of receptor activity.[6] Glycosylation is not necessary in promoting the combination of subunits, as assessed by several criteria including restoration of receptor binding, nor are the carbohydrate components involved in the specificity of the receptor hormone interaction.[6]

These studies highlight several aspects in the study of heterogeneity of gonadotrophins: (1) several measures of FSH activity need to be monitored in order to obtain a clear understanding of the biological role of hormone heterogeneity; (2) since it is likely that partial rather than full deglycosylation will be experienced in hormone heterogeneity studies, subtle differences may exist between various activities, (e.g., immunological and receptor binding) thus requiring close attention in the characterization of respective assays; (3) an assessment of the biological activity both in vivo and in vitro of the various isoforms is essential.

A. Biosynthesis of FSH

The biosynthesis of FSH has not yet been examined in any depth. The following account is largely derived from studies involving the glycosylation and processing of TSH.[7,8] FSH, like LH and TSH, is synthesized, at least in the human, as two presubunits from single separate genes.[9] These observations provide strong evidence that each subunit is a single gene product and that modifications to the protein resulting in its various heterogeneous forms occur posttranslationally. The α- and β-subunits following synthesis are sequestered within the endoplasmic reticulum whereupon the signal (pre-) peptide is cleaved and the remaining polypeptide chain glycosylated by the attachment of a mannose-glucose complex to the amino acid asparagine through a complex enzymatic process involving a lipid carrier (Chapter 2). Studies on the biosynthesis of mouse TSH showed that the presence of this carbohydrate complex protects the protein from intercellular proteolysis.[8] The glycosylated subunits, within vesicle extensions of the endoplasmic reticulum, migrate to the Golgi region where secretory granules are formed. Biochemically, several processes occur; initially the glucose and up to six of the nine mannose sugars are enzymatically removed from the carbohydrate chain of each subunit whereupon the sugars N-acetylglucosamine, galactose, N-acetylgalactosamine, and sialic acid, under the action of various transferases, are attached sequentially to the exposed mannose residues. Meantime, by a presumably nonenzymatic process, the α- and β-subunits interact to form the intact molecule. Is the hormone heterogeneity attributed in part to the presence of various intermediates produced during the glycosylation process? One approach used in determining the change in carbohydrate structure during synthesis has been the use of an enzyme — endoglycosidase H — which specifically cleaves the glucose-mannose oligosaccharide but not the mature oligosaccharide containing N-acetylglucosamine, galactose, and sialic acid. Using this approach it was found that the α- and β-subunits of rat LH[10] and mouse TSH[8] could combine to form the intact hormone containing immature glucose-mannose oligosaccharides.

Only the secreted hormone contained the mature oligosaccharide moieties. Similar conclusions were drawn from studies of FSH synthesis by ovine pituitary cell cultures in which, based on the size of oFSH synthesized and released, it was concluded that the released form was fully processed and glycosylated.[11] Thus it would be anticipated that in the examination of hormone heterogeneity of pituitary extracts a certain proportion of these immature forms may be present. It also would be anticipated that some of these immature forms may be secreted in times of hyperstimulation, for example, following luteinizing hormone-releasing hormone (LHRH) stimulation. From morphological evidence,[12] secretory granules in the pituitary gonadotroph can undergo fusion followed by degradation within lysosomes providing a mechanism to reduce the number of secretory granules in cells which are not actively secreting. The nature of this catabolic process is not understood, particularly in regard to gonadotrophins, however the presence of FSH metabolites in pituitary extracts may be a contributing factor to their apparent heterogeneity.

It has also been observed that the α-subunit from bovine pituitaries can be glycosylated through an O-linkage to the amino acid threonine[13,14] and that this form is secreted as the free subunit.[15] The observation that glycosylation can occur at sites other than asparagine raises the possibility that FSH α- and β-subunits and even the intact molecule may be glycosylated in this manner.

Other chemical changes leading to the formation of heterogeneous forms have been observed in which sulfate has been found, in some cases in addition to sialic acid, attached to the terminal sugar N-acetylglucosamine of human LH and TSH, bovine TSH, FSH, and LH,[16] rat LH[17] but not hCG or hFSH.[16] The biological role of the sulfate moiety is not clear. It is thought to play a protective role comparable with sialic acid although, as noted by Sairam,[6] desialylated bovine LH is biologically inactive in vivo despite the presence of sulfated sugars.

B. Action of Hormones on FSH Synthesis

FSH synthesis and release is controlled by stimulatory effects of LHRH, the direct and indirect effects of gonadal steroids, and the inhibitory effects of a gonadal protein, inhibin. Studies in vitro have shown that LHRH stimulates the incorporation of ^{3}H proline but not ^{3}H glucosamine into immunoprecipitable FSH within 4 hr.[18,19] However, in contrast to others,[20-22] these workers[18,19] also observed a similar effect of LHRH on the incorporation of these precursors into rat LH. The role of steroids on FSH synthesis and release shows species differences. In the sheep a dramatic inhibitory effect of estradiol in vitro on FSH synthesis and FSH mRNA levels has been observed.[23,24] In the rat, estradiol has minimal direct effects in vitro on FSH synthesis,[25] but has a synergistic effect with LHRH[26] while testosterone and progesterone stimulate basal FSH release, cell content, and synthesis but inhibit LHRH-stimulated release of FSH and LH.[27-29] This inhibition is associated with a reduction in the number of LHRH receptors in the pituitary.[30] Castration leads to an increase in pituitary content of FSH and LH and elevated levels of FSH mRNA.[31,32] This increase in mRNA levels in the rat is attributed to elevated production levels of LHRH by the hypothalamus in conjunction with increased pituitary levels of LHRH receptors. The effects of gonadectomy on FSH and LH synthesis are reversed by the addition of estradiol.[23,33,34] The effects of inhibin are less clearly understood. Inhibin suppresses FSH synthesis in vitro without affecting synthesis and release of other pituitary hormones[35-37] and competes with LHRH in the release of FSH and perhaps LH. Furthermore, using granulosa cell culture media or Sertoli cell culture media as inhibin sources, the stimulatory effect of steroids on FSH release in vitro was reversed suggesting that steroids and inhibin interact through a common mechanism.[38,39] These observations have yet to be confirmed with purified inhibin preparations.

A number of studies have shown that (1) the pituitary gonadotrophs are heterogeneous in terms of size, FSH, and LH content,[40-42] (2) they are functionally heterogeneous in their response to LHRH stimulation and androgen treatment in vitro, and (3) this heterogeneity is modified according to the sex and age of the animal. Pituitary gonadotrophs from immature males are larger than those found in the adult with differing ratios of FSH to LH content. Maximum LHRH-stimulated gonadotrophin release was observed with medium sized gonadotrophs in the immature male and with the largest gonadotrophs in the immature female.[43,44] Inhibitory effects of dihydrotestosterone on LHRH stimulated release of FSH and LH depended on the gonadotroph size. These results suggest that the pituitary contains morphologically distinct gonadotrophs which, depending on the age and sex of the animal, respond differently to different stimuli. It would be reasonable to suspect that these classes of gonadotrophs produce their own pattern of gonadotrophins. It has also been observed that not all LH- and FSH-containing cells bound LHRH as assessed using a potent analog of LHRH conjugated to biotin suggesting that not all gonadotrophs are susceptible to LHRH stimulation.[45] Additional studies using a reverse hemolytic plaque assay[46] indicated that the proportion of LHRH-insensitive cells is under hormonal control.

II. HETEROGENEITY OF FSH

A. Fractionation and Assay Methods

The techniques most commonly employed in the study of hormone microheterogeneity consist of gel filtration, various charge separation methods (e.g., electrophoresis, electrofocusing, chromatofocusing), and lectin binding.

1. Gel Filtration

The physicochemical basis for the separation of FSH isoforms on gel filtration is not clear. The weight of evidence suggests that the primary basis for differences between isoforms lies with their sialic acid content. On this basis the separation of FSH isoforms on gel filtration should show small incremental changes in the molecular weight of approximately 1% per sialic acid residue. An examination of gel filtration profiles of the various pituitary FSH isoforms in the rat[47] and castrate and intact monkeys[48,49] shows in the rat a range in Stokes radius (a measure of effective size in solution) between 32 and 36 Å (Table 1), and in the monkey between 27 and 32 Å. When using glycoproteins of known molecular weight with known Stokes radius as molecular weight markers (see legend Table 1), the difference in molecular mass between the most acidic and basic isoforms for rat FSH was 6250 daltons

Table 1

pH DISTRIBUTION OF THE RATIO OF FSH ACTIVITY AS ASSAYED BY IN VITRO BIOASSAY (B), RADIORECEPTOR ASSAY (R), AND RADIOIMMUNOASSAY (I) OF MALE AND FEMALE PITUITARY EXTRACTS AFTER ELECTROFOCUSING

		pH range					
		3.4—3.8	3.8—4.0	4.0—4.2	4.2—4.4	4.4—4.8	4.8—6.0
R/I ratio	Male	1.8[a]	2.0[a]	1.8[a]	1.8[a]	1.2 [b]	1.3[b]
	Female	2.0	2.1	2.2	2.3 [c,*]	1.7[d,*]	1.9*
B/I ratio	Male	1.4	1.7	1.6	1.5	0.9	1.1
	Female	3.1*	2.3	2.4	2.8*	2.5*	2.6*
B/R ratio	Male	0.8	0.9	0.9	0.8	0.8	0.8
	Female	1.5*	1.1	1.1	1.2*	1.5*	1.4*
Stokes radius	Male	35.7	34.5	33.8	32	32	
(Å)	Female			33.2	32		
Molecular mass	Male	37.2	35.1	33.5	30.9	30.9	
(kDa)	Female			33.1	30.9		

Note: *, a vs. b, c vs. d, $p < 0.05$. The Stokes radius (Å) and molecular mass of FSH in each pool as determined by gel filtration are also included.

From Foulds, L. M. and Robertson, D. M., *Mol. Cell. Endocrinol.*, 31, 117, 1983. With permission. The following were used as standards: hFSH 32.2 Å mol wt 32,600, hCG 33 Å mol wt 37,7000, and hLH 29.5 Å mol wt 27,800.[106,107]

and for castrated and intact monkeys, 12,600 daltons which is equivalent to 20 and 40 sialic acid residues per FSH molecule, respectively. Such sialic acid content values are clearly in excess of those known for any gonadotrophin although the sialic acid content for the above preparations is not known. These results suggest that the increase in Stokes radius in addition to that attributable to its sialic acid content is due to either the presence of additional sugars and/or conformational changes whereby the presence of additional sugars increases the hydrodynamic volume of the molecule. Assuming that the differences in FSH sialic acid content noted above are a maximum of eight residues per molecule,[50] although it is most likely less, then rat FSH and monkey FSH have increased their size by 5% maximum due to the addition of sialic acid and 6 and 12% minimum, respectively, by addition of sugars and/or conformational changes. If the observed changes in size between FSH isoforms is attributable to these two (or three) factors then it is not surprising that a wide range in size differences have been reported, varying from no apparent size differences, e.g., human[51] and cynomolgus monkeys,[52] to marked size heterogeneity (rat,[47,53-57] rhesus monkey,[48,49] frog[58]) depending on the respective contributions of these factors to the hydrodynamic size. Furthermore, these size changes will most likely modify the FSH molecule in its interaction with antibodies or receptors which be reflected in changes in affinity and biological activity.

In order to achieve adequate resolution between FSH isoforms, a high resolution gel filtration system is required. This has been achieved[47-49,53-58] using long (1-meter) columns of Sephadex® G-100 Superfine or G-150 at low flow rates. Protein markers are necessary to standardize runs.

2. Electrofocusing, Chromatofocusing, Electrophoresis

These fractionation procedures are the most widely used in assessing the heterogeneity of pituitary hormones. Electrofocusing in sucrose gradients at the preparative level (100- or 400-mℓ columns) has been extensively used in fractionation of large protein loads (100 to 200 mg) of plasma and urinary extracts. The carrier ampholytes and sucrose can be readily removed by gel filtration systems or dialysis. Following electrofocusing of extracts of pituitaries from a number of species, the recovery of FSH bioactivity as determined by an FSH in vitro bioassay and radioreceptor assay was shown to be near quantitative in the human,[59] baboon,[60] and rat.[47] The electrofocusing method is reproducible and no evidence of artifacts attributable to carrier ampholyte-hormone interactions have been observed.[61] In situations where a lower protein loading (10 to 20 mg) is employed an analytical (5 mℓ) electrofocusing system can be employed.[62,63] Recoveries are similar to that observed with the larger column, however resolution is reduced in comparison with the larger IEF system or with the IEF-PAG (isoelectric focusing in polyacrylamide gels) system (vide infra) owing to diffusion and mixing associated with the column emptying process.

To circumvent the loss of resolution in the above procedures and to increase its practicability, electrofocusing in polyacrylamide gel slabs (IEF-PAG) has been employed[64,65] and the isoforms isolated by sectioning the gel and allowing the protein to diffuse into an appropriate buffer. The resolving capacity of this procedure is very high; however, its capacity is limited and some question of uniform recoveries exists.

Chromatofocusing, a more recent technique used to separate hormone isoforms, is based on the column chromatographic separation of proteins according to their pI values on modified ion-exchange columns in which a pH gradient can be established using appropriate amphoteric buffers.[66,67] The method permits higher protein loading and potentially a similar resolving capacity to that observed with electrofocusing in large sucrose gradients. However, the available buffers have a limited pH range (4 to 7 and 6 to 9) which limits the fractionation of FSH isoforms with pI values of <4. The future availability of amphoteric buffer mixtures covering both a larger and narrower pH range used in conjunction with optimized column packings (in terms of particle size, pore diameter, bonded phase composition, etc.) should enable the fractionation of hormone isoforms with higher resolution and capacity within a short time period.

Zone electrophoresis in columns (1.3 or 2.8 × 67 cm) of 0.17% agarose has been successfully employed in separating FSH isoforms in human plasma and pituitary extracts.[65-71] The method is highly reproducible, shows quantitative recoveries of FSH immunoactivity, and is capable of resolving up to 20 different FSH forms from a human pituitary extract (Figure 1). Furthermore, the method has a relatively high capacity such that serum samples of up to 4 mℓ can be fractionated. Its main advantage over other charge separating methods is that it does not employ amphoteric buffers which may cause separation and assay artifacts.

3. FSH Assays

Three types of FSH assays, including in vitro bioassay, radioreceptor (RRA), and radioimmunoassay (RIA), have been used in the study of FSH heterogeneity, of which the most commonly used are RRA and RIA methods. The validity of RRA and RIA procedures rests on their specificity for the intact hormone in question and, since they are competitive assays, on the assumption that tracer and standard are identical to the isoforms under study. It can be realized that if the affinities of the various isoforms under study for the receptor or antibodies are not similar to that of the standard then spurious results must be anticipated. It is generally believed that although isoforms present in purified gonadotrophin preparations used as antigens and RIA tracer may not be representative of what is found in biological extracts, they are likely to be immunologically similar, irrespective of their differences in carbohydrate structure. In contrast, a number of studies suggest that receptor-binding activity and, in particular, in vitro bioactivity are strongly influenced by the carbohydrate composition of the hormone although the contributions of different sugars to the various activities have only been studied with hCG.[72-75] As a consequence, the ratios of bio- and radioreceptor activities to immunological activities have been widely used in the characterization of the

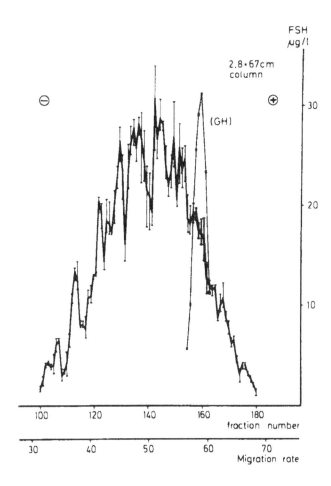

FIGURE 1. Elution pattern of FSH following zone electrophoresis of a pituitary extract from a 21 year-old man. The elution profile growth hormone (GH) is included to indicate the resolution of the system. (From Wide, L., *Acta Endocrinol.*, 109, 181, 1985. With permission.)

biological activity of the various isoforms. However, Weiss,[76] in a study of purified isoforms of hLH, showed a 2.4-fold difference in immunopotency between various isoforms, a corresponding 8.4-fold change in in vitro bioactivity, with a 4-fold change in the ratio of in vitro biological to immunological activities (B/I). Furthermore, significant differences in slope values of the response lines in the radioreceptor assay for the various isoforms were observed suggesting that the various isoforms had different affinities for the receptor. Collectively these results indicate that the various isoforms have inherently different immunological, in vitro biological, and receptor activities and that interpretations of the ratio of these activities must be treated cautiously.

FSH radioreceptor assays (RRA) employed by groups in this review are based on the competition between iodinated FSH and highly purified rat or human FSH preparations for testes receptor preparations from either the rat,[77] hamster,[78] or calf.[79] Since significant differences in slope values of the dose response lines in the RRA have been observed[77] between species for FSH, usual practice is to use a standard pituitary preparation of low purity obtained from the species under study. FSH RRA methods[79] are very sensitive to small changes in salt concentrations, therefore samples are usually gel filtered or dialyzed against the assay buffer prior to assay. Results from this laboratory have shown that carrier ampholytes used in electrofocusing studies can interfere markedly in the RRA and because of their size, separation from FSH on Sephadex® G25 gel filtration columns is not complete. One commercial ampholyte mixture (Ampholine,® LKB Bromma, Sweden) showed minimal interference in the FSH RRA[79] and its effects can be eliminated by gel filtration on Sephadex® G25 and/or dilution.

Results obtained by FSH methods in the assay of human FSH preparations of varying purity[80] show a good correlation with those obtained by FSH RIA and in vitro bioassay although significant differences were observed in FSH activity ratios, including B/R ratios, with some preparations. These differences were attributed to differential losses of biological activity through the purification procedure as either crude pituitary extracts or highly purified preparations gave B/R ratios close to unity. FSH α- and β-subunits and other glycoprotein hormones show negligible cross-reaction in the assay.

Considerable discussion has occurred over the quantitative importance of B/I ratios, particularly in situations where B/I ratios are elevated. In studies with hFSH[80,81] and hLH,[82,83] the choice of standard, quality of tracer, specificity of antiserum, and evidence of parallelism between standard and unknown were some of the variables, with the choice of standard playing the most important role. Many of the gonadotrophin standards contain considerable amounts of immunological activity which were associated with reduced or negligible biological activity. Since the unitage of the standard is described in terms of its biological activity, the use of a standard containing both bioactive and bioinactive immunologically active material will give an underestimation in the assay of unknowns in RIA systems which detect both activities. Thus, for example, the use of the first international reference preparation for LH/FSH for bioassay (which is a crude pituitary extract) as an hFSH standard will result in elevated B/I ratios while the use of a purified hFSH preparation as standard will result in ratios nearer unity.[80] It also follows that the use of antisera with higher specificity for biologically active forms of the hormone will also result in ratios approaching unity. In many of the studies in this review, crude pituitary extracts are employed as standard for the respective assays. Thus R/I or B/I ratios greater than unity have been observed. In this review the absolute ratios of activities described in various reports are largely disregarded with emphasis placed on changes in ratios instead.

In determining recoveries of FSH following IEF in sucrose gradients in the rat,[47] baboon,[60] human pituitary extracts, and highly purified human FSH preparations[59] the recoveries of RRA activity and in vitro bioactivity were uniformly high (88 to 95%). On the other hand the recoveries of immunological activity were much lower (50 to 71%). This observation is surprising as at least one of the RIA systems employed (to human FSH) showed low cross-reaction to the individual subunits and other gonadotrophins, a sign of high specificity. The disappearance of the immunoactive material on electrofocusing suggests either a loss of immunoactivity or more likely its very acidic or very basic pI values results in its migration into the electrode buffer. The nature of this immunoreactive material in pituitary extracts is unknown although the presence of high concentrations of free subunits has not been excluded.

In vitro bioassays which are based on the stimulation by the hormone of a biological end point are largely free of the above limitations associated with competition assays. Two FSH in vitro bioassays have mainly been used. The first is based on the FSH-specific induction of aromatase enzyme in short-term cultures of immature rat Sertoli cells,[84] and the second is based upon the induction of plasminogen activator by rat ovarian granulosa cells in culture.[85] Both methods are sensitive and show minimal cross-reaction with other pituitary hormones, e.g., LH.

B. Species Differences in FSH Heterogeneity

1. Rat and Hamster

Using either gel filtration or IEF techniques, age- and sex-related changes in FSH heterogeneity have been observed. Gel filtration fractionation of rat pituitary extracts under conditions of high resolution have shown that FSH isoforms from male rats are larger on average than those from female rats.[53-57] These isoforms are reduced in size following gonadectomy (Table 1). Androgen treatment of gonadectomized males or females resulted in a larger form of FSH similar in size to that seen with the intact male animal.[53-57] Estrogen treatment was largely ineffective.[53-56] Age-related changes were observed in both sexes with complex changes seen in the female between days 12, 34, and 90. Animals aged 17 and 34 days had similar distribution patterns to those seen in males.[56] Increasing age in the male rat resulted in an increase in FSH size[57] which was reversed by castration. Testosterone administration to the castrate animal reversed the castration effect.[57] The plasma clearance rates of pituitary FSH obtained from rats either undergoing androgen treatment or following castration were longer than that of the intact animal.[54]

Fractionation of pituitary extracts from female rats by IEF-PAG resulted in identification of seven immunological peaks with pI values of 3.8, 4.2, 4.7, 5.1, 5.3, 5.7, and 5.8.[86-88] The pI region 3.8 to 4.3, in comparison with its immunological activity, showed minimal levels of receptor-binding activity. A higher proportion of the more basic forms (pI >5) was found in the female adult compared to the immature animal. In the developing male rat a shift to more basic pI values was also noted with increasing age.[89]

In order to assess if the charge heterogeneity of FSH was attributable to its sialic acid content, a purified rat FSH preparation with a pI value of 4.2 was treated with neuraminidase and the pI profiles were assessed with time.[88] During neuraminidase treatment a large number of peaks of FSH were identified by RRA and RIA, the profiles of which became progressively basic with time. A number of peaks had similar pI values to that found in pituitary extracts suggesting that sialic acid content is a major cause of charge heterogeneity. The FSH R/I ratio increased by two- to threefold during neuraminidase incubation with the highest ratios being found in the more basic fractions.[88]

Fractionation of anterior pituitary extracts of female hamsters using IEF-PAG revealed six species of immunologically active FSH with pI values similar to that observed in the rat[65,87] (4.0, 4.7, 5.1, 5.4, 5.7, and 5.8). Following estradiol treatment, increasing amounts of the more acidic forms were observed[87] although this trend was not quantitatively analyzed. As seen with FSH from the rat[88] and human,[69-71] neuraminidase treatment modified the pI distribution of FSH toward more basic forms. FSH heterogeneity was also examined in the immature, adult, and castrate adult male hamster using both RIA and RRA techniques.[78] pI values of 3.8, 4.2, 4.7, 5.0, 5.3, 5.7, and 6.0 were obtained although the radioreceptor activity in the pH range 3.8 to 4.2 was markedly lower than its corresponding immunological activity. The pI 4.7 isoform was not detected in the immature animal. The R/I ratio for each species of FSH increased with increasing pI value.

A detailed analysis of the pI distribution of FSH as measured by RRA, RIA, and in vitro bioassay was performed in pituitary extracts from male and female rats following electrofocusing in 110-mℓ sucrose gradient columns.[47] In contrast to other studies,[86-94] the pI distribution of FSH was confined to the pH range 3.4 to 4.8 in the male with pI values for peak tubes of 3.99, 4.30, and 4.51 with a broad region <3.8; and pH 3.8 to 4.8 in the female with pI values of 3.91, 4.11, 4.26, and 4.51. Significant but minor increases in FSH R/I ratio (1.5-fold) and B/R ratios (1.2-fold) were observed between the acidic and basic fractions (Figure 2). Of interest was the observation that these ratios as well as the B/R ratios showed differences between sexes within each pI region indicating that while the various isoforms between sexes may have similar physicochemical properties (pI values, Stokes radius) they are not identical. Clearly other factors, e.g., composition of noncharged sugars, are playing a role. When the various FSH isoforms were chromatographed on Sephadex® G100 Superfine a direct relationship between molecular size and pI value of each isoform (Table 1) was observed. Assessment of binding of the various isoforms to the lectin concanavalin A indicated that all isoforms are glycosylated, although significant

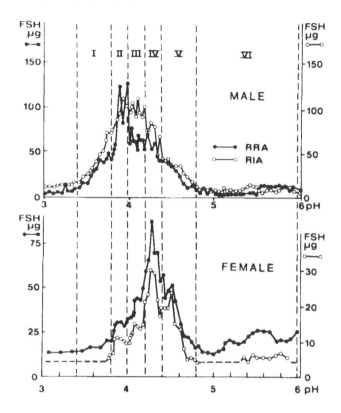

FIGURE 2.   pH distribution of FSH from male and female pituitaries after electrofocusing in 110 mℓ sucrose gradients. FSH was determined by radioimmunoassay (RIA) and radioreceptor assay (RRA) with the NIH-rFSH-RP-1 as standard. The fractions were pooled (I-VI) for further analysis (see Table 1). (Adapted from Robertson, D. M., Foulds, L. M., and Ellis, S., *Endocrinology*, 111, 385, 1982.)

differences in lectin binding between isoforms[47] suggest different carbohydrate composition. Plasma levels of FSH in the rat as determined by RRA and RIA also showed differences in these activities between sexes, following gonadectomy, and after steroid (testosterone and estradiol) treatment; however, the steroids had opposing effects on R/I ratios in plasma in comparison with pituitary extracts.[91]

In order to assess the direct effects of the various hormones known to affect FSH secretion and its heterogeneity, short-term cultures of anterior pituitary cells have been studied in the rat and hamster.[78,92,94] A comparable pI profile of FSH was observed in extracts from either pituitary glands or pituitary cells in culture from either female or male animals[78,92,94] indicating that the pituitary cell culture system does not grossly alter the synthetic pattern of FSH. The addition of LHRH stimulates the release of FSH resulting in the secretion of all the soluble intracellular forms. Insoluble immunoactive forms, some of which are extractable by the detergent Triton® X-100 (unpublished results) were not considered in this study. A clear sex difference in pI profiles of FSH was maintained in culture and following LHRH-stimulated release. Thus, LHRH induces the release of soluble intracellular forms of FSH without modification to their charge profile. However, in other aspects the findings between studies are not complementary. Using pituitary cell cultures from immature male rats a significant decrease in R/I ratio was observed[94] following LHRH stimulation while studies with pituitary cells in culture from mature male and female rats did not show a significant change.[47] Following a 3-day testosterone treatment, a change to more acidic pI values was observed in immature animals[94] while no effect was observed in the adult male[47] which may be attributed to the age difference in the donor animals. The failure to observe any effect of androgens in the adult suggests that either the culture period was too short or the effects of steroids was not a direct action on the pituitary and thus not seen in vitro.

The addition of bovine follicular fluid (a rich inhibin source) which suppresses FSH synthesis and release under the culture conditions used did not lead to a marked change in pI distribution sufficient to account for the differences in pI values between sexes.[92]

2. Human

Electrofocusing of human pituitary extracts[59,96] revealed a heterogeneous pattern of FSH as determined by RIA, RRA, and in vitro bioassay in the pH range 3.5 to 5.5 with common peaks of activity with pI values of 4.00, 4.44, and 5.50. A two- to threefold range in FSH B/I ratios was observed with higher ratios found between pH 4 to 5 than between 5 to 6.

Examination by IEF of eight purified human FSH preparations[96] indicated the presence of 11 FSH isoforms, 5 of which have similar pI values to those observed in crude pituitary extracts, but in different proportions. The in vitro bioactivity of these preparations shows a fourfold range in specific activity (3660 to 14,630 IU/mg). However, the biological activity of the various FSH isoforms in these preparations has not been determined. The B/I ratio of FSH isoforms following electrofocusing of these eight preparations showed a twofold range of values. The preparation with the most basic pI values (pI 5 to 6) also showed the lowest ratio of in vivo/in vitro bioactivity or in vivo bioactivity/immunoactivity suggesting a highly desialylated preparation. The different isoform distribution observed among preparations was attributed in part to the desialylation of FSH which occurred during the purification process and to the selection of particular FSH isoforms for isolation in the purification process.

Using RIA and in vitro and in vivo bioassays in the assay of FSH, a reduced in vivo biological activity but similar immunological activity and in vitro bioactivity in pituitary extracts from young women compared with those from men and elderly women was observed.[97] Electrophoretic fractionation of both pituitary extracts and serum from men and women of various ages[68-71] revealed that (1) the FSH from elderly people was more acidic than from young people, (2) this age-related pattern is much more apparent in women than in men, (3) estrogen treatment can induce secretion of less acidic forms of FSH, and (4) after neuraminidase treatment a marked change to a more basic form was observed in both sexes. The more basic forms of FSH found in serum and pituitary extracts in young women are attributed to a lower sialic acid content and therefore to a shorter circulating half-life. This hypothesis was substantiated by noting that FSH from female, in contrast to male, pituitary extracts is cleared more rapidly from the circulation when tested in mice.[98]

3. Monkey

The gel filtration patterns of FSH from intact and gonadectomized rhesus monkeys are the opposite to that seen in the rat, i.e., larger forms of FSH were observed following gonadectomy which in the case of gonadectomized animals was reversed by estradiol administration.[48,49,99]

Following electrofocusing of pituitary extracts from male and female rhesus monkeys, eight to nine forms of FSH in the pH range 3.75 to 8 have been observed[100] (Figure 3). After gonadectomy in both sexes a similar change in distribution of activity was observed with activity distributed below pH 5, absent between pH 6.5 and 7.5 (in contrast to the intact animal), and the appearance of a new peak with a pI value of 9.0. Overall, the pI values were more acidic following gonadectomy. As seen in the intact animal a twofold increase in B/I ratios was observed between the more basic and acidic isoforms.[100]

Using a chromatofocusing procedure, Chappel et al.[52] separated eight forms of FSH from

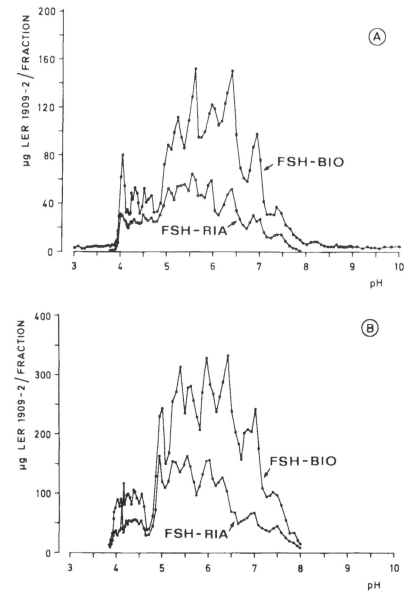

FIGURE 3.   Isoelectrofocusing profiles of FSH in individual pituitary extracts from intact male (A), intact female (B), gonadectomized male (C), and gonadectomized female (D) rhesus monkeys as determined by in vitro bioassay (FSH-BIO) and radioimmunoassay FSH-RIA) procedures. (From Khan, S. A., Sved, V., Fröysa, B., Lindberg, M., and Diczfalusy, E., *J. Med. Primatol.*, 14, 177, 1985. With permission.)

anterior pituitary extracts of cynomolgus monkeys with pI values ranging between 4.1 and 6.4 and an additional peak with a pI value <4. When ovariectomized monkeys were treated with physiological levels of estradiol and progesterone over a 36-hr period a higher proportion (35.5 vs. 18.3%) of the more basic forms (pI >5.3) was observed with a corresponding twofold increase in RRA/RIA ratios. These results support the observations in the rhesus monkey[48,49] that estrogens cause a reduction in the size of FSH associated with a decreased level of sialylation. Neuraminidase treatment of the pituitary extracts resulted in a more

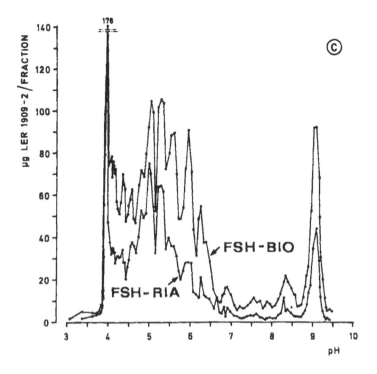

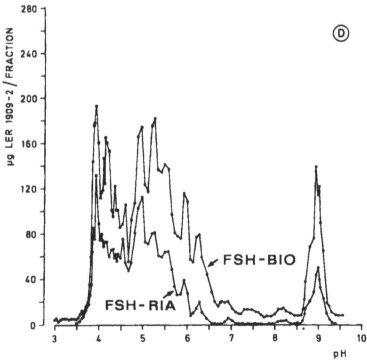

FIGURE 3 continued

basic electrofocusing profile of FSH in support that sialic acid plays a primary role in charge heterogeneity.

Electrofocusing of pituitary extracts from male and female baboons revealed similar pI profiles of FSH bio- and immunoactivity in the pH range 4 to 7 with a two- to fourfold increase in B/I ratio for the more basic forms in comparison with the less basic forms.[60]

4. Other Species

Long-term castration of the male bullfrog *(Rana catesbeiana)* resulted in a marked increase in the size of pituitary and plasma FSH as assessed by gel filtration, with the circulating forms larger than that found in pituitary extracts.[58,101] As expected from studies in other species, the circulating disappearance rates for pituitary FSH from castrate frogs was considerably longer (25.6 hr) than that from intact animals (1 hr).[58]

In the sheep a similar pI profile of pituitary FSH was observed in ewes and rams within the pH range 4 to 7. A close correspondence between radioreceptor and immunological FSH activities was observed although high base-line levels of immunoactivity were observed in the ram in regions devoid of radioreceptor activity.[102] Following electrofocusing of equine pituitary extracts a heterogeneous pI profile of FSH as determined by FSH RRA was observed in the pH range 4.00 to 4.32.[103]

III. CONCLUSIONS

The evidence strongly suggests that the basis for FSH heterogeneity resides in its carbohydrate structure and that the pI values of the various isoforms is largely attributed to its sialic acid content (see also Chapter 7). This evidence is based on (1) the ability of neuraminidase to modify the pI distribution of FSH; (2) a high correlation between sialic acid content and pI values as observed in other glycoprotein hormones, e.g., hCG and LH; and (3) the negligible in vivo although detectable in vitro biological activity of the more basic pI forms of FSH indicating a rapid clearance rate and therefore a reduced sialic acid content.

It is apparent that other factors are involved; the observed differences in R/I or B/I ratios between isoforms would not be anticipated from studies with deglycosylated hormones where radioreceptor and immunological activities, in contrast to in vitro biological activity, are retained following deglycosylation.[6] Furthermore, the observation that significant differences in either R/I or B/I ratios were observed between sexes for isoforms with the same pI values indicates a further level of heterogeneity not resolved by charge separation methods. Similar observations have been made regarding rat pituitary LH.[104,105] Desialylation of purified hormones while rendering them more basic still results in a heterogeneous profile on IEF.[70,71]

Overall, these findings suggest that at least two levels of heterogeneity exist, one related to sialic acid content, the other to factors unknown. However, partial sulfation, phosphorylation, or other modifications to the carbohydrate structure, as discussed above, are possible.

Varying pI profiles of pituitary FSH as assayed by RIA, RRA, and in vitro bioassay have been reported both within and between species. It is generally agreed that the pI values range between 3.5 and 6.0. However, the results of one group (with monkey, hamster, and rat)[86] indicate that the more acidic forms are largely inactive in the RRA and in vitro bioassay while other groups[47,92,102] have reported similar patterns of these activities over the same acidic pH region. There is no ready explanation for these differences. However, this matter is clearly important to resolve for several reasons. First, if differences in B/I ratios do exist between isoforms then these differences may play an important physiological role in conjunction with differences in clearance rates in the biological action of FSH. Second, to argue that different isoforms may exhibit different biological to immunological activities reduces the usefulness of radioimmunoassay methods to reliably measure FSH activity. While radioimmunoassays may be reliable in measuring the levels of FSH protein skeleton, its measurement per se in physiological studies is of limited value unless it also is a measure of biological activity. If it is true that there are differences in B/I ratios between isoforms, then one either employs assays which measure in vitro biological activity or one develops a range of radioimmunoassays specific to each isoform. In the case of LH where differences in B/I ratios have been observed under various endocrine conditions the use of in vitro bioassays is widely used.

The physiological significances of changing populations of FSH isoforms according to the endocrine status of the animal is more readily understood in terms of their differences in FSH clearance rates than in any difference in their (in vitro) biological activity. Differences in FSH clearance rates may reflect the ability of the animal to respond to feedback signals from the gonads or other organs. In the female, for example, where ovarian function is tightly modulated within a limited fixed time frame, the ability to promptly reduce circulating levels of FSH may be highly advantageous, while in the male there is most likely much less need for such sensitive controls. If, however, there is a marked change in FSH in vitro bioactivity, in particular a decrease in activity associated with those isoforms with the longer clearance rates, then the overall advantage of longer clearance rates may be negated.

Much of the confusion surrounding the characterization and physiological importance of the various FSH isoforms will most likely persist until the individual isoforms are isolated and characterized. There is a requirement however that the isolated isoforms must be structurally identical to isoforms native to the pituitary. Based on differences in biological activity and various physicochemical properties of purified preparations of gonadotrophins, the development of mild (nonmodifying) fractionation conditions must be the first requirement. With the advent of new affinity chromatography and reversed phase and ion-exchange HPLC procedures the development of suitable purification procedures may now be possible.

Clear species differences are seen in the distribution of FSH isoforms following endocrine manipulation. In the human, hamster, and rat, pituitary FSH isoforms are either larger and/or more acidic with a longer circulating half-life in the male than in the female. Gonadectomy of either sex results in a smaller, more basic molecule while androgen replacement reverses this trend. In contrast, FSH isoforms in the male rhesus monkey are smaller than in the female and are larger and/or more acidic following gonadectomy. Androgen and estrogen treatment reverses the effect of gonadectomy.

The biochemical basis for these endocrine-related changes in the distribution of FSH isoforms is unclear. These changes reflect differences in carbohydrate structure and thus reflect differences in specific concentrations of the appropriate transferase enzymes. It would seem likely that, since the type and secretory properties of gonadotrophs are modified by steroid treatment, hormone heterogeneity may be a reflection of gonadotroph heterogeneity.

REFERENCES

1. **Pierce, J. G. and Parsons, T. F.,** Glycoprotein hormones structure and function, *Annu. Rev. Biochem.,* 50, 465, 1981.
2. **Saxena, B. B. and Rathnam, P.,** Structure-function relationship of human pituitary FSH and LH, in *Serono Symposium,* Vol. 49, Motta, M., Zanisi, M., and Piva, F., Eds., Academic Press, New York, 1982, 63.
3. **Ryle, H., Chaplan, M. F., Gray, C. J., and Kennedy, J. G.,** The action of a neuraminic acid-free derivative of FSH on mouse ovarian culture *in vitro,* in *Gonadotropins and Ovarian Development,* Butt, W. R., Crooke, A. C., and Ryle, M. Eds., Livingstone, Edinburgh, 1970, 98.
4. **Aggarwal, B. B. and Papkoff, H.,** Relationship of sialic acid residues to in vitro biological and immunological activities of equine gonadotropins, *Biol. Reprod.,* 24, 1082, 1981.

5. **Vaitukaitis, J. L. and Ross, G. T.,** Altered biologic and immunologic activities of progressively desi-alylated human urinary FSH, *J. Clin. Endocrinol. Metab.,* 53, 308, 1971.

6. **Sairam, M. R.,** Gonadotropic hormones: relationship between structure and function with emphasis on antagonists, in *Hormone Protein and Peptides,* Vol. 11, Li, C. H., Ed., Academic Press, New York, 1983, 1.

7. **Weintraub, B. D.,** Biosynthesis and secretion of TSH: relationship to glycosylation, in *Serono Symposium,* Vol. 49, Motta, M., Zanisi, M., and Piva, F., Eds., Academic Press, New York, 1982, 43.

8. **Weintraub, B. D., Siannard, B. S., Magner, J. A., Ronin, C., Taylor, T., Joshi, L., Constant, R. B., Menezes-Ferreira, M. M., Petrick, P., and Gesundheit, N.,** Glycosylation and posttranslational processing of thyroid-stimulating hormone: clinical implications, *Rec. Prog. Horm. Res.,* 41, 577, 1985.

9. **Fiddes, J. C. and Talmadge, K.,** Structure, expression and evolution of the genes for the human gly-coprotein hormones, *Rec. Prog. Horm. Res.,* 40, 43, 1984.

10. **Hoshina, H. and Boime, I.,** Combination of rat lutropin subunits occurs early in the secretory pathway, *Proc. Natl. Acad. Sci. U.S.A.,* 79, 7649, 1982.

11. **Whitfield, G. K. and Miller, W. L.,** Biosynthesis and secretion of FSH beta subunit from ovine pituitary cultures: effect of 17β estradiol treatment, *Endocrinology,* 115, 154, 1984.

12. **Farquhar, M. G.,** Processing of secretory products by cells of the anterior pituitary gland, *Mem. Soc. Endocrinol.,* 19, 79, 1971.

13. **Parsons, T. F., Bloomfield, G. A., and Pierce, J. G.,** Purification of an alternative form of the α subunit of the glycoprotein hormones from bovine pituitaries and identification of its O-linked oligosaccharide, *J. Biol. Chem.,* 258, 240, 1983.

14. **Parsons, T. F. and Pierce, J. G.,** Oligosaccharide moieties of glycoprotein hormones: bovine LH resists enzymatic deglycosylation because of O-sulfated *N*-acetylhexosamines, *Proc. Natl. Acad. Sci. U.S.A.,* 77, 7089, 1980.

15. **Corless, C. L. and Boime, I.,** Differential secretion of O-glycosylated gonadotropin α-subunit and LH in the presence of LHRH, *Endocrinology,* 117, 1699, 1985.

16. **Green, E. D., Baenziger, J. U., and Boime, I.,** Cell-free sulfation of human and bovine pituitary hormones: comparison of the sulfated oligosaccharides of LH, FSH and TSH, *J. Biol. Chem.,* 260, 15631, 1985.

17. **Cozzi, M. G. and Zanini, A.,** Sulfated LH subunits and a tyrosine-sulfated secretory protein (secretogranin II) in female adenohypophyses: changes with age and stimulation of release by LHRH, *Mol. Cell. Endocrinol.,* 44, 47, 1986.

18. **Counis, R., Corbani, M., Poissonnier, M., and Jutisz, M.,** Characterization of the precursors of α and β subunits of FSH following cell-free translation of rat and ovine pituitary mRNAs, *Biochem. Biophys. Res. Commun.,* 107, 998, 1982.

19. **Khar, A., Debeljuk, L., and Jutisz, M.,** Biosynthesis of gonadotropins by rat pituitary cells in culture and in pituitary homogenates: effect of LHRH, *Mol. Cell. Endocrinol.,* 12, 53, 1978.

20. **Liu, T.-C. and Jackson, G. L.,** Effects of synthetic LHRH on incorporation of radioactive glucosamine and amino acids into LH and total protein by rat pituitaries in vitro, *Endocrinology,* 98, 151, 1976.

21. **Liu, T.-C. and Jackson, G. L.,** Modifications of LH biosynthesis and release by LHRH, cycloheximide and actinomycin D, *Endocrinology,* 103, 1253, 1978.

22. **Azhar, S., Reel, J. R., Pastushok, C. A., and Menon, K. M. J.,** LH biosynthesis and secretion in rat anterior pituitary cell cultures: stimulation of LH glycosylation and secretion by LHRH and an agonist analogue and blockade by an antagonist analogue, *Biochem. Biophys. Res. Commun.,* 80, 659, 1978.

23. **Alexander, D. C. and Miller, W. L.,** Regulation of ovine FSH β chain mRNA by 17β-oestradiol *in vivo* and *in vitro, J. Biol. Chem.,* 257, 2282, 1982.

24. **Batra, S. K. and Miller, W. L.,** Progesterone inhibits basal secretion of FSH in ovine pituitary cell culture, *Endocrinology,* 117, 2443, 1985.

25. **Counis, R., Corbani, M., and Jutisz, M.,** Estradiol regulates mRNAs encoding precursors to rat LH and FSH subunits, *Biochem. Biophys. Res. Commun.,* 114, 65, 1983.

26. **Drouin, J. and Labrie, F.,** Interactions between estradiol and progesterone on the control of LH and FSH release in rat anterior pituitary cells in culture, *Endocrinology,* 108, 52, 1981.

27. **Leveque, N. W. and Grotjan, H. E., Jr.,** Interaction of progesterone with testosterone and DHT on FSH hormone release by cultures of rat anterior pituitary cells, *Biol. Reprod.,* 27, 110, 1982.

28. **Legace, L., Labrie, F., Antakly, T., and Pelletier, G.,** Sensitivity of rat adenohypophyseal cells to estradiol and LHRH during long term culture, *Am. J. Physiol.,* 240, E602, 1981.

29. **Legace, L., Massicotte, J., and Labrie, F.,** Acute stimulatory effects of progesterone on LH and FSH release in rat anterior pituitary cells in culture, *Endocrinology,* 106, 684, 1980.

30. **Giguere, V., Lefebvre, F.-A., and Labrie, F.,** Androgens decrease LHRH binding in rat pituitary cells in culture, *Endocrinology,* 108, 350, 1981.

31. **Corbani, M., Counis, R., Starzec, A., and Jutisz, M.,** Effect of gonadectomy on pituitary levels of mRNA encoding gonadotropin subunits and secretion of LH, *Mol. Cell. Endocrinol.,* 35, 83, 1984.

32. **Godine, J. E., Chin, W. W., and Habener, J. F.,** LH and FSH: cell-free translations of mRNAs coding for subunit precursors, *J. Biol. Chem.,* 255, 8780, 1980.
33. **Miller, W. L., Alexander, D. C., Wu, J. C., Huang, E. S., Whitfield, G. K., and Hall, S. H.,** Regulation of β-chain mRNA of ovine FSH by 17β-oestradiol, *Mol. Cell. Biochem.,* 53/54, 187, 1983.
34. **Landefeld, T., Kepa, J., and Karsch, F.,** Estradiol feedback effects on the α-subunit mRNA in the sheep pituitary gland: correlation with serum and pituitary LH concentrations, *Proc. Natl. Acad. Sci. U.S.A.,* 81, 1322, 1984.
35. **Bourguignon, J. P., Demoulin, A., Hoyoux, C., and Franchimont, P.,** Modulation of hypothalamo-pituitary-gonadal axis by inhibin, in *Serono Symposium,* Vol. 42, Flamigni, C. and Givens, J. R., Eds., Academic Press, New York, 1982, 143.
36. **Scott, R. S. and Burger, H. G.,** Mechanism of action of inhibin, *Biol. Reprod.,* 24, 541, 1981.
37. **Robertson, D. M., Giacometti, M., and de Kretser, D. M.,** The effects of inhibin purified from bovine follicular fluid in several *in vitro* pituitary cell culture systems, *Mol. Cell. Endocrinol.,* 46, 29, 1986.
38. **Massicotte, J., Legace, L., Godbout, M., and Labrie, F.,** Modulation of rat pituitary gonadotrophin secretion by porcine granulosa cell inhibin, LHRH and sex steroids in rat anterior pituitary cells in culture, *J. Endocrinol.,* 100, 133, 1984.
39. **Massicotte, J., Legace, L., Labrie, F., and Dorrington, J. H.,** Modulation of gonadotropin secretion by Sertoli cell inhibin, LHRH and sex steroids, *Am. J. Physiol.,* 247, E495, 1984.
40. **Moriarty, G. C.,** Immunocytochemistry of the pituitary glycoprotein hormones, *J. Histochem. Cytochem.,* 24, 846, 1976.
41. **Dada, M. O., Campbell, G. T., and Blake, C. A.,** A quantitative immunocytochemical study of the luteinizing hormone and follicle-stimulating hormone cells in the adenohypophysis of adult male rats and adult female rats throughout the estrous cycle, *Endocrinology,* 113, 970, 1983.
42. **Childs, G. V., Hyde, C., Naor, Z., and Catt, K.,** Heterogeneous luteinizing hormone and follicle-stimulating hormone storage patterns in subtypes of gonadotropes separated by centrifugal elutriation, *Endocrinology,* 113, 2120, 1983.
43. **Denef, C.,** Functional heterogeneity of separated dispersed gonadotropic cells, in *Synthesis and Release of Adenohypophyseal Hormones,* Jutisz, M. and McKern, K., Eds., Plenum Press, New York, 1980, 659.
44. **Denef, C., Hautekeete, E., Dewals, R., and De Wolf, A.,** Differential control of LH and FSH secretion by androgens in rat pituitary cells in culture: functional diversity of subpopulations separated by unit gravity sedimentation, *Endocrinology,* 106, 724, 1980.
45. **Tibolt, R. E. and Childs, G. V.,** Cytochemical and cytophysiological studies of LHRH target cells in the male rat pituitary: differential effects of androgens and corticosterone on GnRH binding and gonadotropin release, *Endocrinology,* 117, 396, 1985.
46. **Smith, P. F., Frawley, L. S., and Neill, J. D.,** Detection of LH release from individual pituitary cells by the reverse hemolytic plaque assay: estrogen increases the fraction of gonadotropes responding to LHRH, *Endocrinology,* 115, 2484, 1984.
47. **Foulds, L. M. and Robertson, D. M.,** Electrofocusing fractionation and characterization of pituitary follicle-stimulating hormone from male and female rats, *Mol. Cell. Endocrinol.,* 31, 117, 1983.
48. **Peckham, W. D. and Knobil, E.,** Qualitative changes in the pituitary gonadotropins of the male rhesus monkey following castration, *Endocrinology,* 98, 1061, 1976.
49. **Peckham, W. D. and Knobil, E.,** The effects of ovariectomy, estrogen replacement and neuraminidase treatment on the properties of the adenohypophyseal glycoprotein hormones of the rhesus monkey, *Endocrinology,* 98, 1054, 1976.
50. **Roos, P.,** Follicle stimulating hormone, *Acta Endocrinol. Suppl.,* 131, 1968.
51. **Wide, L. and Roos, P.,** Pleomorphism of human FSH, in *Serono Symposium,* Vol. 49, Motta, M., Zanisi, M., and Piva, F., Eds., Academic Press, New York, 1982, 75.
52. **Chappel, S. C., Bethea, C. L., and Spies, H. G.,** Existence of multiple forms of FSH within the anterior pituitaries of Cynomolgus monkeys, *Endocrinology,* 115, 452, 1984.
53. **Bogdanove, E. M., Nolin, J. M., and Campbell, G. T.,** Qualitative and quantitative gonad-pituitary feedback, *Rec. Prog. Horm. Res.,* 31, 567, 1975.
54. **Bogdanove, E. M., Campbell, G. T., Blair, E. D., Mula, M. E., Miller, A. E., and Grossman, G. H.,** Gonad-pituitary feedback involves qualitative change: androgens alter the type of FSH secreted by the rat pituitary, *Endocrinology,* 95, 219, 1974.
55. **Bogdanove, E. M., Campbell, G. T., and Peckham, W. D.,** FSH pleomorphism in the rat-regulation by gonadal steroids, *Endocr. Res. Commun.,* 1, 87, 1974.
56. **Blum, W. and Gupta, D.,** Age and sex-dependent nature of the polymorphic forms of rat pituitary FSH: the role of glycosylation, *Neuroendocrinol. Lett.,* 6, 357, 1980.
57. **Conn, M., Cooper, R, McNamara, C., Rodgers, D. C., and Shoenhardt, L.,** Qualitative change in gonadotropin during normal aging in the male rat, *Endocrinology,* 106, 1549, 1980.
58. **McCreery, B. R. and Licht, P.,** Effects of gonadectomy on polymorphism in stored and circulating gonadotropins in the bullfrog, *Rana catesbeiana.* I. Clearance profiles *Biol. Reprod.,* 29, 637, 1983.

59. **Zaidi, A. A., Robertson, D. M., and Diczfalusy, E.**, Studies on biological and immunological properties of human FSH: profiles of two international reference preparations and of an aqueous extract of pituitary glands after electrofocusing, *Acta Endocrinol.*, 97, 157, 1981.

60. **Khan, S. A., Katzija, G., Fröysa, B., and Diczfalusy, E.**, Characterization of various molecular species of follicle-stimulating hormone in baboon pituitary preparations, *J. Med. Primatol.*, 13, 295, 1984.

61. **Robertson, D. M. and Diczfalusy, E.**, Biological and immunological characterization of human LH. II. A comparison of the immunological and biological activities of pituitary extracts after electrofocusing using different standard preparations, *Mol. Cell. Endocrinol.*, 9, 57, 1977.

62. **Katsumata, M. and Goldman, A. S.**, Separation of multiple dihydrotestosterone receptors in rat ventral prostate by a novel micromethod of electrofocusing: blocking action of cyproterone acetate and uptake by nuclear chromatin, *Biochim. Biophys. Acta*, 359, 112, 1974.

63. **Robertson, D. M., Foulds, L. M., and Ellis, S.**, Heterogeneity of rat pituitary gonadotropins on electrofocusing: differences between sexes and after castration, *Endocrinology*, 111, 385, 1982.

64. **Chappel, S. C.**, The presence of two species of FSH within hamster anterior pituitary glands as disclosed by Concanavalin A chromatography, *Endocrinology*, 109, 935, 1981.

65. **Chappel, S. C., Coutifaris, C., and Jacobs, S. J.**, Studies on the microheterogeneity of FSH present within the anterior pituitary gland of ovariectomized hamsters, *Endocrinology*, 110, 847, 1982.

66. **Sluyterman, L. A. E. and Elgersma, O.**, Chromatofocusing: isoelectric focusing on ion-exchange columns. I. General principles, *J. Chromatogr.*, 150, 17, 1978.

67. **Sluyterman, L. A. E. and Wijdenes, J.**, Chromatofocusing: isoelectric focusing on ion-exchange columns. II. Experimental verification, *J. Chromatogr.*, 150, 31, 1978.

68. **Wide, L.**, Electrophoretic and gel chromatographic analysis of FSH in human serum, *Upsala J. Med. Sci.*, 86, 249, 1981.

69. **Wide, L.**, Male and female forms of human FSH in serum, *J. Clin. Endocrinol. Metab.*, 55, 682, 1982.

70. **Wide, L.**, Median charge and charge heterogeneity of human pituitary FSH, LH and TSH. I. Zone electrophoresis in agarose suspension, *Acta Endocrinol.*, 109, 181, 1985.

71. **Wide, L.**, Median charge and charge heterogeneity of human pituitary FSH, LH and TSH. II. Relationship to sex and age, *Acta Endocrinol.*, 109, 190, 1985.

72. **Moyle, W. R., Bahl, O. P., and Marz, L.**, Role of the carbohydrate of hCG in the mechanism of hormone action, *J. Biol. Chem.*, 250, 9163, 1975.

73. **Bahl, O. P. and Moyle, W. R.**, Role of carbohydrate in the action of gonadotropins, in *Receptors and Hormone Action, Vol. 3*, Birnbaumer L. and O'Malley, B., Eds., Academic Press, New York, 1978, 261.

74. **Channing, C. P. and Bahl, O. P.**, Role of carbohydrate residues of hCG in stimulation of progesterone secretion by cultures of monkey granulosa cells, *Biol. Reprod.*, 17, 707, 1978.

75. **Channing, C. P., Sakai, C. N., and Bahl, O. P.**, Role of carbohydrate residues of hCG in binding and stimulation of adenosine 3, 5 monophosphate accumulation by porcine granulosa cells, *Endocrinology*, 103, 341, 1978.

76. **Weise, H. C., Graesslin, D., Lichtenberg, V., and Rinne, G.**, Polymorphism of human LH. Isolation and partial characterisation of seven isoforms, *FEBS Lett.*, 159, 93, 1983.

77. **Reichert, L., Jr. and Bhalla, V. K.**, A comparison of the properties of FSH from several species as determined by a rat testis tubule receptor assay, *Gen. Comp. Endocrinol.*, 23, 111, 1974.

78. **Ulloa-Aguirre, A. and Chappel, S. C.**, Multiple species of FSH within the anterior pituitary gland of male golden hamsters, *J. Endocrinol.*, 95, 257, 1982.

79. **Cheng, K.-W.**, A RRA for FSH, *J. Clin. Endocrinol. Metab.*, 41, 581, 1975.

80. **Marana, R., Robertson, D. M., Suginami, H., and Diczfalusy, E.**, The assay of human follicle-stimulating hormone preparations: the choice of a suitable standard, *Acta Endocrinol.*, 92, 599, 1979.

81. **Storring, P. L., Zaidi, A. A., Mistry, Y. G., Fröysa, B., Stenning, B., and Diczfalusy, E.**, A comparison of preparations of highly purified human pituitary FSH: differences in FSH potencies as determined by in vivo bioassay, in vitro bioassay and immunoassay, *J. Endocrinol.*, 91, 353, 1981.

82. **Robertson, D. M., Puri, V., Lindberg, M., and Diczfalusy, E.**, Biologically active LH in plasma. V. A re-analysis of the differences in ratio of biological to immunological LH activities during the menstrual cycle, *Acta Endocrinol.*, 92, 615, 1979.

83. **Burstein, S., Schaff-Blass, E., Blass, J., and Rosenfield, R. L.**, The changing ratio of bioactive to immunoreactive LH through puberty principally reflects changing LH RIA dose response characteristics, *J. Clin. Endocrinol. Metab.*, 61, 508, 1985.

84. **Van Damme, M.-P., Robertson, D. M., Marana, R., Ritzen, E. M., and Diczfalusy, E.**, A sensitive and specific in vitro bioassay method for the measurement of FSH activity, *Acta Endocrinol.*, 91, 224, 1979.

85. **Beers, W. H. and Strickland, S.**, A cell culture assay for FSH, *J. Biol. Chem.*, 253, 3877, 1978.

86. **Chappel, S. C., Ulloa-Aguirre, A., and Coutifaris, C.**, Biosynthesis and secretion of FSH, *Endocr. Rev.*, 4, 179, 1983.

87. Chappel, S. C., Ulloa-Aguirre, A., and Ramaley, J. A., Sexual maturation in female rats: time related changes in the isoelectric focusing pattern of anterior pituitary FSH, *Biol. Reprod.*, 28, 196, 1983.
88. Ulloa-Aguirre, A., Miller, C., Hyland, L., and Chappel, S., Production of all FSH isoforms from a purified preparation by neuraminidase digestion, *Biol. Reprod.*, 30, 382, 1984.
89. Chappel, S. C. and Ramaley, J. A., Changes in the isoelectric focusing profile of pituitary FSH in the developing male rat, *Biol. Reprod.*, 32, 567, 1985.
90. Galle, P. C., Ulloa-Aguirre, A., and Chappel, S. C., Effects of oestradiol, phenobarbitone and LHRH upon the isoelectric profile of pituitary FSH in ovariectomised hamsters, *J. Endocrinol.*, 99, 31, 1983.
91. Minegishi, T., Igarashi, M., and Wakabayashi, K., Effect of gonadectomy and steroid treatment on the receptor-binding activity and immunoreactivity of serum and pituitary FSH in the adult rats of both sexes, *Endocrinol. Jpn.*, 28, 347, 1981.
92. Foulds, L. M. and Robertson, D. M., Electrofocusing fractionation of FSH in pituitary cell culture extracts from male and female rats, *Mol. Cell. Endocrinol.*, 41, 129, 1985.
93. Ulloa-Aguirre, A., Coutifaris, C., and Chappel, S. C., Multiple species of FSH are present within hamster anterior pituitary cells in vitro, *Acta Endocrinol.*, 102, 343, 1983.
94. Kennedy, J. and Chappel, S., Direct pituitary effects of testosterone and luteinizing hormone-releasing hormone upon follicle-stimulating hormone: analysis by radioimmuno- and radioreceptor assay, *Endocrinology*, 116, 741, 1985.
95. Miller, C., Ulloa-Aguirre, A., Hyland, L., and Chappel, S., Pituitary FSH heterogeneity: assessment of biologic activities of each FSH form, *Fertil. Steril.*, 40, 242, 1983.
96. Zaidi, A. A., Fröysa, B., and Diczfalusy, E., Biological and immunological properties of different molecular species of human FSH: electrofocusing profiles of 8 highly purified preparations, *J. Endocrinol.*, 92, 195, 1982.
97. Wide, L. and Hobson, B. M., Qualitative difference in FSH activity in the pituitaries of young women compared to that of men and elderly women, *J. Clin. Endocrinol. Metab.*, 56, 371, 1983.
98. Wide, L. and Wide, M., Higher plasma disappearance rate in the mouse for pituitary FSH of young women compared to that of men and elderly women, *J. Clin. Endocrinol. Metab.*, 58, 426, 1984.
99. Peckham, W. D., Yamaji, T., Dierschke, D. J., and Knobil, E., Gonadal function and the biological and physiochemical properties of FSH, *Endocrinology*, 92, 1669, 1973.
100. Khan, S. A., Sved, V., Fröysa, B., Lindberg, M., and Diczfalusy, E., Influence of gonadectomy on IEF profiles of pituitary gonadotrophins in rhesus monkeys, *J. Med. Primatol.*, 14, 177, 1985.
101. Licht, P., McCreery, B. R., and Papkoff, H., Effect of gonadectomy on polymorphism in stored and circulating gonadotropins in the bullfrog, *Rana catesbeiana*. II. Gel filtration chromatography, *Biol. Reprod.*, 29, 646, 1983.
102. Robertson, D. M., Ellis, S., Foulds, L. M., Findlay, J. K., and Bindon, B. M., Pituitary gonadotrophins in Booroola and control Merino sheep, *J. Reprod. Fert.*, 71, 189, 1984.
103. Irvine, C. H. G., Kinetics of gonadotrophins in the mare, *J. Reprod. Fert. Suppl.*, 27, 131, 1979.
104. Keel, B. A. and Grotjan, H. E., Jr., Characterization of rat pituitary luteinizing hormone charge microheterogeneity in male and female rats using chromatofocusing: effects of castration, *Endocrinology*, 117, 354, 1985.
105. Dufau, M. L., Nozu, K., Dehejia, A., Garcia Vela A., Solano, A. R., Fraioli, F., and Catt, K. J., Biological activity and target cell actions of LH, in *Serono Symposium*, Vol. 49, Motta, M., Zanisi, M., and Piva, F., Eds., Academic Press, 1982, 117.
106. Bishop, W. H. and Ryan, R. J., Human luteinizing hormone and its subunits. Physical and chemical characterization, *Biochemistry*, 12, 3076, 1973.
107. Rathnam, P. and Saxena, B. B., Isolation and physiochemical characterization of luteinizing hormone from human pituitary glands, *J. Biol. Chem.*, 245, 3725, 1970.

Chapter 7

FOLLICLE-STIMULATING HORMONE STRUCTURE-FUNCTION RELATIONSHIPS

Werner F. Blum and Derek Gupta

TABLE OF CONTENTS

I. INTRODUCTION

Follicle-stimulating hormone (FSH) is an essential factor for the regulation of reproductive functions. It belongs to the group of glycoprotein hormones — FSH, luteinizing hormone (LH), thyroid-stimulating hormone (TSH), and choriogonadotropin (CG) — which share a number of structural features. They are composed of two dissimilar, noncovalently complexed subunits, α and β. Within a species, the α-subunit of these hormones is identical, having a chain length of 92 (human) or 96 amino acids (ovine, bovine, equine, and rat).[1,2] Two asparagine residues at positions 52 and 82 carry oligosaccharide chains of the N-linked complex type[1] (Figure 1). The β-subunits, which are thought to determine the hormone-specific biological effects, differ between the various glycoprotein hormones on an intraspecies level, although considerable homology exists in some portions of the sequence. They comprise about 120 to 130 amino acids with asparagine-linked complex oligosaccharide chains at positions 7 and 24 for FSH.[1] The biological activity of glycoprotein hormones is bound to their integral structure. The individual subunits are biologically inactive.[3]

The earliest attempts to isolate FSH made it clear that this hormone possesses heterogeneity like many other glycoproteins. This phenomenon was most obvious when FSH preparations were subjected to isoelectric focusing (IEF).[4,5]

The polymorphism of gonadotropins attracted particular interest when it became clear that it was not just a mere unpleasant peculiarity for the protein chemist, who wanted to isolate the pure hormone, but the molecular and biological properties of the different forms varied according to the physiological status. Sex, age, gonadectomy, and sex steroid substitution exerted characteristic influences in primates[6-8] as well as in the rat.[9-15] It could be shown that the in vivo bioactivity, the metabolic clearance, and the apparent molecular weight by exclusion chromatography all depend on the endocrine status of the experimental animal.

The development of chromatofocusing (CF), a chromatographic technique analogous to IEF, prompted us to utilize this new method for separation of various FSH forms and to investigate their structural differences as well as their biological properties (in vitro bioactivity, metabolic clearance rates). The results may give new insights into structure-function relationships and point to a new dimension of FSH regulation, namely qualitative control.

II. SEPARATION OF FSH COMPONENTS

FSH components can be separated on the basis of charge differences by IEF.[4,5,16-22] Analogously, when rat pituitary extracts were subjected to CF, a number of FSH peaks could be resolved with apparent isoelectric points (pI) between 5.1 and 3.1 (Figure 2).[23] Ten components could clearly be distinguished in this pH range. They were termed FSH-I to -X. A small amount of immunoreactive material passed through the column unretarded at pH 6.2 and eluted in the basic range at about pH 9.4 after refocusing.

III. SEX- AND AGE-DEPENDENT NATURE OF FSH POLYMORPHISM

When the elution patterns of male and female rats were compared, marked differences emerged (Figure 2, Table 1).[23] There is a clear preponderance of more acidic components in males. In females there is a broad peak of FSH at pH 5.1 which is insignificant in males. This peak, however, is also present in prepubertal animals of both sexes (unpublished data). Moreover, in prepubertal animals, the relative contribution of the more basic forms to total pituitary FSH is significantly higher than in pubertal or adult animals. At the time of the prepubertal FSH peak there is a shift of the componental pattern to more acidic forms which is partially reversed at a later stage.

A number of reports clearly document that experimental manipulations of the endocrine

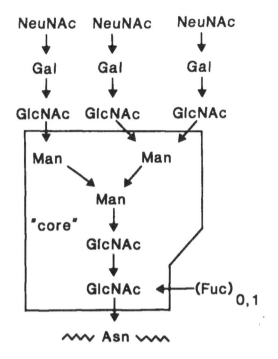

FIGURE 1. Structure of a triantennary N-linked complex oligosaccharide chain. NeuNAc: D-N-acetylneuraminic acid, Gal:
D-galactose, GlcNAc: D-N-acetylglucosamine, Man: D-mannose, Fuc: L-fucose, Asn: L-asparagine.

milieu characteristically influence the polymorphic pattern of pituitary FSH in IEF or
CF.[18,20,24,25]

Determination of the apparent molecular weight of the various FSH components by exclusion chromatography revealed an inverse relationship between the molecular weight and
the pI value[21,23] (Table 1). That is, a shift of the elution pattern in CF to more acidic FSH
forms is paralleled by a shift to a higher apparent molecular weight in exclusion chromatography. Therefore, the findings on the dependence of FSH polymorphism on the endocrine
status obtained by CF or IEF support previous results by exclusion chromatography.[11,12,15]

IV. THE STRUCTURAL CAUSE OF FSH POLYMORPHISM

An answer to the question, how the endocrine status can influence the polymorphic pattern
of FSH, presumes information on the structural differences between FSH components.
However, any suggestion on the structural differences between FSH forms must provide a
rationale for the dependence of FSH polymorphism on the endocrine status.

In analogy to other glycoproteins the polymorphism of FSH has been assumed to be due
to heterogeneity of the carbohydrate moiety.[7] Clear evidence for this hypothesis stems from
lectin-binding studies (Table 2).[15] The fact that pituitary FSH only partially binds to a number
of immobilized lectins indicates that its oligosaccharide chains are heterogenous. The results
demonstrate the presence of the following accessible monosaccharides in a fraction of pituitary FSH: D-N-acetylglucosamine and/or N-acetylneuraminic acid (binding to WGA), β-
D-galactose (binding to RCA I), and α-D-mannose and/or α-D-glucose (binding to LcH and
ConA). Only very little FSH binds to UEA I which makes the presence of accessible L-
fucose residues questionable. Except for D-glucose, the indicated monosaccharides are typical
constituents of N-linked complex oligosaccharide chains.[26] Further, the differential binding

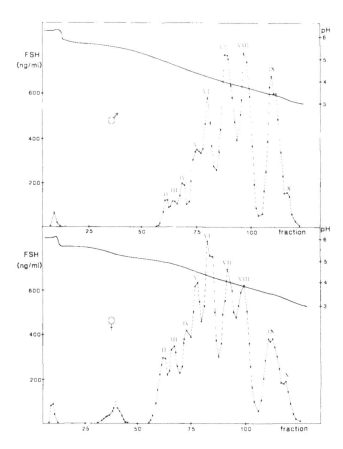

FIGURE 2.   Chromatofocusing of extracts from a pool of 20 pituitary glands from 90-day-old intact male (upper panel) and female (lower panel) rats in the range of pH 6 to 3. Pituitaries were homogenized in 25 mM L-histidine-HCl buffer, pH 6.2, containing 1 mM phenylmethylsulfonyl-fluoride. The extract was applied to a column (0.9 × 55 cm) of Polybuffer® exchanger PBE 94 (Pharmacia, Freiburg, F.R.G.). Elution was carried out at 4°C with Polybuffer® PB 74 (Pharmacia) at a dilution of 1:12, adjusted to pH 3.0 with HCl. Fractions of 3.3 mℓ were collected at a rate of 12 mℓ/hr. FSH of the various fractions was quantitated by radioimmunoassay utilizing the NIADDK FSH RIA kit. The concentration is expressed in terms of NIADDK-rat-FSH-RP-1 (From Blum, W. F. P., Riegelbauer, G., and Gupta, D., *J. Endocrinol.*, 105, 17, 1985. With permission.)

of FSH obtained from male and female rats to some lectins, particularly to RCA I, indicates differences in the carbohydrate moiety of FSH between sexes. Similar observations were made with ConA, where the binding of FSH varied according to the endocrine status.[18,19,27]

The information, however, that may be gained from these kinds of studies is limited. More conclusive were experiments with exoglycosidases. Gradual release of sugar residues and consecutive analysis of the modified FSH forms yielded a more detailed insight into the problem of structural differences between FSH components.[23]

Of particular interest in this context was neuraminic acid, as it introduces negative charge into the molecule, thus influencing its pI. Therefore, the hypothesis was tested that the FSH components as defined by CF on a charge basis differ by their neuraminic acid content. For this purpose FSH components were separated by CF and treated with a highly purified preparation of neuraminidase varying the enzyme concentration and length of incubation. The modified components were then refocused by CF. The results of this experiment are

Table 1
APPARENT ISOELECTRIC POINTS (pI) BY CF AND PARTITION COEFFICIENT (K_{av}) VALUES OF DIFFERENT FSH COMPONENTS TOGETHER WITH THEIR RELATIVE PROPORTIONS IN PITUITARY EXTRACTS FROM INTACT MALE AND FEMALE RATS

Component	Apparent pI (n = 7)	K_{av}	FSH recovered (%) Males (n = 4)	FSH recovered (%) Females (n = 3)
I	5.12 ± 0.06	N.D.		5.8 ± 2.1
II	4.81 ± 0.04	0.278	1.6 ± 0.8	6.5 ± 1.1
III	4.66 ± 0.03	0.276	1.5 ± 0.8	6.2 ± 1.3
IV	4.51 ± 0.03	0.269	2.2 ± 1.2	6.0 ± 1.5
V	4.36 ± 0.04	0.262	8.9 ± 2.1	11.6 ± 2.4
VI	4.22 ± 0.03	0.252	13.9 ± 2.5	17.9 ± 3.1
VII	3.94 ± 0.04	0.244	25.5 ± 2.2	17.7 ± 4.0
VIII	3.70 ± 0.02	0.227	24.5 ± 2.9	13.4 ± 3.2
IX	3.42 ± 0.03	0.224 }	22.0 ± 2.6[a]	14.4 ± 3.4[a]
X	3.14 ± 0.05	N.D. }		

Note: Partition coefficients were determined by exclusion chromatography on Sephadex® G-100 Superfine; values of pI and relative proportions are given as means ± SD from n experiments; N.D. = not determined.

[a] Because of poor resolution in CF, components IX and X were estimated together.

From Blum, W. F. P., Riegelbauer, G., and Gupta, D., *J. Endocrinol.,* 105, 17, 1985. With permission.

shown in Figure 3. Incubation of the individual FSH components II to IX in the presence of increasing amounts of neuraminidase produced new FSH peaks in the more basic range. The newly generated immunoreactive FSH forms exhibited pI values identical to those of their endogenously produced counterparts. Depending on the neuraminidase activity (or time of incubation) they were produced in a consecutive manner. Mild neuraminidase treatment produced components adjacent to the original position, while higher neuraminidase activity resulted in a more pronounced shift. At pH 5.3 and 5.8 new components emerged which were not present in untreated pituitary extracts. As the only parameter being varied was neuraminidase activity, it seems unlikely that the progressive generation of less acidic components is due to the interference of endogenous glycosidases or proteinases.

In a similar experiment Ulloa-Aguirre et al.[22] could produce FSH components with more basic pI values from a highly purified FSH preparation (NIADDK-rat-FSH-I-5) having a pI of about 4.2. Chappel et al.[25] reported the generation of more basic forms from highly acidic FSH (pI <4) by ncurmainidase treatment. A progressive shift of the FSH pattern by IEF to higher pI values could be demonstrated by incubation of a pituitary homogenate with neuraminidase.

Therefore, since all the FSH components present in pituitary extracts can be generated from more acidic forms by gradual release of neuraminic acid, it may be concluded that the essential structural parameter by which they differ from each other is a varying degree of

Table 2
BINDING OF FSH FROM MALE AND FEMALE PITUITARY GLANDS TO LECTINS

Lectin	Specificity	Eluting sugar	% FSH		
			Unbound	Eluted	
ConA	α-D-Glc	α-D-methylglucopyranosid	18.5	68.7	Male
	α-D-Man		15.9	70.4	Female
	α-D-GlcNAc				
	α-D-ManNAc				
WGA	D-GlcNAc	D-GlcNAc	11.9	73.3	Male
	D-NeuNAc		5.9	74.2	Female
RCA I	β-D-Gal	D-Gal	28.5	27.5	Male
			74.5	25.1	Female
LcH	α-D-Man	D-Man	88.5	9.6	Male
	α-D-Glc		79.3	18.1	Female
UEA I	α-L-Fuc	L-Fuc	96.3	2.0	Male
			89.4	5.5	Female
HPA	α-D-GalNAc	D-GalNAc	96.3	2.4	Male
			93.3	0	Female

Note: Pituitary extracts were passed over 1 mℓ columns of various agarose-immobilized lectins; bound FSH was eluted with a 0.1 *M* solution of the indicated sugars; unbound and eluted FSH was quantitated by radioimmunoassay. The results are given as percent of applied hormone. For abbreviations of the sugars see Figure 1. Glc: glucose; GalNAc: *N*-acetylgalactosamine. ConA: *Canavalia ensiformis* agglutinin; WGA: wheat germ agglutinin; RCA I: *Ricinus communis* agglutinin I; LcH: *Lens culinaris* hemagglutinin; UEA I: *Ulex europaeus* agglutinin; HPA: *Helix pomatia* agglutinin.

From Blum, W. F. P. and Gupta, D., *Neuroendocrinol. Lett.*, 2, 357, 1980. With permission.

sialylation. An inverse relationship exists between the sialic acid content and the pI value. In our experiments with neuraminidase we found 12 FSH components in the acidic range between pH 6 and 3. Assuming a triantennary structure of the oligosaccharide chains, the maximum possible number of neuraminic acid residues is 12. Thus, it is tempting to speculate that from one component to the next the neuraminic acid content increases by one residue thereby decreasing the pI.

If the FSH polymorphism is exclusively due to heterogeneity of the carbohydrate moiety, complete removal of the oligosaccharide chains should theoretically produce a single peak in CF. This, however, was not the case.[23] Incubation of pituitary extracts with a mixture of various exoglycosidases containing high enzymatic activities except neuraminidase hardly altered the polymorphic pattern. Addition of high neuraminidase activity produced a dramatic shift of total FSH to the basic range with a major peak at pH 9.4 and three minor peaks (Figure 4). A similar pattern was obtained if pituitary FSH was treated exclusively with neuraminidase. Variation of the incubation time (1 and 24 hr) demonstrated that incomplete removal of sialic acid may not be the cause of heterogeneity of the modified FSH. As the hydrolytic release of neutral sugars by exoglycosidases is largely inhibited by terminal O-sulfated *N*-acetylhexosamines,[28] the presence of such residues may be a further source of heterogeneity. So far, however, sulfated carbohydrates have only been demonstrated in LH and TSH.[1] Further, the possibility must be envisaged that the occurrence of the three minor peaks is an experimental artifact due to the action of endogenous or exogenous proteases in spite of the fact that high amounts of protease inhibitors (aprotinin and phenylmethylsulfonylfluoride) were present during incubation.

Finally, primary heterogeneity of the protein moiety must be taken into consideration.

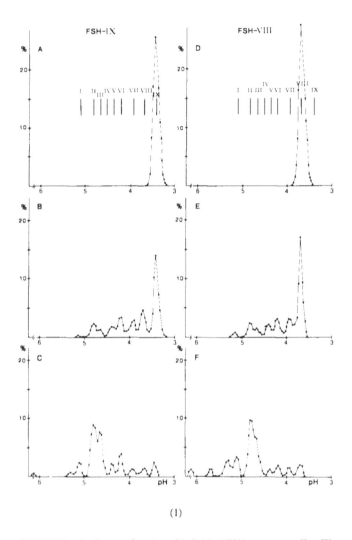

(1)

FIGURE 3. Rechromatofocusing of individual FSH components II to IX after partial desialylation. FSH components were separated by CF (see Figure 2) in a preparative scale and diafiltrated with Amicon® centriflo cones CF 25 and 0.05 M sodium acetate buffer, pH 5.5, containing 1 mM phenylmethylsulfonylfluoride, 1% bovine serum albumin, and 0.004% chlorohexidine diacetate. After addition of 100 U aprotinin, incubation of the samples was carried out at 37°C for 1 hr with either 60 or 500 μU of highly purified neuraminidase from *Cl. perfringens* (Sigma, type X). Analysis of the modified FSH was performed as described in the legend to Figure 2. The ordinate gives percent of total FSH recovered. A and D: unmodified components; B and E: incubation with 60 μU neuraminidase; C and F: incubation with 500 μU neuraminidase. The vertical lines in panels A and D indicate the positions of FSH components I to IX from untreated pituitary extracts (From Blum, W. F. P., Riegelbauer, G., and Gupta, D., *J. Endocrinol.*, 105, 17, 1985. With permission.)

The genome of various species including rat contains one gene for the α-subunit.[2] In humans there is also only one gene coding for the β-subunit of FSH.[29] Therefore, genetic variants due to gene multiplicity do not seem to play a role. It is, however, conceivable that differential processing of the primary genetic products may produce heterogeneity of the FSH molecule.

In conclusion, these experiments demonstrate that the various FSH components as defined

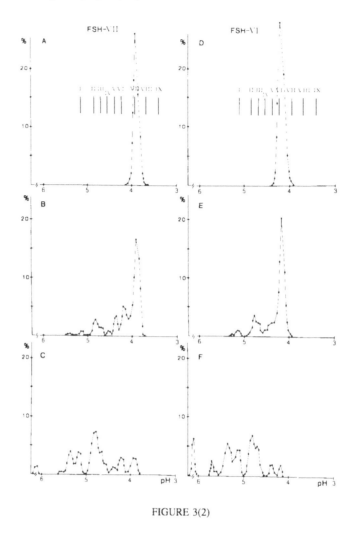

FIGURE 3(2)

by CF or IEF differ with respect to their sialic acid content. The more acidic the components are, the more neuraminic acid residues are attached to the molecule. Since the release of neutral sugars did not influence the pI of the components, it cannot be excluded that apparently homogeneous components by CF are still heterogeneous with regard to these residues. The heterogeneous pattern of FSH in the basic range obtained after treatment with high activities of exoglycosidases including neuraminidase points to the possibility that besides the varying degree of sialylation another cause for microheterogeneity may exist which, however, may not be revealed by CF in unmodified FSH preparations.

V. BIOACTIVITY OF FSH COMPONENTS

In vivo bioactivity of FSH is largely destroyed by desialylation[30,31] due to rapid clearance of the asialo hormone from the circulation.[32] Therefore, the question, whether the various FSH components differ with respect to their activity at the target cell must be investigated in vitro. In general, the biological response of a target cell to a hormonal stimulus is the result of two independent parameters. First, the affinity of a hormone to its receptor determines the range of concentration where hormonal stimulation may take place. Second, the potency of a receptor-bound hormone to trigger further postreceptor events, its intrinsic

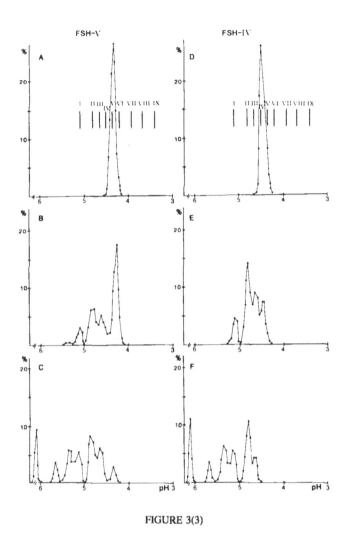

FIGURE 3(3)

bioactivity, determines the maximum response to a hormonal stimulus. Hence, high receptor affinity is not necessarily associated with high bioactivity which is evident in the case of hormone antagonists.

In vitro bioassay systems for FSH are based on its potency to stimulate cyclic AMP synthesis in granulosa or Sertoli cells,[33] estrogen synthesis,[34] or plasminogen activator synthesis in granulosa cells.[35] For the study of the biological activity of FSH components we utilized an assay developed by Beers and Strickland[35] measuring the FSH-dependent plasminogen activator production of rat granulosa cells. The determination of dose-response curves for individual FSH components (FSH-II to -IX) exhibited two important results which would have been missed by one- or two-point measurements[23] (Figures 5 and 6): (1) the dose for half-maximum response (ED_{50}) of FSH-II to -VI is identical to that of NIADDK-rat-FSH-RP-1. For the more acidic components FSH-VII, -VIII, and -IX, however, the ED_{50} increases considerably as the pI decreases; and (2) the maximum response to FSH-II to -VI is almost identical with the response to NIADDK-rat-RP-1. Again, for the acidic components it increases with decreasing pI up to a threefold value.

These differences in bioactivity cannot be explained by the presence of LH, as its concentration was far below the range where LH stimulates plasminogen activator production. Interference of low molecular weight compounds (prostaglandins, cyclic AMP, gonadotropin-releasing hormone, or ampholytes)[36-38] can also be excluded, as these possibly contam-

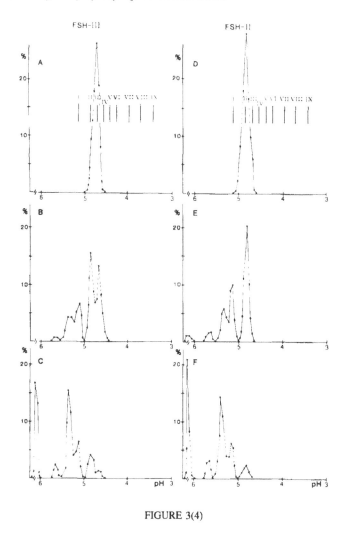

FIGURE 3(4)

inating substances were removed by exclusion chromatography prior to testing. Therefore, it may be concluded that the difference in bioactivity is a consequence of structural differences between FSH molecular components.

In the previous section evidence has been provided that the essential structural parameter by which the FSH components as defined by CF differ from each other is the number of sialic acid residues. Consequently, it is tempting to speculate that the degree of sialylation is also the essential parameter being responsible for the observed differences in bioactivity at the target cell.

The question remains, how the shift of the ED_{50} and of the maximum effect paralleling the decrease of pI may be interpreted. Theoretically, a shift of the ED_{50} to larger dose values would be expected if the acidic components had a higher immunopotency than the alkaline ones, as the dose was established by RIA and not on a weight basis. From several studies on FSH from various species, however, it appears that a high sialic acid content rather decreases immunopotency.[39-41]

Although the formation of a glycoprotein hormone-receptor complex and subsequent triggering of a biological response is a most complex process which cannot simply be regarded as a bimolecular reaction, the increase of the ED_{50} might be interpreted as a decrease of receptor affinity paralleling the decrease of pI. This hypothesis, indeed, is directly substan-

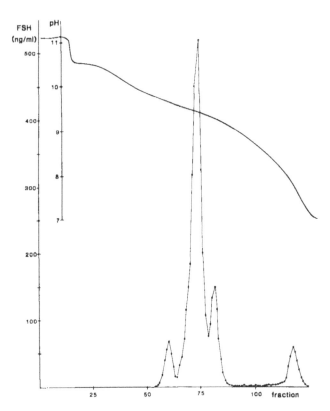

FIGURE 4. Chromatofocusing of a neuraminidase-treated extract of pituitary glands from 90-day-old intact male rats. The extract was incubated with 5 units neuraminidase at 37°C for 24 hr. Chromatofocusing was performed on a 0.9 × 55 cm column of PBE 118 (Pharmacia) equilibrated with 0.025 M triethylamine-HCl, pH 11.2. Elution was carried out at 4°C with Pharmalyte® pH 8 to 10.5 at a dilution of 1:45, adjusted to pH 7.0, and the FSH content of each fraction was determined by radioimmunoassay. (From Blum, W. F. P., Riegelbauer, G., and Gupta, D., *J. Endocrinol.*, 105, 17, 1985. With permission.)

tiated by a number of studies, where the receptor binding of different FSH components was determined.[18,22,25,42] Radioreceptor assay activity considerably decreased with decreasing pI of the components.

That it is in fact the neuraminic acid content which is of decisive importance for receptor binding has been shown by Ulloa-Aguirre et al.[22] They not only produced the various less acidic FSH components from NIADDK-rat-FSH-I-5 by mild neuraminidase digestion, but they were also able to demonstrate that radioreceptor assay activity increases with successive loss of neuraminic acid utilizing homogenates of seminiferous tubules. Moreover, the newly generated FSH forms exhibited a radioreceptor assay to radioimmunoassay ratio similar to that of the corresponding components from pituitary extracts. Studies with pure FSH preparations on the biological role of the carbohydrate moiety support this conception. Deglycosylation significantly enhances the affinity of FSH to its receptor.[41,43]

The maximum biological response to the very acidic FSH components is significantly higher than the response of the more basic components. This phenomenon may be interpreted as an increase of the "intrinsic biological activity" which means that a high sialic acid content of FSH may amplify its potency to trigger a postreceptor signal.

This interpretation is supported by studies on the in vitro bioactivity of pure desialylated FSH.[40] The maximum stimulation of cyclic AMP production in Sertoli cells by asialo FSH is significantly reduced as compared to the unmodified FSH preparation.

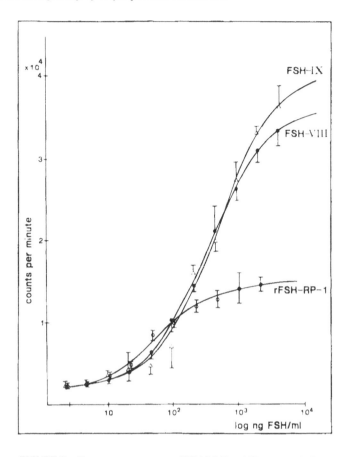

FIGURE 5.   Dose-response curves of NIADDK-rat-FSH-RP-1, pituitary
FSH-VIII, and FSH-IX in the in vitro bioassay. FSH components were
separated by chromatofocusing as described in the legend to Figure 2.
Before bioassay they were chromatographed on a Sephadex® G-75 column
to remove ampholytes and other interfering small molecular weight compounds. The assay was carried out according to Beers and Strickland
determining the FSH-dependent production of plasminogen-activator by
rat granulosa cells. The radioactivity released from ^{125}I-fibrinogen-coated
wells was measured and taken as the biological response. Total releasable
radioactivity after trypsinization was 104,000 c.p.m. The logistic curves
were fitted by the method of least squares. (From Blum, W. F. P., Riegelbauer, G., and Gupta, D., *J. Endocrinol.*, 105, 17, 1985. With permission.)

In conclusion, evidence is accumulating that the carbohydrate moiety and in particular
neuraminic acid affects the interaction of FSH with its target cells. Obviously, an increase
of the sialic acid content of FSH diminishes its receptor affinity, in contrast, however, it
enhances the intracellular transmission of the FSH signal. The molecular mechanisms by
which these adverse effects on bioactivity may be produced remain obscure. One may
conceive an influence of neuraminic acid on FSH conformation or directly on the interaction
with the receptor. However, the question still remains a challenge to further research.

VI. THE METABOLIC CLEARANCE OF FSH COMPONENTS

The clearance of FSH like other glycoprotein hormones follows a multiexponential
law.[6-8,10,14,44] The semilogarithmic presentation shows a curvilinear decay with a rapid decline

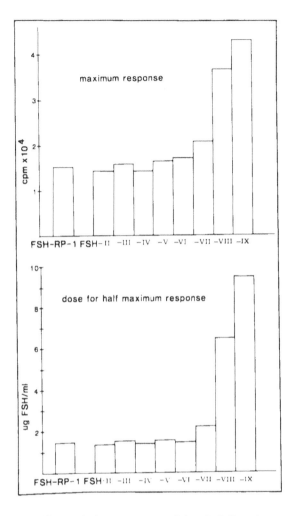

FIGURE 6. Maximum response and dose for half-maximum
response of NIADDK-rat-FSH-RP-1 and FSH components II
to IX in the in vitro bioassay. For details see legend to Figure
5. The dose was established by radioimmunoassay.

at the beginning and progressive flattening of the curve at a later stage. The disappearance
of injected pituitary FSH is more rapid than the clearance of ''native'' FSH (FSH secreted
into the blood). Moreover, the disappearance rate varies according to the endocrine status
of the donor animal.[6-8,10]

A number of hypotheses have been put forward to explain the experimentally observed
multiexponential decay: (1) distribution between multiple compartments,[14,44] (2) structural
alterations in the circulation,[14] and (3) heterogeneity of secreted or injected hormone with
different clearance rates of the hormonal components.[14]

In order to test the third hypothesis, individual FSH components separated by CF were
injected into adult male rats. Their disappearance from the circulation was followed up by
withdrawing blood samples at regular time intervals and measuring serum FSH by RIA.[45]
Figure 7 shows representative semilogarithmic decay curves of FSH-II, -III, and -VI to
-IX. the injected doses were so chosen that 5 min after the injection, the FSH concentration
was about 20-fold of the basal FSH level. Two aspects of the results are worth considering:
(1) the semilogarithmic decay curves were linear until more than 85% of the injected FSH

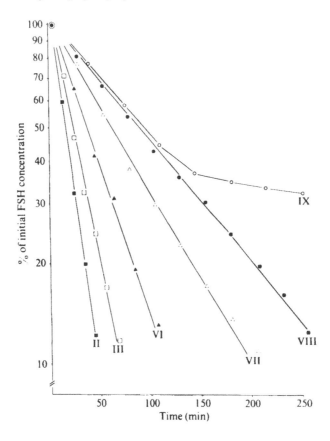

FIGURE 7.   Representative semilogarithmic disappearance curves of FSH components from the circulation of adult rats. Components of FSH from pituitary glands of 90-day-old intact male rats were separated by CF. They were concentrated with Amicon® centriflo cones CF 25 and diafiltrated with Tris-HCl buffer, pH 7.4, containing 0.14 *M* NaCl to remove ampholytes. The solutions of FSH components were adjusted to 60 to 70 μg/mℓ and 0.6 mℓ aliquots were injected into the circulation of 120-day-old male rats. For this purpose the vena femoralis was cannulated under ether anesthesia. After application of 20 IU heparin per 100 g body weight, 0.6 mℓ blood was taken for determination of the basal FSH concentration. Then FSH (10 μg/100 g body weight) was injected and blood samples of 0.6 mℓ were collected at regular time intervals. After each bleeding the volume was substituted with 0.6 mℓ 0.9% NaCl containing 3 IU heparin. FSH in the plasma samples was measured by radioimmunoassay. Values obtained 5 min after injection were taken as 100%. The lines for components IV and V were almost identical with FSH-III. (From Blum, W. F. P., and Gupta, D., *J. Endocrinol.*, 105, 29, 1985. With permission.)

was cleared from the circulation except component IX which obviously was heterogeneous. Therefore, the hypotheses 1 and 2 — distribution between various compartments and structural alterations in the circulation — can evidently be discarded, as in these cases the decay curves would still be curvilinear; and (2) the more acidic the FSH components the more slowly they disappeared from the circulation (Table 3). These results definitely confirm hypothesis 3 according to which FSH is composed of a heterogeneous population of molecules differing with respect to their metabolic clearance rates. The dependence of FSH metabolism on the endocrine status is therefore exclusively the consequence of the variable FSH polymorphism described above. The slower disappearance of endogenous FSH (secreted FSH) as compared to injected pituitary FSH may be explained by the accumulation of long-living components in the circulation.

Table 3

APPARENT ISOELECTRIC POINTS (pI), DISAPPEARANCE RATE CONSTANTS (k), HALF-TIMES ($t_{0.5}$), DISTRIBUTION VOLUMES (V_d), AND METABOLIC CLEARANCE RATES (MCR) OF DIFFERENT FSH COMPONENTS (II-IX) SEPARATED BY CF

Component	pI	$10^2 \times k^a$ (min^{-1})	$t_{0.5}^b$ (min)	V_d^c (ml/100 g body wt)	$10^2 \times MCR^d$ (ml/min 100 g body wt)
II	4.81	5.33 ± 0.45	13.0 ± 1.1	3.4 ± 0.4	18.1
III	4.66	3.74 ± 0.24	18.5 ± 1.2	3.0 ± 0.2	11.2
IV	4.51	3.59 ± 0.17	19.3 ± 0.9	3.0 ± 0.1	10.8
V	4.36	5.56 ± 0.27	19.5 ± 1.5	3.3 ± 0.3	11.9
VI	4.22	2.08 ± 0.29	33.3 ± 4.6	3.0 ± 0.2	6.2
VII	3.94	1.22 ± 0.16	56.7 ± 7.4	3.0 ± 0.2	3.7
VIII	3.70	0.83 ± 0.03	83.9 ± 2.7	3.1 ± 0.1	2.6
IXe a) b)	3.42	0.79 ± 0.04	88.1 ± 4.2	3.5 ± 0.4	2.8
		<0.39	>3 hr		<1.4

Note: The means ± SD from three experiments are given.

[a] k: slope of the semilogarithmic decay curve.

[b] $t_{0.5} = \ln 2/k$.

[c] $V_d = m_i/c_0$ where m_i is the total amount of FSH injected and c_0 is the extrapolated FSH concentration at t = 0 minus the basal concentration.

[d] MCR = $k \times V_d$.

[e] Values for IX a) and b) were calculated from the steep and flat part of the decay curve, respectively.

From Blum, W. F. P. and Gupta, D., *J. Endocrinol.*, 105, 29, 1985. With permission.

The structural parameter which determines the circulatory half-life of the FSH molecule seems again to be associated with the degree of sialylation. As the number of neuraminic acid residues progressively increases with decreasing pI of the components, it may be concluded that a direct causality exists between neuraminic acid content and metabolic half-life. The number of neuraminic acid residues appears to serve as a recognition marker for the clearance of FSH molecules from the circulation.

The mechanism being probably involved in this type of FSH metabolism is via the so-called asialo-glycoprotein receptor on hepatocytes.[32,46] This well-characterized integral membrane protein binds to galactose residues of glycoproteins not being masked by terminal neuraminic acid. From quantitative studies it appears that a critical density of exposed galactose is required. Internalization of bound glycoprotein occurs via coated pits. Coated vesicles then deliver the glycoprotein-receptor complex to the lysosomes, where the glycoprotein is degraded. The receptor is finally recycled to the cell surface.

Morrell et al.[32] showed that 30 min after injection of radioiodinated asialo FSH into rats, 72% of the radioactivity was taken up by the liver. Presuming that the liver is the major site of asialo-FSH catabolism one can calculate from these data a metabolic half-life of 16.3 min. This value is in the range of the half-lives of the less acidic FSH components II to V. Assuming that the half-life of asialo FSH is similar to the half-lives of FSH-II to -V it follows that the liver is the major site of metabolism at least for the FSH components with a low degree of sialylation.

The binding to the asialo-glycoprotein receptor decreases with a decreasing number of exposed galactose residues.[47] The clearance of highly sialylated FSH components therefore requires either a different mechanism or, alternatively, involves partial desialylation in the circulation. The polymorphic pattern of serum FSH in fact suggests that this may be the

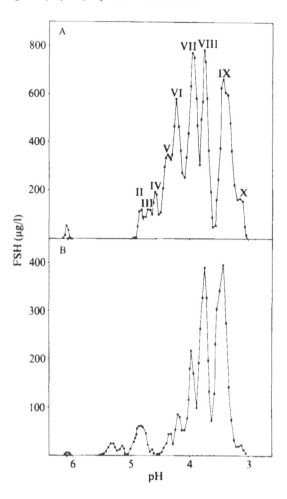

FIGURE 8.   Chromatofocusing of (A) pituitary FSH and (B) serum FSH of 90-day-old intact male rats. Serum FSH was extracted from a pool of 400 mℓ serum by immunoaffinity chromatography utilizing an antibovine FSH column. About 70% of bound FSH could be eluted with 0.02 *M* glycine-HCl, pH 3.0, at 4° C. Under the conditions of desorption less than 5% of immunoassayable FSH was lost. (From Blum, W. F. P. and Gupta, D., *J. Endocrinol.*, 105, 29, 1985. With permission.)

case. It contains a considerable amount of FSH eluting at the pI of components I and II which have a low degree of sialylation (Figure 8, Table 4). The site of desialylation in the rat may be the kidney where neuraminidase activity has been demonstrated.[48] The possible participation of other mechanisms in FSH metabolism (i.e., renal filtration or the mannose/*N*-acetylglucosamine receptor on Kupffer cells[46]) remains unclear.

VII. POLYMORPHISM OF SECRETED FSH

The dependence of pituitary FSH on the endocrine status together with the different biological properties of the FSH components is of physiological significance only if the various components are secreted into the circulation likewise. In order to examine this important question, two experiments were performed. In the first experiment rat pituitary pieces were incubated in vitro. Thereafter, tissue homogenates and incubation medium were

Table 4
RELATIVE AMOUNTS OF FSH COMPONENTS IN POOLED PITUITARY GLANDS AND SERUM OF 90-DAY-OLD INTACT MALE RATS

Component	App. pI	Pituitary (n = 4)	Serum	
			Experimental	Calculated[d]
Unbound[a]	>6.2	0.6 ± 0.2	0.1	0
I	5.1—5.3	0	2.9	0
II	4.81	1.6 ± 0.8	7.6	0.3
III	4.66	1.5 ± 0.8	0.5	0.4
IV + V[b]	4.51, 4.36	11.1 ± 2.8	1.6	3.1
VI	4.22	13.9 ± 2.5	5.4	6.5
VII	3.94	25.5 ± 2.2	19.3	20.1
VIII	3.70	24.5 ± 2.9	28.5	29.2
IX + X[c]	3.42, 3.14	22.0 ± 2.6	33.9	39.3

Note: Values are given as % of total FSH recovered. For details, see legend to Figure 7.

[a] FSH not bound to Polybuffer exchanger PBE 94 at pH 6.2.
[b] Because separation was poor and kinetic parameters were almost identical, IV + V were treated as one component.
[c] As separation was poor, IX + X were estimated together. For calculation the ratio of the fast and slowly disappearing component was assumed to be 55:45.
[d] Calculation of relative amounts of FSH components in serum was performed as follows: $\%s_i = \%p_i \, t_{0.5i} \times 100/\Sigma \, \% \, p_i \, t_{0.5i}$, where $\%s_i$ is the relative amount of component i in serum and $\%p_i$ is the corresponding value for the pituitary gland. $t_{0.5i}$ denotes the metabolic half-time of component i.

From Blum, W. F. P. and Gupta, D., *J. Endocrinol.*, 105, 29, 1985. With permission.

subjected to CF. In a second experiment, FSH was extracted from rat serum by immunoaffinity chromatography and further analyzed by CF.[45]

Figure 9 shows the polymorphic pattern of intracellular FSH and FSH secreted in vitro in the absence and presence of GnRH. It resembles the pattern obtained with frozen pituitary glands without in vitro incubation (Figure 2). No significant difference between the stored and the secreted FSH was evident either with respect to the apparent pI values of FSH components or their relative amounts, albeit FSH-II and -III as well as FSH not bound to the column at pH 6.2 seemed to be somewhat reduced in the medium. Although GnRH stimulated FSH secretion about fourfold, it did not significantly alter the composition of the intracellular or the secreted FSH.

Analogous findings were reported by Ulloa-Aguirre et al.[42] who analyzed FSH secreted by hamster pituitary cell cultures utilizing IEF for separation and by Chappel et al.[25] with primate pituitary cell cultures using CF for analysis.

If secretion of polymorphic FSH does not only occur in vitro but also in vivo, one has to expect that serum FSH is also composed of multiple components. However, as the metabolic clearance rates of the components differ, there should be a relative abundance of the more acidic FSH forms. This, in fact, has been demonstrated by the second experiment (Figure 8, Table 4). The apparent pI values of corresponding components in pituitary and serum were identical. Estimates of the expected relative amounts of FSH components in serum utilizing the experimentally determined metabolic half-lives of FSH components for

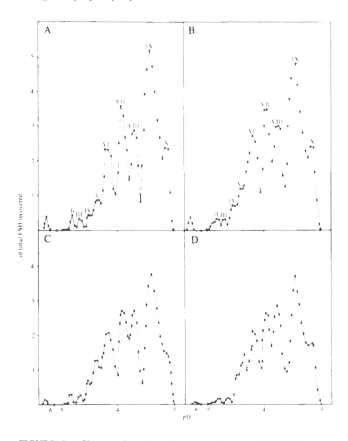

FIGURE 9.   Chromatofocusing of stored and secreted FSH. Pituitary glands of adult male rats were quartered and incubated in vitro in Leibovitz medium L 15 for 4 hr at 37°C with and without 1 μg gonadotropin-releasing hormone (GnRH) per mℓ. Tissue extracts and incubation medium were chromatofocused between pH 6 and 3 as described above. Fractions were assayed for FSH by radioimmunoassay. (A) Intracellular FSH in the absence of GnRH; (B) intracellular FSH in the presence of GnRH; (C) secreted FSH in the absence of GnRH; (D) secreted FSH in the presence of GnRH. The small peaks before pH 6 are immunoreactive FSH not bound to the column (From Blum, W. F. P. and Gupta, D., *J. Endocrinol.*, 105, 29, 1985. With permission.)

calculation exhibited good agreement between the theoretical and empirical pattern in the intermediate pH range. FSH-X at the acidic margin and FSH-I and -II at the basic margin of the elution profile, however, differed significantly from the calculated values (Table 4). This discrepancy could be due to partial desialylation during the extraction procedure. Alternatively, it is possible that FSH is exposed to weak neuraminidase activity in the circulation. This would cause a decrease of the most acidic component and an increase of the most basic components. Intermediate FSH forms would be produced from more acidic forms at about a similar rate as they were further converted to components with higher pI values. Although hypothetical, such a mechanism would also meet the requirements for FSH clearance via the hepatic asialo-glycoprotein receptor as discussed above.

In conclusion, there is convincing evidence that secreted FSH is polymorphic. Its composition at the moment of secretion is almost identical to the composition of pituitary FSH. Hence, regulatory changes of pituitary FSH polymorphism reflect changes of the polymorphism of circulating FSH, albeit the componental pattern in these compartments is not the same. Having longer half-lives, the acidic components accumulate in the blood.

VIII. POSSIBLE MECHANISMS INFLUENCING THE STRUCTURAL HETEROGENEITY OF FSH

An answer to the question, how the endocrine environment exerts its influence on the polymorphic pattern of FSH, needs still more information on FSH regulatory mechanisms. The following somewhat speculative ideas, however, may outline possible mechanisms being involved in the problem of variable FSH polymorphism.

The synthesis of glycoprotein hormones follows the general pathway of glycoprotein synthesis.[1,26,49] The subunits are separately translated and transferred into the cisternal space of the endoplasmatic reticulum as prepeptides. After cleavage of the signal peptide, the mannose-rich oligosaccharide complexes are transferred from a dolichol intermediate *en bloc* to the nascent protein chain (Chapter 2). Then, trimming of the carbohydrate moiety, association of the subunits, and transport to the Golgi apparatus occurs where further processing takes place. This includes the sequential attachment of the peripheral sugars N-acetylglucosamine, galactose, and sialic acid by specific glycosyl transferases. It may be hypothesized that the various FSH components are thereby produced from each other successively by progressive sialylation. The elongation of the oligosaccharide chains appears to be a slow process as compared to the earlier events in the biosynthesis of FSH, as about 99% of the intracellular immunoreactive FSH does already contain sialic acid. Obviously, the completion of the carbohydrate moiety is not a prerequisite for secretion.

The biosynthetic pathway suggests that the core region of the oligosaccharide chains is not involved in FSH heterogeneity, as it is attached to the molecule as part of a presynthesized complex. Heterogeneity rather seems to be the consequence of incomplete synthesis of the peripheral region. Therefore, the question, how the endocrine environment influences FSH polymorphism, may be addressed to an activation (or deactivation) of the glycosyltransferases necessary for the attachment of the peripheral sugar residues. Alternatively, mechanisms providing substrate for these reactions may be involved as well.

A second conceivable mechanism could be associated with the secretion rate. If the secretion rate is increased, the FSH molecule is exposed to the action of glycosyltransferases for a shorter period of time. Consequently, the relative amount of less glycosylated (and hence less sialylated) components should increase and vice versa. A similar albeit transient shift to less sialylated components should be observed, if the rate of neosynthesis of FSH is increased, as further glycosylation of the newly produced FSH takes time. In fact, when we described such a model, where FSH components are produced from each other in a sequential mode, by a linear kinetic model utilizing empirical data for calculations, we were able to demonstrate by computer simulation that the polymorphic pattern of FSH depends on the rates of neosynthesis, sialylation, and secretion in the above-mentioned manner (unpublished data). Thus, the influence of the endocrine milieu does not necessarily presume the existence of mechanisms which specifically alter the FSH structure.

IX. CONCLUSION

Rat FSH (and FSH from other species) is polymorphic. Its polymorphism is subject to the endocrine status of the animal. The structural basis of this phenomenon appears to be heterogeneity of the carbohydrate moiety. The essential difference between FSH components as defined by CF or IEF is the number of neuraminic acid residues. Other possible structural variations which could be a source of heterogeneity, e.g., the type or number of neutral sugars or variations of the protein moiety are not revealed by the currently applied analytical methods on a charge basis.

The maximum biological response of granulosa cells to FSH in vitro (plasminogen activator production) increases progressively with decreasing pI of the components, while receptor affinity decreases. These findings suggest that an increasing sialic acid content of the FSH molecule hinders its binding to the FSH receptor at the target cell. In contrast, however, it enhances the potency of FSH to trigger postreceptor events.

The inverse relationship between the apparent pI of the FSH components and their metabolic half-life suggests that the degree of sialylation is a major structural determinant concerning the survival time in the circulation.

Secreted FSH is a heterogeneous mixture of FSH molecules differing with respect to their pI values. It reflects the polymorphism of pituitary FSH, although the acidic components predominate due to longer half-lives. Any shift of the polymorphic pattern to more sialylated components causes an increase of serum FSH and alters the biological activity at the target cell. This adds a novel aspect to the problem of FSH regulation: modulation of serum FSH levels and FSH bioactivity by qualitative control.

REFERENCES

1. **Pierce, J. G. and Parsons, T. F.**, Glycoprotein hormones: structure and function, *Annu. Rev. Biochem.*, 50, 465, 1981.
2. **Chin, W. W.**, Organization and expression of glycoprotein hormone genes, in *The Pituitary Gland*, Imura, H., Ed., Raven Press, New York, 1985, 103.
3. **Dufau, M. L. and Catt, K. J.**, Gonadotrophin binding and activation of testicular steroidogenesis, in *Recent Progress in Reproductive Endocrinology*, Crosignani, P. G. and James, V. H. T., Eds., Academic Press, New York, 1974, 581.
4. **Graesslin, D., Trautwein, A., and Bettendorf, G.**, Gel isoelectric focusing of glycoprotein hormones, *J. Chromatogr.*, 63, 475, 1971.
5. **Reichert, L., Jr.**, Electrophoretic properties of pituitary gonadotropins as studied by electrofocusing, *Endocrinology*, 88, 1029, 1971.
6. **Peckham, W. D., Yamaji, T., Dierschke, D. J., and Knobil, E.**, Gonadal function and the biological and physicochemical properties of follicle-stimulating hormone, *Endocrinology*, 92, 1660, 1973.
7. **Peckham, W. D. and Knobil, E.**, The effect of ovariectomy, estrogen replacement, and neuraminidase treatment on the properties of the adenohypophysial glycoprotein hormones of the rhesus monkey, *Endocrinology*, 98, 1054, 1976.
8. **Peckham, W. D. and Knobil, E.**, Qualitative changes in the pituitary gonadotropins of the male rhesus monkey following castration, *Endocrinology*, 98, 1061, 1976.
9. **Diebel, N. D., Yamamoto, M., and Bogdanove, E. M.**, Discrepancies between radioimmunoassays and bioassay for rat FSH: evidence that androgen treatment and withdrawal can alter bioassay-immunoassay ratios, *Endocrinology*, 92, 1065, 1973.
10. **Bogdanove, E. M., Campbell, G. T., Blair, E. D., Mula, M. E., Miller, A. E., and Grossman, G. H.**, Gonad-pituitary feedback involves qualitative change: androgens alter the type of FSH secreted by the rat pituitary, *Endocrinology*, 95, 219, 1974.
11. **Bogdanove, E. M., Campbell, G. T., and Peckham, W. D.**, FSH pleomorphism in the rat — regulation by gonadal steroids, *Endocr. Res. Commun.*, 1, 87, 1974.
12. **Bogdanove, E. M., Nolin, J. M., and Campbell, G. T.**, Qualitative and quantitative gonad-pituitary feedback, *Rec. Prog. Horm. Res.*, 31, 567, 1975.
13. **Weick, R. F.**, A comparison of the disappearance rates of luteinizing hormone from intact and ovariectomized rats, *Endocrinology*, 101, 157, 1977.
14. **Bogdanove, E. M. and Nansel, D. D.**, Biological and immunological distictions between pituitary and serum LH in the rat, in *Structure and Function of the Gonadotropins*, McKerns, K. W., Ed., Plenum Press, New York, 1978, 415.
15. **Blum, W. F. P. and Gupta, D.**, Age- and sex-dependent nature of the polymorphic forms of rat pituitary FSH: the role of glycosylation, *Neuroendocrinol. Lett.*, 2, 357, 1980.
16. **Zaidi, A. A., Robertson, D. M., and Diczfalusy, E.**, Studies on the biological and immunological properties of human follitropin: profiles of two international reference preparations and of an aqueous extract of pituitary glands after electrofocusing, *Acta Endocrinol. (Copenhagen)*, 97, 157, 1981.
17. **Zaidi, A. A., Fröysa, B., and Diczfalusy, E.**, Biological and immunological properties of different molecular species of human follicle-stimulating hormone: electrofocusing profiles of eight highly purified preparations, *J. Endocrinol.*, 92, 195, 1982.

18. **Ulloa-Aguirre, A. and Chappel, S. C.**, Multiple species of follicle-stimulating hormone exist within the anterior pituitary gland of male golden hamsters, *J. Endocrinol.*, 95, 257, 1982.
19. **Chappel, S. C., Coutifaris, C., and Jacobs, S. J.**, Studies on the microheterogeneity of follicle-stimulating hormone present within the anterior pituitary gland of ovariectomized hamsters, *Endocrinology*, 110, 847, 1982.
20. **Robertson, D. M., Foulds, L. M., and Ellis, S.**, Heterogeneity of rat pituitary gonadotropins on electrofocusing; difference between sexes and after castration, *Endocrinology*, 111, 385, 1982.
21. **Foulds, L. M. and Robertson, D. M.**, Electrofocusing fractionation and characterization of pituitary follicle-stimulating hormone from male and female rats, *Mol. Cell. Endocrinol.*, 31, 117, 1983.
22. **Ulloa-Aguirre, A., Miller, C., Hyland, L., and Chappel, S.**, Production of all follicle-stimulating hormone isohormones from a purified preparation by neuraminidase digestion, *Biol. Reprod.*, 30, 382, 1984.
23. **Blum, W. F. P., Riegelbauer, G., and Gupta, D.**, Heterogeneity of rat FSH by chromatofocusing: studies on in-vitro bioactivity of pituitary FSH forms and effect of neuraminidase treatment, *J. Endocrinol.*, 105, 17, 1985.
24. **Galle, P. C., Ulloa-Aguirre, A., and Chappel, S. C.**, Effects of oestradiol, phenobarbitone and luteinizing releasing hormone upon the isoelectric profile of pituitary follicle-stimulating hormone in ovariectomized hamsters, *J. Endocrinol.*, 99, 31, 1983.
25. **Chappel, S. C., Bethea, C. L., and Spies, H. G.**, Existence of multiple forms of follicle-stimulating hormone within the anterior pituitaries of cynomolgus monkeys, *Endocrinology*, 115, 452, 1984.
26. **Kornfeld, R. and Kornfeld, S.**, Assembly of asparagine-linked oligosaccharides, *Annu. Rev. Biochem.*, 54, 631, 1985.
27. **Chappel, S. C.**, The presence of two species of follicle-stimulating hormone within hamster anterior pituitary glands as disclosed by Concanavalin A chromatography, *Endocrinology*, 109, 935, 1981.
28. **Parsons, T. F. and Pierce, J. G.**, Oligosaccharide moieties of glycoprotein hormones: bovine lutropin resists enzymatic deglycosylation because of terminal O-sulfated N-acetylhexosamines, *Proc. Natl. Acad. Sci. U.S.A.*, 77, 7089, 1980.
29. **Watkins, P., Eddy, R., Beck, A., Vellucci, V., Gusella, J., and Shows, T.**, Assignment of the human gene for the beta subunit of follicle stimulating hormone (FSHB) to chromosome 11, *Cytogenet. Cell Genet.*, 40, 773, 1985.
30. **Gottschalk, A., Whitten, W. K., and Graham, E. G.**, Inactivation of follicle-stimulating hormone by enzymic release of sialic acid, *Biochim. Biophys. Acta*, 38, 183, 1960.
31. **Papkoff, H.**, Some biological properties of a potent follicle-stimulating hormone preparation, *Acta Endocrinol. (Copenhagen)*, 48, 439, 1965.
32. **Morell, A. G., Gregoriadis, G., Scheinberg, I. H., Hickman, J., and Ashwell, G.**, The role of sialic acid in determining the survival of glycoproteins in the circulation, *J. Biol. Chem.*, 246, 1461, 1971.
33. **Farmer, S. W. and Papkoff, H.**, PMSG and FSH stimulation of cyclic AMP production in rat seminiferous tubule cells, *J. Endocrinol.*, 76, 391, 1978.
34. **Van Damme, M.-P., Robertson, D. M., Marana, R., Ritzen, E. M., and Diczfalusy, E.**, A sensitive and specific in vitro bioassay method for the measurement of follicle-stimulating hormone activity, *Acta Endocrinol. (Copenhagen)*, 91, 224, 1979.
35. **Beers, W. H. and Strickland, S.**, A cell culture assay for follicle-stimulating hormone, *J. Biol. Chem.*, 253, 3877, 1978.
36. **Strickland, S. and Beers, W. H.**, Studies on the role of plasminogen activator in ovulation. In vitro response of granulosa cells to gonadotropins, cyclic nucleotides, and prostaglandins, *J. Biol. Chem.*, 251, 5694, 1976.
37. **Wang, C.**, Luteinizing hormone releasing hormone stimulates plasminogen activator production by rat granulosa cells, *Endocrinology*, 112, 1130, 1983.
38. **Wang, C. and Leung, A.**, Gonadotropins regulate plasminogen activator production by rat granulosa cells, *Endocrinology*, 112, 1201, 1983.
39. **Vaitukaitis, J. L. and Ross, G. T.**, Altered biologic and immunologic activities of progessively desialylated human urinary FSH, *J. Clin. Endocrinol.*, 33, 308, 1971.
40. **Aggarwal, B. B. and Papkoff, H.**, Relationship of sialic acid residues to in vitro biological and immunological activities of equine gonadotropins, *Biol. Reprod.*, 24, 1082, 1981.
41. **Manjunath, P., Sairam, M. R., and Sairam, J.**, Studies on pituitary follitropin. X. Biochemical, receptor binding and immunological properties of deglycosylated ovine hormone, *Mol. Cell. Endocrinol.*, 28, 125, 1982.
42. **Ulloa-Aguirre, A., Coutifaris, C., and Chappel, S. C.**, Multiple species of FSH are present within hamster anterior pituitary cells cultured in vitro, *Acta Endocrinol. (Copenhagen)*, 102, 343, 1983.
43. **Sairam, M. R. and Manjunath, P.**, Studies on pituitary follitropin. XI. Induction of hormonal antagonistic activity by chemical deglycosylation, *Mol. Cell. Endocrinol.*, 28, 139, 1982.

44. **Yen, S. S. C., Llerena, L. A., Pearson, O. H., and Littell, A. S.,** Disappearance rates of endogenous follicle-stimulating hormone in serum following surgical hypophysectomy in man, *J. Clin. Endocrinol.,* 30, 325, 1970.

45. **Blum, W. F. P. and Gupta, D.,** Heterogeneity of rat FSH by chromatofocusing: studies on serum FSH, hormone released in vitro and metabolic clearance rates of its various forms, *J. Endocrinol.,* 105, 29, 1985.

46. **Ashwell, G. and Harford, J.,** Carbohydrate-specific receptors of the liver, *Annu. Rev. Biochem.,* 51, 531, 1982.

47. **Kawaguchi, K., Kuhlenschmidt, M., Roseman, S., and Lee, Y. C.,** Synthesis of some cluster galactosides and their effect on the hepatic galactose-binding system, *Arch. Biochem. Biophys.,* 205, 388, 1980.

48. **Cohen-Forterre, L., Mozere, G., Andre, J., and Sternberg, M.,** Cleavage of oligosaccharides by rat kidney sialidase, *FEBS Lett.,* 173, 191, 1984.

49. **Weintraub, B. D., Stannard, B. S., Magner, J. A., Ronin, C., Taylor, T., Joshi, L., Constant, R. B., Menezes-Ferreira, M. M., Petrick, P., and Gesundheit, N.,** Glycosylation and posttranslational processing of thyroid-stimulating hormone: clinical implications, *Rec. Prog. Horm. Res.,* 41, 577, 1985.

Chapter 8

LUTEINIZING HORMONE MICROHETEROGENEITY

Brooks A. Keel and H. Edward Grotjan, Jr.

TABLE OF CONTENTS

I. INTRODUCTION

The pituitary hormones luteinizing hormone (LH, lutropin), follicle-stimulating hormone (FSH, follitropin), and thyroid-stimulating hormone (TSH, thyrotropin), as well as the placental hormone human chorionic gonadotropin (hCG) are glycoproteins which are composed of two distinct polypeptide chains. Their alpha subunits are identical in amino acid sequence within a species, while their beta subunits possess a distinct sequence which is specific for the respective hormone. It is the beta subunit, in combination with the common alpha subunit, which determines hormonal specificity (reviewed in Chapter 1). These four hormones have carbohydrate moieties attached to the protein core. Both of the subunits are glycosylated, containing one or two asparagine-linked oligosaccharide units per subunit. The beta subunit of hCG also possesses four serine-linked oligosaccharide residues (reviewed in Chapter 2). The carbohydrate moieties very often contribute to the expression of molecular pleomorphism. Glycoprotein hormones which exhibit microheterogeneity exist as a series or family of isohormones which may differ in molecular weight, isoelectric point (pI), circulatory half-life, receptor-binding activity and/or biological activity. It is becoming clear that subtle changes in the carbohydrate structure of glycoprotein hormones can dramatically alter their respective biological activities. Thus, qualitative as well as quantitative changes in the production of the glycoprotein hormones may occur.

LH is synthesized by the anterior pituitary, secreted into the circulation and is carried to its site of action by the blood. Its target organs are the testis and ovary where it primarily functions to regulate steroidogenesis.[1] LH interacts with specific membrane receptors on granulosa, luteal, and Leydig cells resulting in the stimulation of adenylyl cyclase, formation of cyclic AMP, activation of enzyme-specific protein kinase, and subsequent phosphorylation of hormone-specific cellular proteins.[2] The culmination of this cascade of events is a dramatic increase in the steroidogenic capacity of the gonad. Any alteration in the circulating levels of biologically active LH can profoundly affect gonadal steroidogenic efficiency. Changes which occur in the quality, as well as the quantity, of LH may play a critical role in the regulation of gonadal function. Therefore, the study of LH microheterogeneity is important to our overall understanding of the control of gonadal function.

Discrepancies between the biological and immunological potencies of several polypeptide hormones have been known for some time (reviewed in Reference 3). With the advent of sensitive in vitro bioassays for LH, measurements of the biologically active concentrations of pituitary and circulating LH have become possible. These in vitro bioassays, utilizing isolated interstitial cells from rat[4-17] or mouse[18-24] testes, have allowed accurate determination of low levels of bioactive LH from several species and in a variety of normal and abnormal endocrine states. The results from these studies have shown that the ratio of bioactive to immunoreactive LH (B/I ratio) is lower in cycling women than in men, postmenopausal women, or girls with gonadal dysgenesis,[5,6,9] suggesting that the bioactivity of LH might be under steroidal regulation. Estrogen administration to patients with Turner's syndrome lowered the B/I ratio of circulating LH while testosterone was found to prevent the castration-associated decrease in the B/I rato of rat LH,[13] adding further evidence that the biological potency of LH is dependent upon the steroidal milieu. Thus, qualitative changes in the synthesis and/or secretion of LH by the pituitary could represent a highly sensitive fine-tuning mechanism for controlling gonadal function.

In this chapter we will examine the biological characterization of LH microheterogeneity. Emphasis will be placed on potential relationships between the presence of LH isohormones and the overall bioactivity of LH and on the possible endocrine control mechanism involved in the biosynthesis of isohormones.

II. SIZE MICROHETEROGENEITY

Early attempts to isolate, purify, and characterize pituitary LH led to the observation that the hormone existed as a heterologous population of molecules. Reichert and Jiang[25] purified LH from bovine (bLH), porcine (pLH), equine (eLH), ovine (oLH), and human (hLH) pituitaries and compared their migration on columns of Sephadex® G-100. Two major forms of biologically active LH were observed; one having an apparent molecular weight estimated to be about 45,000 daltons and the other eluting in the void volume (Vo) of the column.[25] In addition, bLH and hLH displayed another minor peak of biological activity eluting just after Vo. These investigators postulated that, among other things, a variety of biologically active LH molecules are extractable from pituitary tissue or that purified LH is capable of forming aggregates.[25] Further attempts at the isolation and characterization of human pituitary LH revealed several electrophoretic variants.[26-30] Taken together, these early observations strongly suggested that LH exists within the pituitary as a series of molecular variants. Whether the molecular heterogeneity associated with purified LH preparations occurred naturally or was a consequence of the isolation procedures could not be determined.

Bogdanove and co-workers[31-33] were among the first to describe gonadotropin heterogeneity in crude pituitary extracts and to relate changes in heterogeneity with the endocrine status of the animal. Peckham, Knobil, and colleagues[34-36] extended these observations by demonstrating that LH in sera and pituitary extracts from ovariectomized monkeys is larger in molecular weight, as judged by gel filtration, than LH from intact female animals.[35] The effect of ovariectomy could be reversed by chronic estradiol administration. In fact, the LH produced in these estrogen-treated ovariectomized animals was even smaller than that produced by intact female monkeys. Pituitary LH obtained from intact male monkeys was found to be smaller than LH from intact females.[36] Orchidectomy resulted in identical changes in the size of LH[36] as did ovariectomy,[35] that is, an increase in molecular weight. The effects of steroid replacement in male animals was not examined.[36] These investigators went on to demonstrate that the ability of LH to survive in circulation depended upon the endocrine status of the animal from which the LH was isolated. The relative circulatory half-lives of LH was found to be intact male LH $<$ intact female LH $<$ gonadectomized male or female LH.[35,36] Because the ability of a glycoprotein to survive in circulation has been related to the degree of sialylation,[37,38] Peckham, Knobil, et al. speculated[34,35] that the differences in circulatory half-life and molecular size of LH from animals of differing endocrine states might be the result of varying degrees of sialylation. These investigators observed that the differences in apparent molecular size of LH from intact and ovariectomized monkeys were essentially eliminated by digestion with neuraminidase.[35] Taken together, these results strongly suggest that qualitative differences in monkey LH exist within the pituitary and that these differences are dependent upon the endocrine status of the animal. Further, these qualitative differences appear to be at least due to subtle alterations in the carbohydrate moiety of the molecule. Because the half-life of LH is dependent upon endocrine status,[35,36] the observed qualitative changes in LH could affect stimulation of the gonad.

Although Bogdanove et al.[33] found no significant differences in the gel filtration elution profiles of rat (r) LH as a result of castration, other investigators described multiple forms.[39-43] Conn et al.[39] observed that the aging process in rat may be associated with qualitative changes in LH. These investigators reported that the pituitary and circulating forms of rLH prepared from old animals or from young castrated rats had a higher molecular weight, as judged by gel filtration, than rLH preparations from younger intact animals.[39] Testosterone administration reversed the observed effects in both castrates and old animals. The qualitative differences observed by Conn and co-workers[39] were apparently related to the carbohydrate moiety of the LH in that removal of sialic acid residues by neuraminidase abolished the apparent size differences. Reddy and Menon,[40] using gel filtration on ultragel columns, showed two closely migrating but distinct peaks of immunoreactive pituitary rLH, corresponding to molecular weights of 40,000 and 26,000 daltons. Both peaks were active in an in vitro bioassay, with the lower molecular weight form being twice as biologically active as the larger form. Based on these observations, these investigators discounted the possibility that their findings might have resulted from dissociation of the native hormone into subunits.[40] However, partial dissociation cannot be ruled out entirely. Large molecular weight rLH-like substances, eluting near the Vo following gel filtration of pituitary extracts, have also been reported.[41-43] These large molecular weight forms, termed "big" LH (BLH), displayed immunoreactive specificity towards specific anti-LH beta antisera, appeared early in biosynthesis, contained little carbohydrate, possessed little biological activity, and resisted dissociation.[41,42] Further studies have suggested that BLH is a precursor of native rLH and may represent a pool of immature rLH.[42] Others have observed that the "big" LH strongly interacts with concanavalin A suggesting that it was glycosylated and that processing of the carbohydrate moiety may not have been complete.[43]

Using gel filtration, we have investigated the molecular forms of rLH produced and released by pituitary cells in culture.[44] In that study, large molecular weight rLH-like substances were not detected. If present, BLH was of insufficient cross-reactivity or existed in quantities too low to be detected by the radioimmunoassays employed.[44] However, all samples were subjected to ultracentrifugation prior to gel filtration[44] which substantially reduces or totally eliminates large molecular weight gonadotropin-like materials.[45,46] More recently, we have examined the gel filtration profiles of oLH obtained from rams, castrated sheep (wethers), and in wethers treated with estradiol (E_2), dihydrotestosterone (DHT), or combinations of the two steriods (DHT + E_2).[47] In contrast to the earlier work of Reichert and Jiang,[25] gel filtration profiles consistently exhibited a single peak of oLH immunoreactivity corresponding to an average apparent molecular weight of 47,300 (Figure 1). Neither castration nor steroid implantation caused a change in the apparent molecular size of pituitary oLH. In this series of experiments the extraction of the pituitaries was performed in such a way as to minimize large molecular weight gonadotropin-like materials.[45,46] Therefore, the size heterogeneity observed by others may be species-specific and may vary depending upon the procedures used for the extraction of the LH or upon the techniques utilized for determination of the molecular weight of the hormone. It is entirely possible that large molecular weight forms of LH represent aggregated forms of the hormone.

III. CHARGE MICROHETEROGENEITY

LH exists within the pituitary as a family or series of charged isomers termed "isohormones". Each of these isohormones possesses a distinct and reproducible isoelectric point (pI). Interestingly, these isohormones differ not only in pI but also in circulatory half-life, target cell receptor binding, and biological activity. Results from several laboratories suggest that at least a portion of this charge microheterogeneity may result from subtle differences in oligosaccharide structure. Other investigators have shown that the endocrine status of the animal alters the pattern of the isohormones. Thus, changes in the animal's endocrine environment not only alters the quantity of pituitary gonadotropins but also induces qualitative changes in the hormone which may be important in the normal control of gonadal function. In the next section we will provide evidence for the existence of LH isohormones, discuss the relationships between charge and biological activity, and examine the effects of varying endocrine states on the profile of LH isohormones.

A. Characterization of Crude LH Isohormones

The charge characteristics of LH from a variety of species including the human,[48-63] monkey,[64] equine,[48,65,66] bovine,[48,67] ovine,[48,68-72] porcine,[48] rabbit,[48] rat,[73-82] avian,[83,84] and

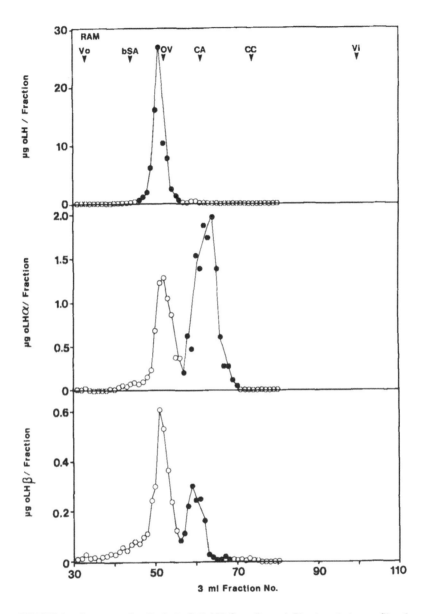

FIGURE 1. Representative Sephadex® G-100 Superfine gel filtration elution profile of immunoreactive oLH, oLHα, and oLHβ in the extract (100 mg tissue equivalents) of an anterior pituitary from a ram. The 1.5 × 185 cm column was eluted at 3 mℓ/hr with 0.15 M ammonium bicarbonate. Three-mℓ fractions were collected. (From Keel, B. A., Schanbacher, B. D., and Grotjan, H. E., Jr., *Biol. Reprod.*, 36, 1114, 1987. With permission.)

whale[85] have been described. The apparent pIs of these isoelectric variants are depicted in Table 1. These values were obtained using LH preparations of differing degrees of purity and were characterized by different isoelectric focusing techniques. It is readily apparent from Table 1 that with the exception of equine LH, the isohormones of LH focus primarily as basic species. This is in contrast to the isohormones of FSH,[86] TSH,[87] and prolactin[88] (see also Chapters 5, 6, 7, and 10, and Figure 8 below) which focus predominately in the acidic region.

With the exception of highly purified preparations, LH from several species can be separated into as many as five to eight isohormones. Robertson and co-workers[53,54,59] sub-

Table 1

REPORTED APPARENT ISOELECTRIC POINTS (pI) OF LH ISOHORMONES FROM VARIOUS SPECIES

Species	Ref.	Number of isohormones	Apparent pI	Comments
Human	Reichert[48]	4	9.18, 8.18, 7.62, 7.18	Purified preparation
	Robertson et al.[53]	7	9.55, 9.17, 8.81, 8.23, 7.80, 7.33, 6.75	Pituitary or plasma LH
	Van Damme et al.[55]	4	8.14, 7.72, 7.26, 6.71	Postmenopausal urine
	Weise et al.[60,61]	7	8.80, 8.40, 7.80, 7.40, 6.90, 6.50, 5.90	Pituitary LH
	Suginami et al.[63]	6	9.13, 8.60, 8.16, 7.67, 7.24, <7.00	Serum LH
	Graesslin et al.[51]	3	8.10, 7.60, 6.90	Urine
	Maffezzoli et al.[50]	3	6.20, 4.20, 2.80	Urine
	Strollo et al.[58]	4	8.54, 8.10, 7.58, 7.10	Plasma LH
Monkey	Hamada and Suginami[64]	6	9.29, 8.88, 8.34, 7.84, 7.38, <7.00	Plasma LH
Equine	Braselton and McShan[65]	4	7.30, 6.60, 5.90, 4.50	Purified preparation
	Reichert[48]	1	4.24	Purified preparation
	Matteri and Papkoff[66]	5	6.60, 6.10, 5.70, 5.20, 4.80	Purified preparation
Bovine	Reichert[48]	1	9.55	Purified preparation
	Yoshida and Ishii[67]	3	8.80, 8.30, 4.00	Partially purified
Ovine	Reichert[48]	1	10.47	Purified preparation
	Keel et al.[69-72]	8	>9.80, 9.26, 9.14, 9.07, 8.98, 8.91, 8.81, <7.0	Pituitary extracts
Porcine	Reichert[48]	1	9.80	Purified preparation
Rabitt	Reichert[48]	2	9.32, 8.88	Purified preparation
Rat	Wakabayashi et al.[74,76,81]	7	9.80, 9.60, 9.30, 9.10, 8.80, 8.50, 7.90	Pituitary extracts
	Dufau et al.[14]	6	9.44, 9.20, 9.02, 8.76, 8.64, 8.46	Pituitary extracts
	Uchida and Suginami[77]	7	10.30, 9.30, 9.00, 8.70, 8.30 neutral Lh, acidic LH	Pituitary extracts
	Keel and Grotjan[78-80]	7	>9.80, 9.25, 9.23, 9.17, 9.06, 8.97, <7.00	Pituitary extracts
Avian	Hattori and Wakabayashi[83,84]	5	9.80, 9.40, 9.00, 8.50, 8.00	Pituitary extracts
Whale	Tamura-Takahashi and Ui[85]		8.70, 8.50, 8.30, 8.10	Partially purified

jected human pituitaries to column electrofocusing using a broad (pH 3.5 to 10) pH range. These investigators discovered a heterogeneous profile of hLH activity with a large proportion present within the pH range of 10 to 6.5.[53] Seven main regions of hLH activity within this pH range were observed[53] with mean pI values displayed in Table 1. When the pituitary extracts were subjected to a narrow pH gradient (pH 6 to 9) it was shown that the main LH-reactive fractions proved to be heterogeneous, although the average pI value for each peak region was comparable to that observed in the broad pH range studies.[53] These investigators went on to show that considerable variation existed between (with respect to the relative proportions of these isoelectric species) as well as within samples.[53] More recently, Weise et al.[60] demonstrated seven peaks (in the range of 8.8 to 5.9) for pituitary hLH subjected to column IEF. Using analytical gel IEF, refocusing of the different hLH forms revealed homogeneity.[60] The presence of multiple isoelectric variants of LH have also been demonstrated in the serum of humans[53,58,63] and monkeys[64] suggesting that the isohormones of

LH are secreted into circulation. Additional studies by Hattori and co-workers[76,84] have shown that isolation and refocusing of LH isohormones results in homogeneous forms having identical pIs as those observed previously, thus, significant changes in the charge characteristics do not occur during preparation of the components suggesting that LH isohormones are not artifacts of the separation technique.

The rat has been used extensively as an animal model for the characterization of gonadotropin isohormones. To date, IEF has been the method of choice to separate rLH[14,74-77,81,82] as well as rFSH[75,86,89,92] isohormones. Using this technique the laboratories of Wakabayashi[74,76,81] and Dufau[14] have observed six to seven rLH isohormones with pI values in the range of 9.8 to 7.9. In contrast, Robertson et al.,[75] who employed IEF using a broad pH gradient, reported that pituitary rLH focused as two broad immunoreactive peaks in the pH ranges of 9.5 to 7.0 and 6.5 to 3.5, respectively. Of major interest was the finding that the acidic forms of rLH constituted 25 to 48% of the rLH in pituitary extracts.[75] Similar results have been reported by others who also used broad pH gradient IEF.[77,82] Uchida and Suginami[77] observed distinct peaks of rLH in the range of 10.3 to 8.3; two additional subpopulations of rLH migrated in the neutral range (pI = 8.14 to 7.00) and the acidic range (pI >6.99). The acidic species of rLH represented 22 to 32% of the immunoreactive rLH recovered from the IEF column. More recently, Baldwin et al.,[82] using polyacrylamide gel IEF and a pH range of 3 to 10, demonstrated two major isoelectric species of rLH; one broad peak in the alkaline pH range and a second broad peak in the acidic pH range. The acidic species represented 21 to 35% of recovered rLH. The possible significance of the acidic forms of LH is discussed below.

A major disadvantage of IEF on polyacrylamide gels is that proteins which focus outside the defined pH window migrate into the electrode strips (i.e., off the gel) and are lost. Although proteins cannot be run off an IEF column, those which migrate into the electrode solutions may lose activity or may be discarded.[100] Chromatofocusing has several distinct advantages over IEF:[78] (1) it is relatively inexpensive, (2) relatively large sample volumes and protein concentrations can be utilized, (3) the limits of the pH gradient can easily be controlled, and (4) it can be used as a preparative technique. Another major advantage of chromatofocusing is that proteins outside the pH window can be collected and analyzed. Furthermore, base-line resolution of proteins with pI differences of as little as 0.02 can be achieved.[101] Several investigators have noted that advantage of chromatofocusing for separating the isohormones of FSH from several species including rat,[93-96] hamster,[97,98] and monkey[99] (see also Chapter 7).

We have used chromatofocusing to isolate and characterize the isohormones of rat,[78-80] hamster, and ovine[69-72] pituitary LH. As illustrated in Figure 2, chromatofocusing gradients in the pH 10.5 to 7.0 range are very reproducible, which is essential for comparing the isohormones observed in different elution profiles. When extracts of rat anterior pituitaries were subjected to this technique at least seven species of immunoreactive rLH were observed (Figure 3). Five species exhibited apparent pIs of 9.25, 9.23, 9.17, 9.06, and 8.97; similar rLH isohormones have previously been observed using IEF.[14,74,76,81] However, two additional immunoreactive peaks of rLH were observed after chromatofocusing: one in the void volume (pI >9.80) and one which was bound to the column but eluted with 1.0 M NaCl (pI <7.00). The latter two peaks were outside the pH window of the column but were easily detected. The resolution of both IEF and chromatofocusing can be increased by narrowing the pH gradient. However, proteins with pIs outside the pH window of IEF gels are run off the ends and are lost. Thus, it would appear that chromatofocusing is advantageous for studying gonadotropin isohormones because of the flexibility in defining the pH gradient and because materials outside the pH window are not lost.

Anterior pituitaries obtained from morning proestrus hamsters, when subjected to chromatofocusing, displayed similar LH isohormones as seen with the rat (Figure 4). However,

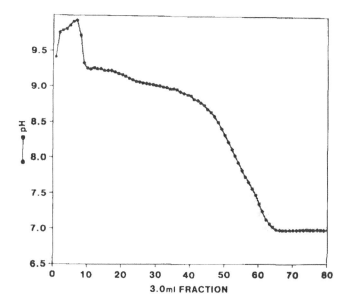

FIGURE 2.   Reproducibility of the pH 10.5 to 7.0 chromatofocusing gradient. The solid line represents the mean of seven profiles and the shaded area illustrates one SD on each side of the line. (From Keel, B. A. and Grotjan, H. E., Jr., *Anal. Biochem.*, 142, 267, 1984. With permission.)

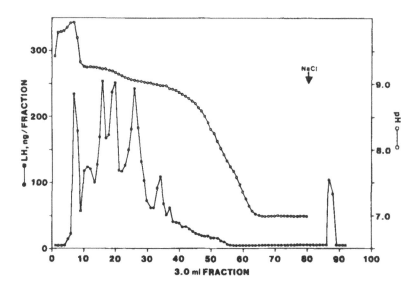

FIGURE 3.   Chromatofocusing elution profile of immunoreactive rLH in the extract of four pituitaries from intact male rats. Note that the first peak elutes before the gradient starts to form (pH 9.8) and that the last peak was bound to the column (pH 7.0) and eluted only after changing the eluant to 1.0 *M* NaCl. (From Keel, B. A. and Grotjan, H. E., Jr., *Anal. Biochem.*, 142, 267, 1984. With permission.)

a markedly different relative distribution profile was obvious. In particular, approximately three times as much hamster LH focused as the most basic isohormone (apparent pI >9.8) as observed for the rat (Table 2). It should be noted, however, that the hamster pituitary extract was obtained from a proestrus female. Other species display cyclic changes in the relative distribution of LH among its isohormones (see below). Uchida and Suginami[77] have

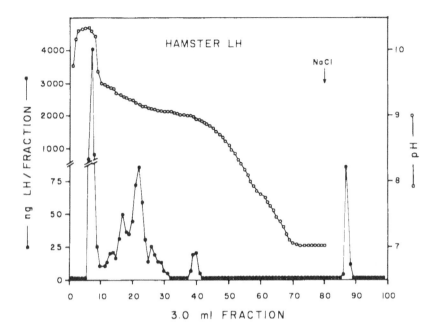

FIGURE 4. Chromatofocusing elution profile of immunoreactive hamster LH in the extract of four pituitaries from intact proestrous female hamsters. Note that the first peak elutes before the gradient starts to form (broken scale) and that the last peak was bound to the column and eluted only changing the eluent to 1.0 M NaCl.

Table 2
RELATIVE DISTRIBUTION OF RAT, HAMSTER, AND OVINE LH ISOHORMONES

Apparent pI	Relative Amount (%)		
	Rat LH	Hamster LH	Ovine LH
>9.80	25	70	3
9.00—9.80	58	24	9
8.00—8.99	10	3	72
<8.00	7	3	16

Note: Values expressed as a percentage of the total amount of immunoreactive LH obtained from the chromatofocusing column.

observed high amounts of rLH migrating in the high alkaline region in pituitary extracts obtained by proestrus rats. Species differences may also be a factor.

When extracts of anterior pituitaries from intact male sheep were chromatofocused, at least eight species of immunoreactive oLH were observed (Figure 5). Six of the species exhibited apparent pIs of 9.26, 9.14, 9.07, 8.98, 8.91, and 8.81. An additional highly basic (pI >9.80) and an acidic (pI <7.00) species were also observed. At first glance, it might appear that the isohormones of oLH are similar to those of rLH. However, close examination reveals significant differences. As shown in Table 2, the major species of rLH have pIs greater than 9.00. In contrast, the predominant forms of oLH are more acidic in nature exhibiting pIs between 8.00 and 9.00. The biochemical basis for these differences remains to be more fully delineated.

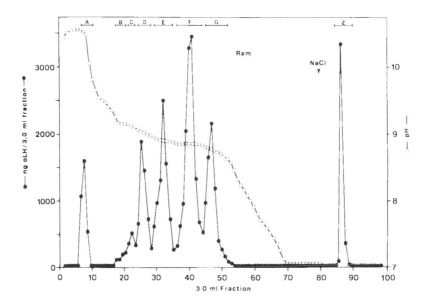

FIGURE 5.   Representative chromatofocusing elution profile of immunoreactive oLH in the extract of an anterior pituitary from a ram. Note that the first peak (isohormone A) eluted before the pH gradient started to form (pI >9.8) and that the last peak (isohormone Z) was bound to the column (pI <7.0) but eluted with 1 *M* NaCl. Brackets correspond to fractions pooled to divide the profile into respective isohormones (designated with letters). (From Keel, B. A., Schanbacher, B. D., and Grotjan, H. E., Jr., *Biol. Reprod.*, 36, 1102, 1987. With permission.)

To assure that the bound materials observed above for rLH indeed represented an acidic LH species, this peak was rechromatofocused (Figure 6). During the rechromatofocusing, all the immunoreactive LH was again bound and none eluted in earlier fractions. The acidic species of rLH and oLH illustrated here is reminiscent of the acidic rLH observed by others.[75,77,82] However, in our hands, acidic forms of rLH consistently constitute less than 20% of the immunoreactive LH in pituitary extracts from intact untreated animals.[78-80] Recent preliminary findings from Baldwin's laboratory[82] using chromatofocusing and narrow pH gradients suggest that the acidic rLH may be composed of at least three peaks in the pH range of 6.8 to 5.4. We have observed that when the acidic forms of rLH isolated by chromatofocusing on a pH 10.5 to 7 gradient were refocused on a pH 7 to 4 gradient, multiple peaks of rLH were observed (Figure 7). Therefore, LH may be a complex composite of 12 or more isoelectric variants spanning a broad pH range. In support of this concept Hartee et al.[62] recently isolated 14 subfractions of hLH by fast protein liquid chromatography; none of the subfractions was homogeneous with respect to charge. Clearly, more work is required before the precise number of LH isohormones in each species is known.

Apparent isohormones detected in crude pituitary extracts by radioimmunoassay (RIA) could result, in part, from the cross-reaction of other pituitary hormones in the LH RIA. To rule out this possibility we examined chromatofocusing elution profiles of rLH, rFSH, rTSH, and rPRL in pituitary extracts prepared from intact, adult male rats.[80] As observed previously, at least seven rLH isohormones were detected using a pH gradient from 10 to 7 (Figure 8). In contrast to rLH, >92% of the immunoreactive rFSH, rTSH, and rPRL remained bound to the column and eluted only in the presence of 1.0 *M* NaCl indicating that these hormones focus as predominately acidic forms as has been demonstrated by others.[86-88] Tests of parallelism were also performed comparing dose response curves for the various LH isohormones with purified LH. As illustrated in Figure 9, all seven oLH

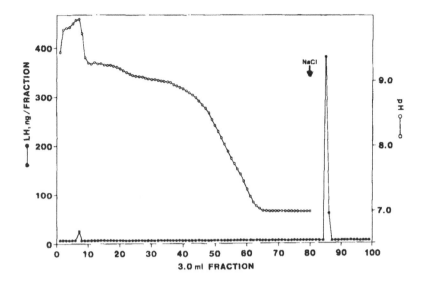

FIGURE 6.   Rechromatofocusing of the bound fraction of rLH from a previous elution profile. Note that all of the immunoreactive rLH again eluted with a pI of less than 7. The arrow indicates changing the eluant to 1.0 *M* NaCl. (From Keel, B. A. and Grotjan, H. E., Jr., *Anal. Biochem.*, 142, 267, 1984. With permission.)

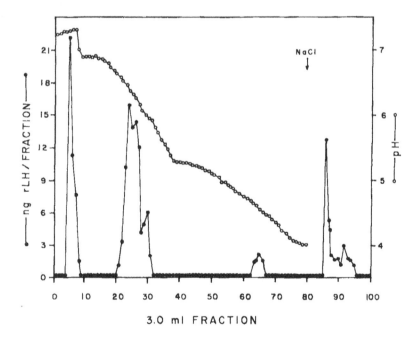

FIGURE 7.   Chromatofocusing elution profile of acidic forms of rLH obtained from a previous elution profile. Pituitary extracts were initially applied to a pH 10.5 to 7.0 gradient. The bound material, eluted with NaCl and representing rLH species with apparent pIs <7.00, was pooled and subjected to a pH 7.40 to 4.00 gradient.

isohormones were parallel to the standard. Identical results have been obtained for the isohormones of rLH.[80] It is apparent from these results that the basic isohormones of LH are not due to immunocrossreactivity of the other pituitary hormones. The acidic forms of LH could result, at least in part, from the cross-reaction of other hormones in the LH RIA. Although it has been suggested that FSH isohormones represent alpha and beta dimers

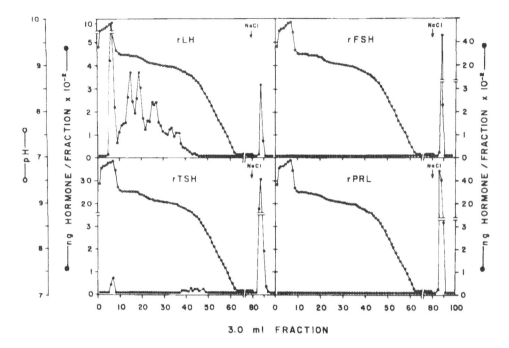

FIGURE 8.   Chromatofocusing elution profile of immunoreactive rLH, rFSH, rTSH, and rPRL in the extracts of pituitaries from intact male rats. The arrow indicates treatment of the column with 1.0 *M* NaCl. (From Keel, B. A. and Grotjan, H. E., Jr., *Endocrinology*, 117, 354, 1985. With permission.)

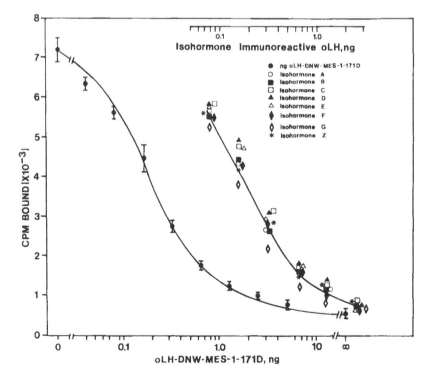

FIGURE 9.   Tests for parallelism of oLH isohormones in the oLH radioimmunoassay. Note that all eight forms observed were parallel to oLH-DNW-MES-1-171D, the standard used. (From Keel, B. A., Schanbacher, B. D., and Grotjan, H. E., Jr., *Biol. Reprod.*, 36, 1102, 1987. With permission.)

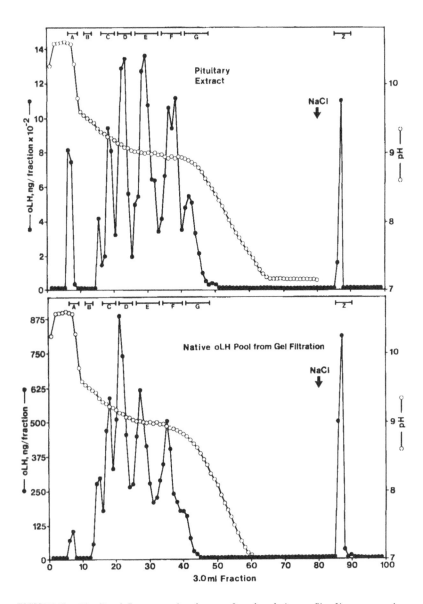

FIGURE 10.   Top Panel: Representative chromatofocusing elution profile of immunoreactive oLH in the extract of an anterior pituitary obtained from a wether. Bottom Panel: Chromatofocusing elution profile of immunoreactive oLH in pools of gel filtration fractions containing native oLH. Components bound to the column were eluted with 1.0 M NaCl. (From Keel, B. A. and Grotjan, H. E., Jr., *Biol. Reprod.*, 36, 1125, 1987. With permission.)

rather than uncombined subunits,[99,102] it is possible that certain LH isohormones detected by RIA could represent uncombined subunits or could be contaminated with uncombined subunits. To test this possibility we have resolved native oLH from its uncombined alpha and beta subunits by gel filtration[44] and subjected the partially purified native hormone and subunits to chromatofocusing.[69,71] Fractions from gel filtration containing native oLH were pooled. When chromatofocused, a pattern of immunoreactive oLH similar to that found in pituitary extracts was observed (Figure 10). Fractions from gel filtration corresponding to uncombined oLH subunits were also pooled and subjected to chromatofocusing and the resulting elution profiles were analyzed for immunoreactive oLH alpha and oLH beta by specific RIAs. Chromatofocusing elution profiles of highly purified oLH alpha and beta

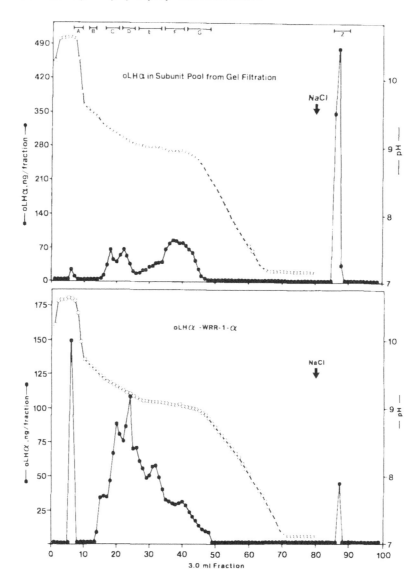

FIGURE 11. Top Panel: Chromatofocusing elution profile (means of six samples) of immunoreactive oLHα in pools of uncombined oLH subunits prepared by subjecting pituitary extracts to gel filtration. Some individual profiles exhibited distinct peaks in the pH 9.5 to 8.7 region but such peaks were not present in all samples. Bottom Panel: Chromatofocusing elution profile of oLHα-WRR-1-α, a highly purified preparation presumably derived by the dissociation of native oLH, as assessed by radioimmunoassay. In both cases, components bound to the column were eluted with 1 *M* NaCl. (From Keel, B. A. and Grotjan, H. E., Jr., *Biol. Reprod.*, 36, 1125, 1987. With permission.)

subunits are also shown for comparison. Approximately 60% of the immunoreactive free oLH alpha resolved from the native oLH by gel filtration eluted in the pH range of 10.5 to 8.7 which corresponds to the basic isohormones of native oLH (Figure 11, top). The remainder of the immunoreactive oLH alpha was bound to the column under the conditions utilized and thus exhibited acidic characteristics (apparent pI <7.00). More than half of the immunoreactive oLH beta was bound to the chromatofocusing column indicating a high percentage of acidic forms (Figure 12, top). In addition, a significant amount of highly basic forms which eluted in the flow through peak (apparent pI >9.80) were observed. The

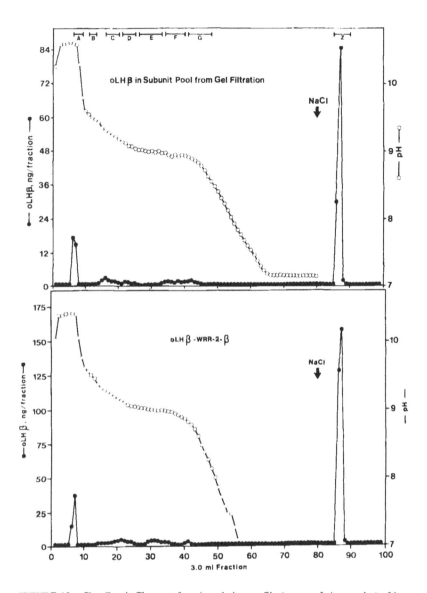

FIGURE 12.   Top Panel: Chromatofocusing elution profile (means of six samples) of immunoreactive oLHβ in pools of uncombined oLH subunits prepared by subjecting pituitary
extracts to gel filtration. Some individual profiles exhibited distinct peaks in the pH 9.5 to
8.7 region but such peaks were not present in all samples. Bottom Panel: Chromatofocusing
elution profile of oLHβ-WRR-1-β, a highly purified preparation presumably derived from
the dissociation of native oLH, as assessed by radioimmunoassay. In both cases, components
bound to the column were released with 1 *M* NaCl. (From Keel, B. A. and Grotjan, H. E.,
Jr., *Biol. Reprod.*, 36, 1125, 1987. With permission.)

remainder of the immunoreactive oLH beta was more or less uniformly distributed across
the gradient in the pH range of 9.70 to 8.70. These results suggest that the small amounts
of uncombined LH subunits present in pituitary extract appear to only minimally affect the
pattern of immunoreactive LH observed during chromatofocusing.

B. Characterization of Isohormones in Purified LH

For the most part, multiple LH isohormones are observed in crude pituitary extracts or
blood plasma. The situation appears to be different for purified preparations of LH, and

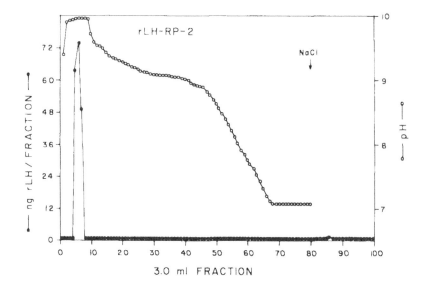

FIGURE 13.   Chromatofocusing elution profile of NIADDK-rLH-RP2. All of the immunoreactive rLH eluted before the pH gradient started to form (pI >9.8). The arrow indicates treatment of the column with 1.0 *M* NaCl. Identical results were obtained using rLH-I-5. (From Keel, B. A. and Grotjan, H. E., Jr., *Endocrinology*, 117, 354, 1985. With permission.)

depends to a great extent upon the degree of purity and possibly the chemical methods utilized in the purification procedures. Reichert[48] reported that highly purified bovine, ovine, and porcine LH existed as a single alkaline species with pIs of 9.55 or greater. In contrast to these results, Yoshida and Ishii,[67] using a much less stringent purification scheme, observed lower and multiple isoelectric points for bLH. These investigators suggested that the more acidic forms of bLH may have been eliminated by the ion-exchange chromatography step employed by Reichert[48] resulting in a preparation of LH consisting of only one component with an alkaline pI. Robertson et al.[56] observed that the hLH isoelectric profile for a partially purified pituitary extract used as a reference preparation (designated 68/104) was considerably more acidic than that obtained with a highly purified preparation (designated 68/40). In a comprehensive study evaluating the electrofocusing profiles of nine different highly purified human pituitary LH preparations, significant differences were observed in the spread and distribution patterns of the various immunoreactive species.[57] Wakabayashi[74] noted that rLH-RP-1, a relatively crude reference preparation, displayed multiple isohormones with pIs ranging from 9.0 to 8.0, while the highly purified rLH-I-4 focused as a highly basic species with an apparent pI of 9.5 to 9.6.

We have examined the chromatofocusing profiles of several highly purified preparations of rat[80] and ovine[71] native LH as well as the subunits of ovine LH.[71] The chromatofocusing elution profile of rLH-RP-2, a highly purified standard, is illustrated in Figure 13. Sixty to ninety percent of the immunoreactive rLH recovered from the column focused as a symmetrical peak which eluted in the void (flow through) volume indicating a highly basic form (pI >9.80). Identical results were obtained when rLH-I-5 was chromatofocused. Although these results are in general agreement with results of Wakabayashi,[79] it is not possible to tell whether the form with a pI >9.8 noted for rLH-I-5 and rLH-RP-2 is identical to the pI 9.5 to 9.6 form observed by Wakabayashi for rLH-I-4.[79] It is generally known[96,103] that the chromatofocusing technique used by us may not always yield the same pI for a given protein as the IEF procedure used by Wakabayashi.[79]

Reichert[48] previously reported that highly purified oLH electrofocused with an apparent pI of about 10.0. Similar results were obtained when oLH-LER-1056-C2, a highly purified oLH preparation obtained from Reichert, was subjected to chromatofocusing (Figure 14, middle). In contrast, two oLH preparations obtained from Dr. D. N. Ward's laboratory, Houston, Tex., exhibited different chromatofocusing patterns. The oLH-DNW-LDB-I-38A contained predominately basic forms with pIs in the 8 to 9 range (Figure 14, bottom) while oLH-DNW-MES-1-171D contained predominately acidic forms (Figure 14, top). Thus, it would appear that purified oLH preparations may each contain a distinct complement or subset of oLH isohormones.

Wakabayashi[74] suggested that, in an effort to eliminate contaminating TSH and FSH from the highly purified preparations of LH, most of the slightly alkaline (or acidic) species were eliminated during purification, leaving the component with the most alkaline pI. It was further suggested that this may be why purified preparations tend to exhibit a single pI or a few forms with pIs very close to each other. Zaidi et al.[57] suggested that the significant differences in electrofocusing profiles observed for a large number of highly purified LH preparations may have resulted from chemical manipulations employed during purifications. Therefore, the disparity between purified preparations and the pattern of LH isohormones in pituitary extracts could reflect selection of a particular form (or forms) during purification, changes in molecular structure during purification, or both. Because the charge microheterogeneity of FSH, hCG, and LH is at least partially due to subtle differences in carbohydrate structure, it is tempting to speculate that the terminal portions of gonadotropin oligosaccharides (which are generally negatively charged) may be lost during some purification procedures resulting in forms with very basic pIs.

C. Biological Activities of LH Isohormones

Several investigators have observed a disparity between the pituitary and plasma levels of immunoreactive LH and the biological activity of the LH. The biochemical basis for this disparity has not been fully elucidated; however, recent studies have attempted to relate the differences between LH immunological and biological activities with its charge microheterogeneity. LH isohormones differ not only in pI but also in receptor binding and biological activity as well as in circulatory half-life. The ability of different isohormones of hLH to bind to Leydig cell receptors, activate cAMP, and stimulate testosterone production has recently been investigated.[61] The results from these studies indicate that the various hLH isohormones have similar activities with respect to maximal cAMP accumulation and Leydig cell testosterone production. However, all seven of the isohormones displayed different relative affinities to the receptors of intact Leydig cells.[61] Dramatic differences were observed when the in vitro biological activities for these isohormones (receptor binding, cAMP accumulation, and testosterone production) were compared with immunoreactivity. Interestingly, the more basic isohormones possessed the greatest B/I ratio, and this ratio decreased linearly with decreasing pI. Similar results have been obtained by a number of investigators[14,56,60,61,63,76-80,84] (Table 3). In addition to these studies, Robertson and Diczfalusy[54] observed that the isoelectric profiles of biological and immunological LH were very similar over the pH range of 7 to 9. However, the B/I ratios were markedly lower in the acidic pH range. These results led to the suggestion that nonbiologically active substances (uncombined subunits, other pituitary hormones, etc.) which cross-react in the immunoassay, might lead to erroneously depressed B/I ratios. However, this is clearly not the case for basic rLH isohormones because they focus devoid of immunoreactive FSH, TSH, and PRL (see Figure 8). Although acidic LH isohormones focus with several other pituitary hormones, they do contain significant LH biological activity. Nonetheless, the lower B/I ratio of these particular isohormones could result, in part, from the cross-reaction of other pituitary hormones or uncombined subunits in the LH RIAs. Additional studies in our laboratory have demonstrated that all of the isohormones of LH are parallel to the standard used in the bioassay,[70] suggesting that the B/I ratios observed (in particular, for the acidic forms) are not artifacts of the bioassay.

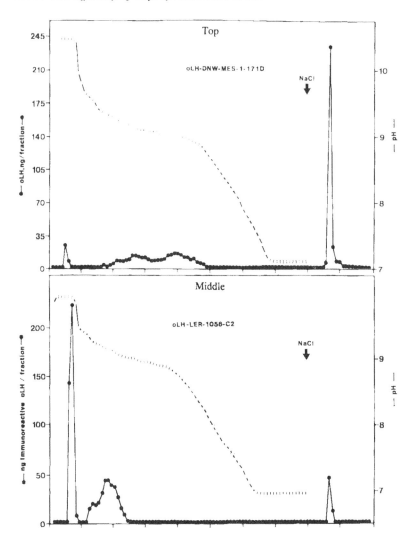

FIGURE 14.   Chromatofocusing elution profiles of purified oLH (Top Panel: oLH-DNW-MES-1-171D; Middle Panel: oLH-LER-1056-C2; Bottom Panel: oLH-DNW-LDB-I-38A) preparations as assessed by radioimmunoassay. In all three cases, components bound to the column were eluted with 1 *M* NaCl. (From Keel, B. A. and Grotjan, H. E., Jr., *Biol. Reprod.*, 36, 1125, 1987. With permission.)

In contrast to the above studies, the relationship between apparent pIs and B/I ratios for oLH isohormones does not appear to be monotonic. The oLH isohormones with apparent pIs slightly less than 9 consistently exhibited the highest B/I ratios.[70] Furthermore, both the most basic form of oLH (apparent pI >9.8) as well as acidic forms (apparent pIs <7.0) exhibited relatively low B/I ratios. We[71] have observed that uncombined oLH subunits predominately chromatofocus in the regions where isohormones A and Z elute (Figures 11 and 12). Thus, it is possible that the B/I ratios for these forms could be slightly reduced as a result of cross-reaction of uncombined subunits in the oLH RIA.

The B/I ratio and absolute amounts of each LH isohormone are important indices of the physiological changes which occur in LH. However, when considered separately they may not adequately reflect subtle changes. Therefore, we also expressed the data as the product of the amount and B/I ratio for each peak to yield a Bio-Index,[70,79,80] which represents the distribution of bioactive LH. Thus the Bio-Index reflects both qualitative (B/I ratio) and

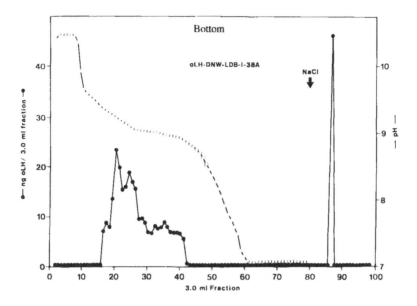

FIGURE 14 (Bottom)

Table 3
CORRELATION BETWEEN B/I RATIOS[a]
AND pIs OF VARIOUS LH
ISOHORMONES

Species	Ref.	Correlation coefficient (r)
Human	Weise et al.[60]	0.80
Rat	Dufau et al.[14]	0.96
	Hattori et al.[76]	0.96
	Uchida and Suginami[77]	0.97
	Keel and Grotjan[80]	0.94
Avian	Hattori and Wakabayashi[84]	0.98

[a] Data based on the B/I ratios obtained using the in vitro Leydig cell testosterone assay.

quantitative (amount of each isohormone) changes. When expressed in this manner, it is readily apparent that the most basic rLH form (apparent pI >9.8) is the most predominant biologically active species (Table 4). However, in the case of oLH, the isohormone having an apparent pI of 8.91 was the predominant biologically active form of oLH. This method of expressing the amount of biologically active LH has proven to be useful when evaluating subtle changes which may occur in the LH isohormones between control and experimental groups (see below).

Robertson and co-workers[54,56,57] pointed out the potential difficulties in interpreting B/I ratio data. B/I ratios obtained using impure standards or standards displaying nonparellelism to the unknowns are particularly susceptible to errors.[54] Liu et al.[104] recently noted that different types of bioassays may give rise to completely different comparisons for the same molecules, while Wide and Hobson[105] have recently shown that the bioassay method chosen to monitor the purification of a hormone has a major influence on the biological properties of the hormone. Burstein et al.[17] demonstrated that the principal cause of variability in B/I

Table 4
BIO-INDEX VALUES OF VARIOUS
OVINE AND RAT LH ISOHORMONES

Rat		Ovine	
pI	Bio-Index	pI	Bio-Index
>9.80	2.0	>9.80	35.2
9.25	0.4	9.26	28.2
9.23	0.8	9.14	36.6
9.17	0.9	9.07	145.1
9.06	0.8	8.98	329.6
8.97	0.4	8.91	1822.5
<7.00	0.1	8.81	508.4
		<7.00	180.3

Note: Bio-Index, the distribution of biologically active
LH, is expressed as the amount of immunoreactive LH (μg/pituitary) multiplied by the B/I ratio
for each isohormone.

ratios is due to changing dose-response characteristics of plasma LH in the RIA. Therefore,
extreme care must be exercised when evaluating changes in the B/I ratios of LH isohormones.
Thus, more work is needed in order to understand the relationships between isoelectric point,
biological activity, and immunological activity of LH isohormones.

In vitro LH bioassays measure the ability of the hormone to bind to the target cell receptor
(for example, the Leydig cell), activate adenylyl cyclase, and ultimately stimulate increased
synthesis of testosterone. These bioassays do not, however, provide any information on the
ability of the hormone to survive in circulation. Little is known concerning the circulatory
half-lives of the LH isohormones. The metabolic clearance rates of FSH isohormones have
been studied in some detail.[95] A decrease in pI of the isohormones was correlated very well
with an increase in the metabolic half-life of the isoelectric variant. Clearly, in vitro bioassay
data only provide partial information concerning the physiological significance of gonadotropin isohormones.

D. Endocrine Alterations of LH Isohormones

Previous investigators have noted that the endocrine status of the animal alters the pattern
of gonadotropin isohormones. Galle et al.[98] demonstrated that endocrine conditions which
decrease gonadotropin-releasing hormone (GnRH) release or the sensitivity of the pituitary
to LHRH cause an increase in the relative proportions of the more acidic, less biologically
active forms of pituitary FSH and a concomitant reduction in the more basic, more biologically active forms of pituitary FSH. Thus, changes in the endocrine status of the animal,
such as castration, cryptorchidism, or exogenous hormone administration, could not only
alter the quantity of pituitary LH but would also induce qualitative changes. These qualitative
changes may be important in the normal control of gonadal function.

1. Changes During the Menstrual Cycle and Effects of GnRH

Robertson and co-workers[53,58,59] demonstrated marked differences in the electrofocusing
pattern of plasma hLH between postmenopausal and cycling females. A significantly higher
proportion of midcycle plasma hLH was found in the pH regions with mean pI values of
8.81, 9.17, and 9.55 than in postmenopausal samples.[53] Conversely, a distinct acidic species
(pI = 5 to 6[56] to 6.87 to 7.36[58]) was detected in postmenopausal samples but was consistently
absent from normally menstruating women. The electrofocusing profile observed in post-

menopausal urine was also composed of acidic forms.[55] Thus, a significant difference exists in the hLH isohormone profile between these two endocrinological states.

Several reports have shown a marked disparity between bioactive and immunoreactive LH levels in peripheral circulation during the midcycle gonadotropin surge in monkeys[8,106,107] and humans.[9,14,19] When LH bioactivity was measured in cycling female rats, a marked increase was observed in both biologically and immunologically active serum LH at 2000 hr of proestrus.[14,108] The serum LH B/I ratio was constant during the cycle except during the LH surge where it was increased.[14,108] These studies have led others to investigate whether these cyclic alterations in B/I ratio reflect changes in the isohormone pattern of LH. Uchida and Suginami[77] observed cyclic changes in both the bioactive and immunoreactive levels of the most basic isohormone of rLH (apparent pI = 10.3) while more alkaline or acidic species remained constant throughout the cycle. The pI 10.3 species was the most predominant during proestrus, was reduced by a factor of two in estrus, and increased during the metestrus and diestrus periods. Similar results were obtained by Strollo et al.[58] in the human. These investigators[58] observed that the less alkaline species of LH were consistently absent from midcycle plasma. Based upon their findings, Uchida and Suginami[77] concluded that the midcycle release of LH may be associated with molecular modifications of the hormone resulting in an alkaline shift in pI and a subsequent increase in biological potency of the hormone.

That this preferential release of a biologically active pool of LH (increased alkalinity) is under GnRH control has recently been demonstrated. Pulsatile release of biologically active LH has been observed in healthy, normal men and postmenopausal women.[15] Although a majority of the episodic peaks of the bioactive and immunoreactive LH occurred concurrently, 13% of the bioactive LH pulses and 23 to 43% of the immunoreactive LH pulses were not associated with increases in immuno- or bioactivity, respectively.[15] Alterations in the B/I ratio of LH were also demonstrated in men when the frequency of the endogenous GnRH signal was amplified by administration of an opiate receptor antagonist.[16,109] Pulsatile secretion of hLH in menstruating women has been observed, and modulation of the frequency of bioactive pulses occurs during the follicular and luteal phases of the cycle. In particular, the pulse frequency of bioactive LH increased significantly during the late follicular phase and then markedly declined during the luteal phase.[15] In contrast to these findings, it has been reported that a single bolus or continuous infusions of GnRH does not result in a preferential release of biologically active LH.[9,110,111] Therefore, the rate and/or degree of GnRH stimulation may be critical in controlling the release of LH with varying biological and immunological activities. Because changes in biological activity of LH are often associated with a concomitant shift in the isohormone profile, it is reasonable to assume that the relative distribution of LH isohormones might also be under GnRH control. This possibility was recently explored by Baldwin et al.[82] who examined changes in the electrofocusing profile of pituitary rLH incubated in vitro in the presence or absence of GnRH. These investigators,[82] using broad pH range gel IEF, found that the rLH migrating in the alkaline pH range was increased by GnRH. Further, the alkaline to acidic rLH ratio of secreted rLH was increased in the presence of GnRH.[83] These observations suggest that the majority of releasable rLH stored in the pituitary possesses alkaline pIs.

It is generally accepted that in response to a sustained stimulation by GnRH, LH is released in a biphasic pattern. An initial increase in LH is observed followed by a plateau or a fall, and a consecutive second rise appears after a latent period. The ''two pool'' theory has been proposed to explain this phenomenon wherein the first acute rise in LH represents the release of stored hormone (pool I) while the second LH release (pool II) involves synthesis of new LH or chemical modification of existing hormone (see references in References 63 and 64). Suginami and colleagues have evaluated the electrofocusing profiles of human[63] and monkey[64] LH during these two phases of exogenous GnRH-stimulated LH release. The dominant hLH species in midcycle plasma pool I had pIs ranging from 8.60 to 9.13 whereas follicular and luteal phase plasma pool I migrated in the pI 7.67 to 8.60 range.[63] In both species, plasma pool II LH consistently contained an increase in the precentage of acidic species; the B/I ratio of these forms was markedly depressed.[63,64] Taken together, these observations suggest[63,64,77] that (1) the isoelectric pattern, and thus the overall bioactivity, of LH changes during the normal menstrual cycle, (2) LH isohormone secretion is under GnRH control, (3) LH isohormones migrating as alkaline forms might represent stored, early releasable forms of LH (pool I), and (4) acidic LH isohormones might represent a newly synthesized generation of hormone requiring chemical modification prior to release (pool II).

2. Differences Between Males and Females

In general, the sex of the animal and its endocrine status does not change the number of LH isohormones present. There may, however, be changes in the distribution of LH among the various isohormones. We observed[80] that pituitary extracts from intact male and proestrus female rats consistently contain the same seven rLH isohormones (Figures 15 and 16). The distributions, B/I ratios, and Bio-Index values between the two sexes were statistically similar (Table 5). In contrast, others found differences in the relative amounts of rLH between sexes. Wakabayashi[74] initially reported that, although corresponding isoelectric species of rLH were observed in intact males and females, the relative amounts of these forms were different. Later reports from that laboratory revealed the relative amounts of components having pIs of 9.8, 9.6, and 9.3 (the three most basic species) were higher in female pituitaries obtained during the morning of proestrus than in their intact male counterparts.[76] Robertson et al.[75] observed significantly more rLH bioactivity in the pH range of 7 to 9.5 in intact males than in either proestrus or diestrus females, while no significant differences between sexes were noted for forms in the acidic pH range. It is conceivable that the varying results may be related to the exact time during proestrus that the females were sacrificed (Keel and Grotjan,[80] 0830 hr; Wakabayashi,[74] 1300 hr; Hattori et al.,[76] "on the morning of proestrus"; Robertson et al.,[75] 1000 to 1500 hr). Although this would seem unlikely, the precise changes which occur in LH isohormones during the day of proestrus remain to be described.

3. Effects of Castration

Significant alterations in the relative amounts and biological activities of LH isohormones as a result of castration have been reported by several laboratories. Wakabayashi[74] and Hattori et al.[76] demonstrated that orchidectomy caused a marked increase in the relative amounts of rLH isohormones having pIs <9.00. A threefold decrease in the relative amount of the most basic forms (pI >9.6) was also observed.[76] Robertson et al.[75] reported that the proportion of alkaline rLH increased from 52 to 57% to 71 to 73% as a result of castration. Hamada and Suginami[64] observed a castration-induced increase in neutral to acidic monkey LH associated with a decrease in forms migrating in the high alkaline region. Dufau et al.[14] evaluated changes in the distribution and B/I ratios of the rLH isohormones as a function of the duration of gonadectomy. They observed that orchidectomy decreased the B/I ratios of all the rLH isohormones by 100%. The effects were maximal by 45 days and returned to mean control values by 60 days past castration.

We have observed[80] a 2.5- to 3.5-fold decrease in the relative amount of the most basic isohormone of rLH as a result of castration in both sexes (Table 6; Figures 15 and 16). No significant differences in B/I ratios were observed between intact and castrate male rats or between intact and castrate female rats.[70] The Bio-Index of rLH isohormones was similar between intact male and female rats (Table 7). The Bio-Index of isohormones with pIs of >9.80, 8.97, and <7.00 were also similar in both sexes and not significantly altered by castration. In contrast, castration in both sexes resulted in a marked increase in the Bio-Index of isohormones focusing in the midalkaline range (pI = 9.25 to 9.06). Thus, the Bio-

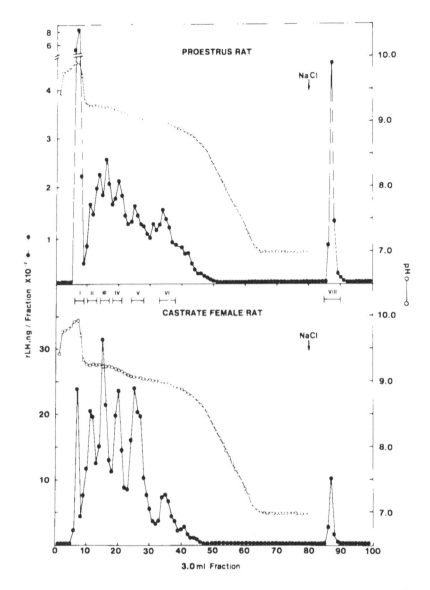

FIGURE 15. Chromatofocusing elution profiles of immunoreactive rLH in the extracts of anterior pituitaries from intact proestrus (top) and castrate (bottom) female rats. Each profile represents the mean obtained from two separate determinations. Note that the ordinate for castrates is 10-fold larger than that for intact rats. Brackets (⊢——⊣) correspond to the fractions pooled to divide the elution profiles into isohormones which are designated with Roman numerals. Peak I (pI >9.8) elutes before the gradient starts to form and peak VII (pI <7.0) elutes only after treatment of the column with 1.0 *M* NaCl (arrow). (From Keel, B. A. and Grotjan, H. E., Jr., *Endocrinology*, 117, 354, 1985. With permission.)

Index data clearly indicate a shift in the pattern of biologically active rLH which is not readily apparent when the data are examined as relative amounts or B/I ratios.

The response to castration in the ovine appears to be much less dramatic[70] than that observed in the rat. As a result of castration there appeared to be a slight shift in the distribution of oLH among its isohormones toward more basic forms and possibly decreased B/I ratios of all forms (compare Figures 5 and 17). In particular there was a two- to fourfold increase in the relative amounts of isohormones ranging from pI 9.26 to 8.98 relative to rams, with a concomitant reduction in the other species. However, no gross redistribution of oLH isohormones as a result of castration was observed.[70]

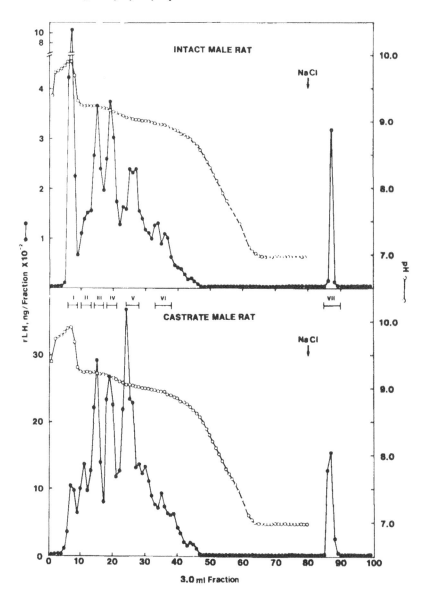

FIGURE 16.   Chromatofocusing elution profiles of immunoreactive rLH in the extracts of anterior pituitaries from intact (top) and castrate (bottom) male rats. Each profile represents the mean obtained from three separate determinations. Note that the ordinate for castrates is 10-fold larger than that for intact rats. Brackets (⊢——⊣) correspond to the fractions pooled to divide the elution profiles into isohormones which are designated with Roman numerals. Peak I (pI >9.8) elutes before the gradient starts to form and peak VII (pI <7.0) elutes only after treatment of the column with 1.0 *M* NaCl (arrow). (From Keel, B. A. and Grotjan, H. E., Jr., *Endocrinology*, 117, 354, 1985. With permission.)

In addition to castration, cryptorchidism has been shown to markedly alter the distribution of rLH isohormones in the pituitary.[79] Identical numbers of rLH isohormones were observed in intact and bilateral cryptorchid rat pituitaries. However, the relative amount of the most basic, most biologically active form was increased from 31 to 72% in the cryptorchid animal. As shown in Table 8, this resulted in a dramatic increase in the amount of biologically active rLH (Bio-Index) compared with intact rats.

Taken together, these results strongly suggest that changes in the gonadal status of the

Table 5
RELATIVE AMOUNTS, B/I RATIOS, AND BIO-INDEX
VALUES OF rLH ISOHORMONES FROM INTACT
MALE AND FEMALE RATS

	Relative amount[a]		B/I Ratio[b]		Bio-Index[c]	
pI	Male	Female	Male	Female	Male	Female
>9.80	26	36	5.8	5.2	7.9	11.1
9.25	8	9	3.0	6.1	1.6	5.1
9.23	19	14	3.5	6.3	3.4	4.6
9.17	18	9	3.2	4.5	3.5	2.7
9.06	15	9	2.8	3.3	3.0	1.9
8.97	9	13	2.8	3.3	1.4	1.9
<7.00	5	10	1.2	2.0	0.5	1.4

[a] Values expressed as a percentage of the total amount of immunoreactive LH obtained from the chromatofocusing column.
[b] Biological to immunological ratio expressed as ng bioactivity per ng immunoreactivity.
[c] Values expressed as the amount of immunoreactive LH (μg/4 pituitaries) multiplied by the B/I ratio for each isohormone.

Table 6
RELATIVE AMOUNTS OF RAT LH
ISOHORMONES SEPARATED BY
CHROMATOFOCUSING

	Relative amount[a]			
	Male		Female	
Isohormone	Intact	Castrate	Intact	Castrate
I	24.9	7.2	26.4	9.7
II	8.8	11.0	12.8	16.0
III	16.0	17.4	14.3	20.9
IV	17.0	20.1	12.2	17.2
V	16.6	25.9	11.2	23.0
VI	9.6	10.5	11.2	8.6
VII	6.8	7.7	11.8	4.6

[a] Mean percentage of immunoreactive rLH obtained as each isohormone.

From Keel, B. A. and Grotjan, H. E., Jr., *Endocrinology*, 117, 354, 1985. With permission.

animal leads to qualitative changes in the gonadotropins. Because gonadal manipulations, such as castration and cryptorchidism, cause marked changes in the hormonal milieu of the animal, it seems reasonable that the observed alterations in LH isohormone distribution may be under gonadal steroid regulation.

4. Effects of Steroid Administration
Although the majority of the studies performed to date would suggest that alterations in the hormonal milieu results in qualitative changes in LH, relatively little is known concerning the effects of steroid administration on the isohormones of LH. Chappel and co-workers[98] observed that long-term (30 hr) exposure of ovariectomized hamsters to estradiol caused an

Table 7
BIO-INDEX VALUES OF rLH IN
INTACT AND CASTRATE MALE AND
FEMALE RATS

pI	Male Intact	Male Castrate	Female Intact	Female Castrate
>9.80	8	8	11	21
9.25	2	26	5	39
9.23	3	43	5	54
9.17	4	41	3	50
9.06	3	39	2	13
8.97	1	16	2	13
<7.00	0.5	5	1	5

Note: Bio-Index, the distribution of biologically active
LH, is expressed as the amount of immunoreactive
LH (μg/4 pituitaries) multiplied by the B/I ratio
for each isohormone.

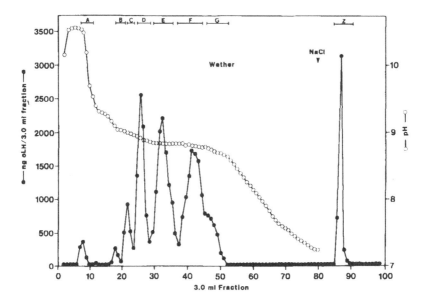

FIGURE 17.   Representative chromatofocusing elution profile of immunoreactive oLH in
the extract of an anterior pituitary from a wether. Brackets correspond to fractions pooled
to divide the profile into respective isohormones (designated with letters). (From Keel, B.
A., Schanbacher, B. D., and Grotjan, H. E., Jr., *Biol. Reprod.*, 36, 1102, 1987. With
permission.)

increase in the relative proportions of the more basic, more biologically active forms of
FSH. Similar results were obtained for FSH isohormones when ovariectomized monkeys
were treated with estradiol and progesterone.[99] With respect to LH, Wakabayashi[74] noted
that estradiol and progesterone administration to orchidectomized rats caused a marked
increase in one of the most acidic, least biologically active forms of rLH.

In an attempt to better understand the endocrine control mechanisms of LH isohormone
biosynthesis, we have examined the effects of steroid administration on the oLH isohormone
profiles. In this study,[70] wethers either received no steroid (castrate controls) or silastic

Table 8
BIO-INDEX VALUES OF
rLH ISOHORMONES
FROM INTACT AND
BILATERALLY
CRYPTORCHID RATS

pI	Intact	Cryptorchid
>9.80	2.68	12.25
9.25	0.34	1.02
9.23	0.79	0.43
9.17	0.93	0.63
9.06	0.62	0.48
8.97	0.36	0.16
<7.00	0.15	0.26

Note: Bio-Index, the distribution of biologically active LH, is expressed as the amount of immunoreactive LH (μg/pituitary) multiplied by the B/I ratio for each isohormone.

implants containing estradiol (E_2), dihydrotestosterone (DHT), or a combination of both steroids. Approximately 10 weeks after steroid implantation, the animals were sacrificed, the pituitaries removed, and the oLH isohormone profiles determined by chromatofocusing. Pituitary extracts yielded eight peaks of immunoreactive oLH with apparent pIs of >9.8, 9.26, 9.07, 8.98, 8.81, and <7.0. These isohormones were designated A to G and Z, respectively. As stated previously, castration in sheep did result in a gross redistribution of oLH isohormones (compare Figures 5 and 17). However, markedly different profiles of oLH emerged when pituitary extracts from steroid-implanted wethers were chromatofocused. DHT treatment resulted in a dramatic increase in the relative amount of isohormone A while isohormones B to D were essentially nondetectable (Figure 18). In contrast, treatment of wethers with E_2 resulted in a marked increase in the more acidic forms of oLH. Approximately 90% of the immunoreactive oLH from the pituitaries of E_2-treated wethers displayed apparent pIs <8.98 (Figure 19). Moreover, E_2 treatment resulted in a two- to fourfold increase in the relative amount of isohormone Z (apparent pI <7.00). Treatment of wethers with a combination of DHT and E_2 resulted in an isohormone profile characteristic of both steroid effects (Figure 20). That is, a large increase was observed in the relative amounts of isohormones A and Z while the relative amounts of isohormones B to D were reduced when compared to rams or untreated wethers.

The B/I ratios of the oLH isohormones present in pituitary extracts of rams are shown in Table 9. All eight isohormones possessed significant biological activity. Overall, castration resulted in a slight decrease in B/I ratios compared with rams. The B/I ratios of oLH isohormones were further reduced when wethers were treated with DHT or E_2, either separately or combined. Regardless of the treatment group, isohormone F consistently exhibited the highest B/I ratio. In general, treatment of wethers with steroids reduced the overall biologically active amounts of LH in isohormones B to G.

These observations, and the findings of Wakabayashi,[74] suggest that gonadal steroids can alter the distribution of pituitary LH among its isohormones. In particular, androgens and estrogens in sheep and progestins and estrogens in rats cause a shift in the distribution towards forms which had relatively low B/I ratios. Taken together, these results suggest that one of the negative feedback effects of gonadal steroids is to induce a qualitative change in

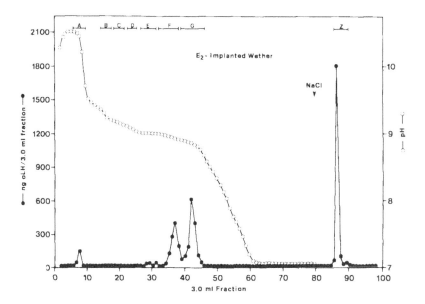

FIGURE 18.   Representative chromatofocusing elution profile of immunoreactive oLH in the extract of an anterior pituitary from a DHT-implanted wether. Brackets correspond to fractions pooled to divide the profile into respective isohormones (designated with letters). (From Keel, B. A., Schanbacher, B. D., and Grotjan, H. E., Jr., *Biol. Reprod.*, 36, 1102, 1987. With permission.)

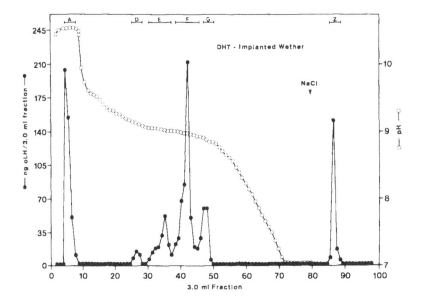

FIGURE 19.   Representative chromatofocusing elution profile of immunoreactive oLH in the extract of an anterior pituitary from an E_2-implanted wether. Brackets correspond to fractions pooled to divide the profile into respective isohormones (designated with letters). (From Keel, B. A., Schanbacher, B. D., and Grotjan, H. E., Jr., *Biol. Reprod.*, 36, 1102, 1987. With permission.).

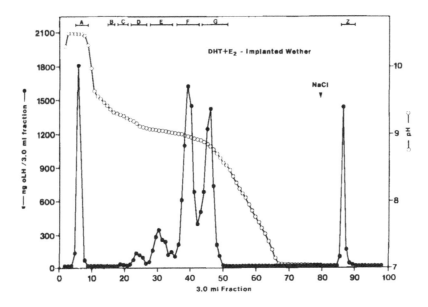

FIGURE 20. Representative chromatofocusing elution profile of immunoreactive oLH in the extract of an anterior pituitary from a DHT + E₂-implanted wether. Brackets correspond to fractions pooled to divide the profile into respective isohormones (designated with letters). (From Keel, B. A., Schanbacher, B. D., and Grotjan, H. E., Jr., *Biol. Reprod.*, 36, 1102, 1987. With permission.)

Table 9
B/I RATIOS OF oLH ISOHORMONES PRESENT IN
ANTERIOR PITUITARY EXTRACTS OF RAMS,
WETHERS, AND STEROID-IMPLANTED WETHERS

			Wethers implanted with		
Isorhomones	Rams	Wethers	DHT	E₂	DHT + E₂
A	0.40	0.29	0.58	0.20	0.42
B	0.94	0.80	ND[a]	0.13	0.19
C	1.00	0.98	ND	0.16	0.19
D	1.21	1.45	0.42	0.48	0.89
E	1.28	1.24	0.64	0.39	0.65
F	2.80	1.57	0.97	1.00	1.34
G	1.22	1.00	0.51	0.51	0.91
Z	0.81	0.43	0.20	0.23	0.32

[a] Not determined because this form was nondetectable.

From Keel, B. A., Schanbacher, B. D., and Grotjan, H. E., Jr., *Biol. Reprod.*, in press. With permission.

pituitary LH by shifting the distribution toward LH isohormones which have reduced B/I ratios.

E. Significance of "Acidic LH"

For the most part, LH is composed of isohormones characterized predominately by basic pIs. However, significant amounts of acidic LH have been described in several species and led to speculation concerning its physiological importance and biochemical basis. Thus, several studies have addressed the importance of acidic LH.

One possible explanation for the presence of acidic forms of LH is that they could partially result from the presence of LH subunits or from the other pituitary glycoproteins which focus in this region (e.g., see Figure 8). These other components could cross-react in the immunoassay but would not be active in a bioassay, resulting in forms with reduced B/I ratios. We have demonstrated that uncombined oLH alpha and beta subunits, partially purified from crude pituitary extracts, are composed predominately of acidic components (see Figures 11 and 12). Recent studies demonstrated an O-glycosylated form of the free alpha subunit, which contains sialic acid, is characteristically acidic, and will not combine with a free beta subunit to form the native hormone[112-115] (also see Chapter 3). Due to its acidic characteristics, it is conceivable that a significant portion of the free alpha subunit observed by us could be of the O-glycosylated form. What contribution this form may have toward the presence of acidic LH remains to be determined. It is important to note, however, that the acidic forms of LH do possess significant biological activity and, therefore, at least partially represent native hormone.

Suginami and co-workers[63,64,77] have proposed that the basic forms of LH might represent a stored, readily releasable form of the hormone. These forms of the hormone, although possessing maximal in vitro biological activity, are characterized by shorter circulatory half-lives. Therefore, under physiological conditions which favor a short-lived, highly biologically active signal (such as during the midcycle LH surge and acute GnRH exposure), the pituitary secretes basic forms of LH. In support of this theory, Suginami et al.[63,64,77] have demonstrated that the stored forms of LH which are readily released in response to acute GnRH exposure are composed of basic forms. The recent work of Baldwin et al.[82] also supports this conclusion. Conversely, acidic forms of LH, which possess a decreased in vitro biological activity but extended circulatory survival, may be released under physiological conditions which favor a long-lived albeit less potent signal.

Uchida and Suginami[77] have proposed an interesting theory of the development of the LH molecule from the comparative biology standpoint. These investigators note that the fish LH-like gonadotrophic hormone, similar to FSH and hCG, possesses pIs in the acidic region, while avian LH migrates as highly basic forms. On the other hand, LH from most developed mammals is distributed more or less over the entire pH range. These findings indicated to Uchida and Suginami[77] that the acidic forms of LH might represent the "primordicity of the hormone in terms of molecular development."

From the above studies and the experiments of others, several general conclusions can be drawn concerning acidic LH, although exceptions to these generalities exist: (1) acidic LH is absent from highly purified preparations and from the circulation of cycling women, (2) acidic forms of LH possess low in vitro biological activities and are characterized by decreased B/I ratios, (3) the isohormones of FSH, TSH, and PRL, as well as the free alpha subunit, coelectrofocus with acidic LH potentially resulting in erroneously elevated immunoreactive levels of LH, (4) acidic forms of LH are reduced during the midcycle surge of the hormone, and 5) steroid administration increases the proportion of acidic LH.

IV. BIOCHEMICAL BASIS FOR LH HETEROGENEITY

A variety of studies have suggested that the gonadotropin microheterogeneity results from differences in oligosaccharide structure. Peckham and Knobil[35] and Conn et al.[39] have shown that treatment of LH with the enzyme neuraminidase, which cleaves sialic acid from the carbohydrate moiety, either abolished or greatly diminished the observed size heterogeneity of the hormone. Treatment of human[60] and rat[81,82] LH with neuraminidase decreased the acidic isohormones of the hormone while concomitantly increasing the proportion of basic forms. These results suggest that rLH and hLH heterogeneity may be related to the number of sialic acid residues on the carbohydrate. The general carbohydrate structure of oLH and bLH suggests an analogous situation where the charge heterogeneity is related to the number of terminal sulfated hexosamines (see Chapter 2).

Grotjan and Cole[116] have recently investigated the role of oligosaccharides in the charge heterogeneity of human and ovine LH. Neuraminidase digestion was found to have a minimal effect on the isohormone profile of oLH suggesting essentially no role for sialic acid.[116] In contrast, the isohormone profile of hLH was markedly shifted from acidic to basic forms as a result of neuraminidase digestion, suggesting that the charge heterogeneity and pI of hLH is primarily due to terminal sialic acid residues.[116] Digestion of oLH with endoglycosidase F (endoF), an enzyme which specifically clips off the N-linked oligosaccharides, yielded only the very basic forms of both oLH and hLH.[116]

Although small deviations in peptide structure cannot be ruled out as an explanation for the existence of LH heterogeneity, the results presented above would strongly suggest that LH oligosaccharides play a primary role in the observed charge heterogeneity. The role of oligosaccharides in the biological activity of LH is not completely understood. Furthermore, the relationships between the animal's endocrine status, oligosaccharide structure, charge heterogeneity, and LH action remain to be elucidated.

V. CONCLUSIONS

LH is one of several pituitary glycoprotein hormones which exhibit both size and charge microheterogeneity. The existence of charge heterogeneity, and the presence of isohormones appears to be the most well documented and reproducible aspect of LH heterogeneity. LH isohormones from a number of species differ not only in charge but also in in vitro and in vivo biological activities. In most cases there is an apparent direct relationship between the charge on the LH variant and its inherent biological activity. Thus, physiological situations which shift the distribution of LH among its isohormones result in an alteration in the overall biological activity of the hormone. Variations in the endocrine status of the animal, such as age, sex, gonadal function, stage of cycle, or exogenous steroid and/or GnRH administration, directly affect the pattern of LH isohormones in the pituitary. Changes in the isohormone profile and thus the overall biological potency of LH can, likewise, dramatically affect gonadal activity. Therefore, it is apparent that the pituitary, in response to the hormonal milieu, alters not only the quantity or amount of LH produced but can also modulate the quality or biological activity of the hormone. The existence of LH microheterogeneity then represents an exquisite fine-tuning mechanism for controlling gonadal activity. With the elucidation of the biochemical basis for LH charge heterogeneity and the precise endocrine control of the biosynthesis of LH isohormones, we will better understand the control mechanisms of the hypothalamic-pituitary-gonadal axis.

ACKNOWLEDGMENTS

We would like to thank Drs. Reichert, Ward, and Niswender and the NIH Pituitary Hormone Distribution Program for radioimmunoassay reagents utilized in these studies. Excellent technical assistance has been provided by Robin Harms, Mame Stark, and Kathy Townley-Wren. Ovine pituitary tissue was generously provided by Pete's Locker Service, Centerville, S.D., Akron Frozen Food Center, Akron, Iowa, and Dr. Bruce D. Schanbacher, USDA Meat Animal Research Center, Clay Center, Neb. Studies from the author's laboratory work supported by Robert A. Welch Foundation Grant No. AU-931, NIH Grant HD-18879, the Parson's Endowment through the University of South Dakota School of Medicine and the Women's Research Institute, Wichita, Kan.

REFERENCES

1. **Payne, A. H., Quinn, P. G., and Stalvey, J. R. D.**, The stimulation of steroid biosynthesis by luteinizing hormone, in *Luteinizing Hormone Action and Receptors*, Ascoli, M., Ed., CRC Press, Boca Raton, Fla., 1985, chap. 4.

2. **Hunzicker-Dunn, M. and Birnbaumer, L.**, The involvement of adenylyl cyclase and cyclic AMP-dependent protein kinases in luteinizing hormone actions, in *Luteinizing Hormone Action and Receptors*, Ascoli, M., Ed., CRC Press, Boca Raton, Fla., 1985, chap. 3.

3. **Rabinowitz, D. and Roth, J.**, *Heterogeneity of Polypeptide Hormones*, Academic Press, New York, 1974.

4. **Dufau, N. L., Mendelson, C., and Catt, K. J.**, A highly sensitive in vitro bioassay of LH in human serum: the rat interstitial cell testosterone (RICT) assay, *J. Clin. Endocrinol. Metab.*, 42, 958, 1976.

5. **Dufau, N. L., Pock, R., Neribauh, A., and Catt, K. J.**, In vitro bioassay of LH in human serum: the rat interstitial cell testosterone (RICT) assay, *J. Clin. Endocrinol. Metab.*, 42, 958, 1976.

6. **Dufau, M. L., Beitins, I. L., McArthur, J. W., and Catt, K. J.**, Effects of luteinizing hormone releasing hormone (LHRH) upon bioactive and immunoactive serum LH levels in normal subjects, *J. Clin. Endocrinol. Metab.*, 43, 658, 1976.

7. **Beitins, I. Z., Dufau, M. L., Laughlin, K. O., Catt, K. J., and McArthur, S. W.**, Analysis of biological and immunological activities in the two pools of LH released during constant infusion of luteinizing hormone-releasing hormone (LHRH) in men, *J. Clin. Endocrinol. Metab.*, 45, 605, 1977.

8. **Dufau, M. L., Hodgen, G. D., Goodman, A. L., and Catt, K. J.**, Bioassay of circulating luteinizing hormone in the rhesus monkey: comparison with radio-immunoassay during physiological changes, *Endocrinology*, 100, 1557, 1977.

9. **Dufau, M., Beitins, I., McArthur, J., and Catt, K.**, Bioassay of serum LH concentrations in normal and LHRH-stimulated subjects, in *The Testis in Normal and Infertile Men*, Troen, P. and Nankin, H. R., Eds., Raven Press, New York, 1977, 309.

10. **Grotjan, H. E., Jr. and Steinberger, E.**, On the relative biological and immunonegative potencies of luteinizing hormone in the oestradial benzoate-treated male rat, *Acta Endocrinol. (Copenhagen)*, 89, 538, 1978.

11. **Lucky, A. W., Rebar, R. W., Rosenfield, R. L., Roche-Bender, B. A., and Helke, J.**, Reduction of the potency of luteinizing hormone by estrogen, *N. Engl. J. Med.*, 300, 1034, 1979.

12. **Solano, A. R., Dufau, M. L., and Catt, K. J.**, Bioassay and radioimmunoassay of serum luteinizing hormone in the male rat, *Endocrinology*, 105, 372, 1979.

13. **Solano, A. R., Garcia-Vela, A,. Catt, K. J., and Dufau, M. L.**, Modulation of serum and pituitary luteinizing hormone bioactivity by androgen in the rat, *Endocrinology*, 106, 1941, 1980.

14. **Dufau, M. L., Nozu, K., Dehejia, A., Garcia-Vela, A., Solano, R. R., Fraioli, F., and Catt, K. J.**, Biological activity and target all actions of luteinizing hormone, in *Pituitary Hormones and Related Peptides*, Motta, N., Zanisi, N., and Piva, F., Eds., Academic Press, New York, 1982, 117.

15. **Dufau, M. L., Veldhuis, J. D., Fraioli, F., Johnson, M. L., and Beitins, I. Z.**, Mode of secretion of bioactive luteinizing hormone in man, *J. Clin. Endocrinol. Metab.*, 57, 993, 1983.

16. **Veldhuis, J. D., Beitins, I. Z., Johnson, M. L., Serabian, M. A., and Dufau, M. L.**, Biologically active luteinizing hormone is secreted in episodic pulsations that vary in relation to stage of the mentrual cycle, *J. Clin. Endocrinol. Metab.*, 58, 1050, 1984.

17. **Burstein, S., Scaff-Blass, E., Blass, J., and Rosenfield, R. L.**, The changing ratio of bioactive to immunoreactive luteinizing hormone (LH) through puberty principally reflects changing LH radioimmunossay dose-response characteristics, *J. Clin. Endocrinol. Metab.*, 61, 508, 1985.

18. **Van Damme, M. P., Robertson, D. M., and Diczfalusy, E.**, An improved in vitro bioassay method for measuring luteinizing hormone (LH) activity using mouse Leydig cell preparations, *Acta Endocrinol.*, 77, 655, 1974.

19. **Mukhopadhyay, A. K., Leidenberger, F. A., and Lichtenberg, V.**, A comparison of bioactivity and immunoactivity of luteinizing hormone stored in and released in vitro from pituitary glands of rats under various gonadal states, *Endocrinology*, 104, 925, 1979.

20. **Storring, P. L., Zaidi, A. A., Misty, Y. G., Lindberg, M., Stemming, B. E., and Diczfalusy, E.**, A comparison of preparations of highly purified human pituitary luteinizing hormone: differences in the luteinizing hormone potencies as determined by in vivo bioassays, in vitro bioassay and immunoassay, *Acta Endocrinol.*, 101, 339, 1982.

21. **Celani, M. F., Montanini, V., Baraghini, G. F., Carani, C., and Marrama, P.**, Effects of acute stimulation with gonadotropin releasing hormone (GnRH) on biologically active serum luteinizing hormone (LH) in elderly men, *J. Endocrinol. Invest.*, 7, 589, 1984.

22. **Celani, M. F., Montanini, V., and Marrama, P.**, Effects of luteinizing hormone-releasing hormone (LRH) upon bioactive and immunoreactive serum LH in patients with Turner's syndrome before and after oestrogen treatment, *Acta Endocrinol.*, 109, 304, 1985.

23. **Schenken, R. S., Williams, R. F., Zowan, B. D., and Hodgen, G. D.,** Progesterone and 17α-hydroxy-progesterone advance the estrogen-induced bioassayable luteinizing hormone surge in castrate monkey, *Fertil. Steril.*, 43, 301, 1985.

24. **Cha, K. Y., Barnes, R. B., Marrs, R. P., and Lobo, R. A.,** Correlation of the bioactivity of luteinizing hormone in follicular fluid with oocyte maturity in the spontaneous cycle, *Fertil. Steril.*, 43, 338, 1986.

25. **Reichert, L. E., Jr. and Jiang, N. S.,** Comparative gel filtration and density gradient centrifugation studies on heterologous pituitary luteinizing hormones, *Endocrinology*, 77, 78, 1965.

26. **Ryan, R. J.,** On obtaining luteinizing hormone and follicle stimulating hormone from human pituitaries, *J. Clin. Endocrinol. Metab.*, 28, 886, 1968.

27. **Shome, B., Parlow, A. F., Ramirez, V. D., Elrick, H., and Pierce, J. G.,** Human and porcine thyrotropins: a comparison of electrophoretic and immunological properties with the bovine hormone, *Arch. Biochem. Biophys.*, 103, 444, 1968.

28. **Peckham, W. D. and Parlow, A. F.,** Isolation from human pituitary glands of three discrete electrophoretic components with high luteinizing hormone activity, *Endocrinology*, 85, 618, 1969.

29. **Rantham, P. and Saxena, B. B.,** Subunits of luteinizing hormone from human pituitary glands, *J. Biol. Chem.*, 246, 7087, 1971.

30. **Roos, P., Nyberg, L., Wide, L., and Gemzell, C.,** Human pituitary luteinizing hormone. Isolation and characterization of four glycoproteins with luteinizing activity, *Biochem. Biophys. Acta*, 405, 363, 1975.

31. **Deibel, N. D., Yamamoto, M., and Bogdanove, E. M.,** Discrepancies between radioimmunoassays and bioassays for rat FSH; evidence that androgen treatment and withdrawal can alter bioassay-immunoassay ratios, *Endocrinology*, 92, 1065, 1973.

32. **Bogdanove, E. M., Campbell, G. T., Blair, E. D., Mula, M. E, and Grossman, G. H.,** Gonad-pituitary feedback involved qualitative change: androgen alters the type of FSH secreted by the rat pituitary, *Endocrinology*, 95, 219, 1974.

33. **Bogdanove, E. M., Campbell, G. T., and Peckham, W. D.,** FSH pleomorphism in the rat-regulation by gonadal steroids, *Endocr. Res. Commun.*, 1, 87, 1974.

34. **Peckham, W. D., Yamaji, J., Dierschke, D. J., and Knobil, E.,** Gonadal function and the biological and physicochemical properties of follicle-stimulating hormone, *Endocrinology*, 92, 1660, 1973.

35. **Peckham, W. D. and Knobil, E.,** The effects of ovariectomy, estrogen replacement, and neuraminidase treatment on the properties of the adenohypophysial glycoprotein hormones of the rhesus monkey, *Endocrinology*, 98, 1054, 1976.

36. **Peckham, W. D. and Knobil, E.,** Qualitative changes in the pituitary gonadotropins of the male rhesus monkey following castration, *Endocrinology*, 98, 1061, 1976.

37. **Vaitukitis, J. L. and Ross, G. T.,** Altered biologic and immunologic activities of progressively desialylated human urinary FSH, *J. Clin. Endocrinol. Metab.*, 33, 308, 1971.

38. **Morell, A. G., Gregoriadis, G., Sheinberg, I. H., Hickman, J., and Ashwell, G.,** The role of sialic acid in determining the survival of glycoproteins in the circulation, *J. Biol. Chem.*, 246, 1461, 1971.

39. **Conn, P. M., Cooper, R., McNamara, C., Rogers, D. D., and Shoenhardt, L.,** Qualitative change in gonadotropin during normal aging in the male rat, *Endocrinology*, 106, 1549, 1980.

40. **Reddy, P. V. and Menon, K. M. J.,** Existence of multiple molecular forms of luteinizing hormone in rat: differences in immunological and biological activities between stored and circulating forms, *Acta Endocrinol.*, 97, 33, 1981.

41. **Lui, T. C., Ax, R. L., and Jackson, G. L.,** Characterization of luteinizing hormone synthesized and released by rat pituitaries in vitro: dissociation of immunological and biological activities, *Endocrinology*, 105, 10, 1979.

42. **Lui, T. C. and Jackson, G. L.,** Big luteinizing hormone (BLH): possible precurser of native LH (NLH) in anterior pituitary glands of rats, *Biol. Reprod.*, 24, 380, 1981.

43. **Chowdhury, M., Grotjan, H. E., Jr., and Steinberger, E.,** Further characterization of the molecular species formed during the biosynthesis of rat luteinizing hormone, *J. Endocrinol.*, 93, 169, 1982.

44. **Grotjan, H. E., Jr., Leveque, N. W., Berkowitz, A. S., and Keel, B. A.,** Quantitation of LH subunits released by rat anterior pituitary cells in primary culture, *Mol. Cell. Endocrinol.*, 35, 121, 1984.

45. **Kourides, I. A., Landon, M. B., Hoffman, B. J., and Weintraub, B. D.,** Excess free alpha relative to beta subunits of the glycoprotein hormones in normal and abnormal human pituitary *Clin. Endocrinol.*, 12, 407, 1980.

46. **Grotjan, H. E., Jr.,** unpublished data, 1986.

47. **Keel, B. A., Schanbacher, B. D., and Grotjan, H. E., Jr.,** Ovine luteinizing hormone. II. Effects of castration and steroid administration on the levels of uncombined subunits within the pituitary, *Biol. Reprod.*, 36, 1114, 1987.

48. **Reichert, L. E., Jr.,** Electrophoretic properties of pituitary gonadotropins as studied by electrofocusing, *Endocrinology*, 88, 1029, 1971.

49. **Graesslin, D., Trantwein, A., and Bettendorf, G.,** Gel electrofocusing of glycoprotein hormones, *J. Chromatogr.*, 63, 475, 1971.

50. **Maffezzoli, R. D., Kaplan, G. N., and Chrambach, A.**, Fractionation of immunoreactive human chorionic gonadotropin and luteinizing hormone by isoelectric focusing in polyacrylamide gel, *J. Clin. Endocrinol.*, 34, 361, 1972.
51. **Graesslin, D., Spies, A., Weise, H. C., and Bettendorf, G.**, Properties of human pituitary and urinary LH, *Acta Endocrinol. Suppl.*, 173, 56, 1973.
52. **Loeber, J. G.**, Human luteinizing hormone: structure and function of some preparations, *Acta Endocrinol. Suppl.*, 210, 1, 1977.
53. **Robertson, D. M., Van Damme, M. P., and Diczfalusy, E.**, Biological and immunological characterization of human luteinizing hormone. I. Biological profile in pituitary and plasma samples after electrofocusing, *Mol. Cell. Endocrinol.*, 9, 45, 1977.
54. **Van Damme, M. P., Robertson, D. M., and Diczfalusy, E.**, Biological and immunological characterization of human luteinizing hormone. III. Biological and immunological profiles of urine preparations after electrofocusing, *Mol. Cell. Endocrinol.*, 9, 69, 1977.
55. **Robertson, D. M. and Diczfalusy, D. M.**, Biological and immunological characterization of human luteinizing hormone. II. A comparison of the immunological and biological activities of pituitary extracts after electrofocusing using different standard preparations, *Mol. Cell. Endocrinol.*, 9, 57, 1977.
56. **Robertson, D. M., Fröysa, B., and Diczfalusy, E.**, Biological and immunological characterization of human luteinizing hormone. IV. Biological and immunological profile of two international reference preparations after electrofocusing, *Mol. Cell. Endocrinol.*, 11, 91, 1978.
57. **Zaidi, A. A., Qazi, M. H., and Diczfalusy, E.**, Molecular composition of human luteinizing hormone: biological and immunological profiles of highly purified preparations after electrofocusing, *J. Endocrinol.*, 94, 29, 1982.
58. **Strollo, F., Harlin, J., Hernandez-Montes, H., Robertson, D. M., Zaidi, A. A., and Diczfalusy, E.**, Qualitative and quantitative differences in the isoelectrofocusing profile of biologically active lutropin in the blood of normally menstruating and post-menopausal women, *Acta Endocrinol.*, 97, 166, 1981.
59. **Reader, S. D. J., Robertson, W. R., and Diczfalusy, E.**, Microheterogeneity of luteinizing hormone in pituitary glands from women of pre- and postmenopausal age, *Clin. Endocrinol.*, 19, 355, 1983.
60. **Weise, H. C., Graesslin, D., Lichtenberg, V., and Rinne, G.**, Polymorphism of human pituitary lutropin (LH): isolation and partial characterization of seven isohormones, *FEBS Lett.*, 159, 93, 1983.
61. **Lichtenberg, V., Weise, H. C., Graesslin, D., and Bettendorf, G.**, Polymorphism of human pituitary lutropin (LH): effect of the seven isohormones on mouse Leydig cell functions., *FEBS Lett.*, 169, 1, 1984.
62. **Hartree, A. S., Lester, J. B., and Shownkeen, R. C.**, Studies of the heterogeneity of human pituitary LH by fast protein liquid chromatography., *J. Endocrinol.*, 105, 405, 1985.
63. **Suginami, H., Yano, M., Hamada, K., Ito, T., Yano, K., and Matsuura, S.**, Qualitative and quantitative differences in hLH release induced by continuous stimulation with synthetic LHRH in normal menstrual cycle as assessed by isoelectrofocusing, *Endocrinol. Jpn.*, 32, 583, 1985.
64. **Hamada, K. and Suginami, H.**, Qualitative and quantitative changes in plasma luteinizing hormone (LH) under stimulation by intravenous infusion of synthetic luteinizing hormone-releasing hormone (LHRH) in Japanese monkeys *(Macaca fuscata)* as assessed by electrofocusing., *Endocrinol. Jpn.*, 30, 101, 1983.
65. **Braselton, W. E. and McShan, W. H.**, Purification and properties of follicle-stimulating and luteinizing hormones from horse pituitary glands. *Arch. Biochem. Biophys.*, 139, 45, 1970.
66. **Matteri, R. L. and Papkoff, H.**, Analysis of equine LH isoforms by chromatofocusing, *Endocrinology*, 118 (Suppl.), 158, 1986.
67. **Yoshida, T. and Ishii, S.**, Electrofocusing of partially purified bovine luteinizing hormone, *Endocrinol. Jpn.*, 6, 629, 1973.
68. **Sherwood, O. D., Grimek, H. J., and McShan, W. H.**, Purification of luteinizing hormone from sheep pituitary glands and evidence for several physicochemically distinguishable active components, *Biochim. Biophys. Acta*, 221, 87, 1970.
69. **Keel, B. A. and Grotjan, H. E., Jr.**, Charge heterogeneity of partially purified ovine LH subunits and the native hormone, in *Proc. 18th Miami Winter Symp.*, *Advances in Gene Technology: Molecular Biology of the Endocrine System*, Vol. 4, Puett, D., Ahmad, F., Black, S., Lopez, D., Melner, M., Scott, W., and Whelan, W., Eds., Cambridge University Press, Cambridge, 1986, 224.
70. **Keel, B. A., Schanbacher B. D., and Grotjan, H. E., Jr.**, Ovine luteinizing hormone. I. Effects of castration and steroid administration on the charge heterogeneity of pituitary luteinizing hormone, *Biol. Reprod.*, 36, 1102, 1987.
71. **Keel, B. A. and Grotjan, H. E., Jr.**, Ovine luteinizing hormone. III. Relationships between the charge heterogeneity of partially purified subunits and the native hormone, *Biol. Reprod.*, 36, 1125, 1987.
72. **Keel, B. A. and Grotjan, H. E., Jr.**, unpublished observations.
73. **Kercret, H. and Duval, J.**, Etude par electrofocalisation des gonadotropins hypophysaires du rat, *Biochimie*, 57, 85, 1975.
74. **Wakabayashi, K.**, Heterogeneity of rat luteinizing hormone revealed by radioimmunoassay and electrofocusing studies, *Endocrinol. Jpn.*, 24, 473, 1977.

75. **Robertson, D. M., Foulds, L. M., and Ellis, S.,** Heterogeneity of rat pituitary gonadotropins on electrofocusing; differences between sexes and after castration, *Endocrinology,* 111, 385, 1982.
76. **Hattori, M., Sakamoto, K., and Wakabayashi, K.,** The presence of LH components having different ratios of bioactivity to immunoreactivity in the rat pituitary glands, *Endocrinol. Jpn.,* 30, 289, 1983.
77. **Uchida, H. and Suginami, H.,** Subpopulations of luteinizing hormone (LH) possessing various ratios of bioactivity to immunoreactivity in the female rat pituitary glands and their changes during the estrous cycle, *Endocrinol. Jpn.,* 31, 605, 1984.
78. **Keel, B. A. and Grotjan, H. E., Jr.,** Characterization of rat lutropin charge microheterogeneity using chromatofocusing, *Anal. Biochem.,* 142, 267, 1984.
79. **Keel, B. A. and Grotjan, H. E., Jr.,** Influence of bilateral cryptorchidism on rat pituitary luteinizing hormone charge microheterogeneity, *Biol. Reprod.,* 32, 83, 1985.
80. **Keel, B. A. and Grotjan, H. E., Jr.,** Characterization of rat pituitary luteinizing hormone charge microheterogeneity in male and female rats using chromatofocusing: effects of castration, *Endocrinology,* 117, 354, 1985.
81. **Hattori, M, Ozawa, K., and Wakabayashi, K.,** Sialic acid moiety is responsible for the charge heterogeneity and the biological potency of rat lutropin, *Biochem. Biophys. Res. Commun.,* 127, 501, 1985.
82. **Baldwin, D. M., Highsmith, R. F., Ramey, J. W., and Krummen, L. A.,** An in vitro study of LH release, synthesis and heterogeneity in pituitaries from proestrous and short-term ovariectomized rats, *Biol. Reprod.,* 34, 304, 1986.
83. **Hattori, M. and Wakabayashi, K.,** Isoelectric focusing and gel filtration studies on the heterogeneity of avian pituitary luteinizing hormone, *Gen. Comp. Endocrinol.,* 39, 215, 1979.
84. **Hattori, M. and Wakabayashi, K.,** Different profiles of isoelectric avian luteinizing hormone components in biological activity and immunoreactivity, *Endocrinol. Jpn.,* 30, 551, 1983.
85. **Tamura-Takahashi, H. and Ui, N.,** Purification and properties of four biologically active components of whale luteinizing hormone, *J. Biochem.,* 81, 1155, 1977.
86. **Chappel, S. C., Ulloa-Aguirre, A., and Coutifaris, C.,** Biosynthesis and secretion of follicle-stimulating hormone, *Endocr. Rev.,* 4, 179, 1983.
87. **Mori, M., Ohshima, K., Fukuda, H., Kobayashi, I., and Wakabayashi, K.,** Changes in the multiple components of rat pituitary TSH and TSH β subunit following thyroidectomy, *Acta Endocrinol.,* 105, 49, 1984.
88. **Ishikawa, J., Wakabayashi, K., and Igarashi, M.,** Charge heterogeneity of rat pituitary prolactin in relation to the estrous cycle, gonadectomy, and estrogen treatment, *Endocrinol. Jpn.,* 32, 725, 1985.
89. **Chappel, S. C. and Ramaley, J. A.,** Changes in the isoelectric focusing profile of pituitary follicle-stimulating hormone in the developing male rat, *Biol. Reprod.,* 32, 567, 1985.
90. **Foulds, L. M. and Robertson, D. M.,** Electrofocusing fractionation and characterization of pituitary follicle-stimulating hormone from male and female rats, *Mol. Cell. Endocrinol.,* 31, 117, 1983.
91. **Foulds, L. M. and Robertson, D. M.,** Electrofocusing fractionation of follicle-stimulating hormone in pituitary cell culture extracts from male and female rats, *Mol. Cell. Endocrinol.,* 41, 129, 1985.
92. **Chappel, S. C., Ulloa-Aguirre, A., and Ramaley, J. A.,** Sexual maturation in female rats: time-related changes in the isoelectric focusing pattern of anterior pituitary follicle-stimulating hormone, *Biol. Reprod.,* 28, 196, 1983.
93. **Ulloa-Aguirre, A., Miller, C., Hyland, L., and Chappel, S.,** Production of all follicle-stimulating hormone isohormones from a purified preparation by neuraminidase digestion, *Biol. Reprod.,* 30, 382, 1984.
94. **Kennedy, J. and Chappel, S.,** Direct pituitary effects of testosterone and luteinizing hormone-releasing hormone upon follicle-stimulating hormone: analysis of radioimmuno- and radioreceptor assay, *Endocrinology,* 116, 741, 1985.
95. **Blum, W. F. P. and Gupta, D.,** Heterogeneity of rat FSH by chromatofocusing: studies on serum FSH, hormone released in vitro and metabolic clearance rates of its various forms, *J. Endocrinol.,* 105, 29, 1985.
96. **Blum, W. F. P., Riegelbauer, G., and Gupta, D.,** Heterogeneity of rat FSH by chromatofocusing: Studies on in vitro bioactivity of pituitary FSH forms and effect of neuraminidase treatment, *J. Endocrinol.,* 105, 17, 1985.
97. **Cameron, J. L. and Chappel, S. C.,** Follicle-stimulating hormone within and secreted from anterior pituitaries of female golden hamsters during the estrous cycle and after ovariectomy, *Biol. Reprod.,* 33, 132, 1985.
98. **Galle, P. C., Ulloa-Aguirre, A., and Chappel, S. C.,** Effects of oestradiol, phenobarbitone and luteinizing hormone releasing hormone upon the isoelectric profile of pituitary follicle-stimulating hormone in ovariectomized hamsters, *J. Endocrinol.,* 99, 31, 1983.
99. **Chappel, S. C., Bethea, C. L., and Spies, H. G.,** Existence of multiple forms of follicle-stimulating hormone with the anterior pituitaries of cynomolgus monkeys, *Endocrinology,* 115, 452, 1984.
100. **Jackson, J. H., Jr. and Russel, P. J., Jr.,** Characteristics of the isoelectric focusing procedure: importance of column size, pH, and protein-protein interactions, *Anal. Biochem.,* 137, 41, 1984.

101. Chromatofocusing with Polybuffer and PBE, Pharmacia Fine Chemicals, Uppsala.
102. **Ulloa-Aguirre, A., Countifaris, C., and Chappel, S. C.**, Multiple species of FSH are present within hamster anterior pituitary cells cultured in vitro, *Acta Endocrinol.*, 102, 343, 1983.
103. **Fagerstam, L. G., Lizana, J., Axio-Fredriksson, U. B., and Wahlstrom, L.**, Fast chromatofocusing of human serum proteins with special reference to α_1-Antitrypsin and Gc-Globulin, *J. Chromatogr.*, 266, 523, 1983.
104. **Liu, W. K., Bousfield, G. R., Moore, W. T., Jr., and Ward, D. N.**, Priming procedure and hormone preparations influence rat granulosa cell response, *Endocrinology*, 116, 1454, 1985.
105. **Wide, L. and Hobson, B.**, Influence of the assay method used on the selection of the most active forms of FSH from the human pituitary, *Acta Endocrinol.*, 113, 17, 1986.
106. **Marut, E. L., Williams, R. F., Cowan, B. D., Lynch, A., Lerner, S. P., and Hodgen, G. D.**, Pulsatile pituitary gonadotropin secretion during maturation of the dominant follicle in monkeys: estrogen positive feedback enhances the biological activity of LH, *Endocrinology*, 109, 2270, 1981.
107. **Schenken, R. S., Werlin, L. B., Williams, R. F., Prihoda, T. J., and Hodgen, G. D.**, Periovulatory hormonal dynamics: relationship of immunoassayable gonadotropins and ovarian steroids to the bioassayable luteinizing hormone surge in Rhesus monkeys, *J. Clin. Endocrinol. Metab.*, 60, 886, 1985.
108. **Solano, A. R., Garcia-Vela, A., Catt, K. J., and Dufau, M. L.**, Regulation of ovarian gonadotropin receptors and LH bioactivity during the estrous cycle, *FEBS Lett.*, 122, 184, 1980.
109. **Veldhuis, J. D., Rogol, A. D., Johnson, M. L., and Dufau, M. L.**, Endogenous opiates modulate the pulsatile secretion of biologically active luteinizing hormone in man, *J. Clin. Invest.*, 72, 2031, 1983.
110. **Beitins, I. Z., Dufau, M. L., O'Loughlin, D., Bercu, B. B., McArthur, J. W., Crawford, J. D., and Catt, K. J.**, Biological and immunological activities of serum LH in normal women during LH-RH infusion and in Turner's syndrome during estrogen treatment, in *The LH-Releasing Hormone*, Beling, C. G. and Wentz, A. C., Eds., Masson, New York, 1980, 135.
111. **Sawyer-Steffan, J. E., Lasley, B. L., Hoff, J. D., and Yen, S. S. D.**, Comparison of in vitro bioactivity and immunoactivity of serum LH in normal cyclic and hypogonadal women treated with low doses of LH-RH, *J. Reprod. Fertil.*, 65, 45, 1982.
112. **Parsons, T. F., Bloomfield, G. A., and Pierce, J. G.**, Purification of an alternative form of the subunit of the glycoprotein hormones from bovine pituitaries and identification of its O-linked oligosaccharide, *J. Biol. Chem.*, 258, 240, 1983.
113. **Parsons, T. F. and Pierce, J. G.**, Free α-like material from bovine pituitaries: removal of its O-linked oligosaccharide permits combination with lutropin-β, *J. Biol. Chem.*, 259, 2662, 1984.
114. **Cole, L. A., Perini, F., Birken, S., and Ruddon, R. W.**, An oligosaccharide of the O-linked type distinguishes the free from the combined form of hCG a subunit, *Biochem. Biophys. Res. Commun.*, 122, 1260, 1984.
115. **Corless, C. L. and Boime, I.**, Differential secretion of O-glycosylated gonadotropin α-subunit and luteinizing hormone (LH) in the presence of LH-releasing hormone, *Endocrinology*, 117, 1699, 1985.
116. **Cole, L. A., Metsch, L. S., and Grotjan, H. E., Jr.**, The steroidogenic activity of LH is maintained following enzymatic removal of oligosaccharides, *Mol. Endocrinol.*, 1, 621, 1987.

Chapter 9

EQUINE LUTEINIZING HORMONE MICROHETEROGENEITY

Robert L. Matteri and Harold Papkoff

TABLE OF CONTENTS

I. INTRODUCTION

The basic biochemical properties of equine pituitary gonadotropins have been well established.[1-6] As in other species, these hormones, as well as thyrotropin (TSH), are glycoproteins, composed of two dissimilar, noncovalently bound alpha and beta subunits. The alpha subunits of luteinizing hormone (LH), follicle-stimulating hormone (FSH), and TSH have identical peptide structures within a single species.[7] In several species, only one gene coding for the alpha subunit of gonadotropins and TSH has been found.[8-10] Although the alpha subunit is required for biological activity, it is the beta subunit which varies and confers the hormonal specificity of these hormones.

It is well known that glycoprotein hormones such as the gonadotropins possess considerable microheterogeneity. Several forms of FSH with varying properties have been reported in rats,[11-13] hamsters,[14] donkeys,[15] horses,[16] monkeys,[17,18] baboons,[19] and humans.[20,21] Several variants of equine[22-24] and human[25-27] chorionic gonadotropin are also known (eCG and hCG, respectively). Likewise, heterogeneity of LH is found to occur in amphibians,[28] birds,[29] rats,[13,30,31] sheep,[32] monkeys,[33] and humans.[34-37]

This microheterogeneity reflects the existence of multiple forms (isohormones, isoforms) of glycoprotein hormones, some of which may differ in isoelectric properties. The occurrence of glycoprotein hormone isoforms may be of physiological importance. Gonadotropin isohormones have been reported to differ from one another in their biological potencies[30,38-40] although in no case have the isohormones been isolated in pure form for evaluation. Furthermore, the relative abundance of one isohormone over another appears to be under endocrine regulation. The isohormone forms of gonadotropins in the pituitary can vary between sexes[12,13,41] and are affected by treatment with sex steroids,[11,33,42] gonadectomy,[13,18,33,43] cryptorchidism,[44] stage of sexual development,[35,45,46] and phase of the reproductive cycle.[37] The existence of such regulation may enable an animal to selectively vary the biopotency of secreted gonadotropins according to reproductive needs.

II. MICROHETEROGENEITY OF EQUINE LH

The biochemical and biological properties of equine luteinizing hormone (eLH) have been of great interest in the study of the gonadotropic hormones. Characteristic of this hormone is its high carbohydrate content, which includes sialic acid, a sugar not present in LH from many species.[6] Additionally, the beta subunit of eLH has a C-terminal amino acid extension which is similar to that found in hCG and eCG.[2] In nonequid species, eLH is extremely potent in LH activity[5,6] and exerts considerable intrinsic FSH activity.[47-49] In light of these interesting properties, the characterization of naturally occurring variants of eLH, such as different isohormone forms, could provide basic information pertaining to the general relationship between the biochemical structure and the biological function of gonadotropins.

Only very recently have studies documented the characterization of specific isohormones of LH from the horse. The microheterogeneity of eLH, however, has long been known. Early reports describing the purification of eLH showed that the prepared hormone did not focus into a discrete band by isoelectric focusing. Isoelectric ranges of 4.5 to 7.3[50] and 3.8 to 8.9[51] were reported in these early studies. As might be expected, fractionation of equine pituitary glycoprotein extracts by ion-exchange chromatography shows a wide spread of LH activity rather than resolution into a single, discrete peak.[52] These data are reflective of the charge microheterogeneity of this hormone and indicate that various forms of eLH can be separated by fractionation methods which discriminate between isoelectric properties.

The charge heterogeneity of eLH has recently been used to prepare three highly purified eLH preparations by anion-exchange chromatography, which differed in apparent negative charge.[52] Figure 1 shows the DEAE elution pattern of the equine pituitary glycoprotein extract

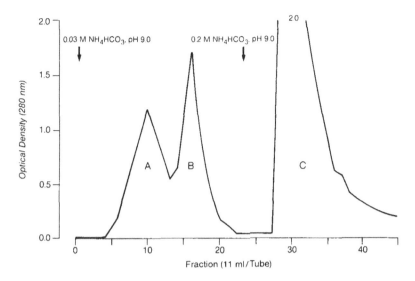

FIGURE 1. DEAE chromatography of glycoprotein extract obtained from horse pituitary glands. Protein fractions A, B, and C were subsequently purified to yield LH preparations eLH-A, -B, and -C. (From Matteri, R. L., Papkoff, H., Ng, D. A., Swedlow, J. R., and Chang, Y.-S., *Biol. Reprod.*, 34, 571, 1986. With permission.)

Table 1
PROPERTIES OF eLH PREPARATIONS

LH preparation	eLH-A	eLH-B	eLH-C
Hexose[a]	9.0	9.2	9.0
Hexosamine[a]	8.8	9.1	11.2
Sialic acid[a]	4.5	6.7	6.6
Total carbohydrate[a]	22.3	25.0	26.8
Ve/Vo[b]	1.72	1.54	1.47
R_F[c]	0.14	0.19	0.26

[a] Percent carbohydrate by weight.
[b] Ve/Vo ratios on Sephadex® G-100.
[c] Relative electrophoretic mobilities on polyacrylamide disc gels.

From Matteri, R. L., Papkoff, H., Ng, D. A., Swedlow, J. R., and Chang, Y.-S., *Biol. Reprod.*, 34, 571, 1986. With permission.

obtained in this study. The three protein peaks obtained from this column were further purified by cation exchange (sulfopropyl-Sephadex®) and size exclusion (Sephadex® G-100) chromatography. The final products of this purification were designated eLH-A, -B, and -C, corresponding to protein peaks A, B, and C from the DEAE chromatography (Figure 1).

Chemical analysis of these hormone fractions revealed relatively minor differences in carbohydrate content (Table 1). All three preparations contained similar amounts of hexose. Equine LH-B and -C possess an equivalent amount of sialic acid, which is greater than that detected in eLH-A. Preparation eLH-C has slightly more hexosamine than the other two fractions. These results are in agreement with the general consensus that charge heterogeneity may reflect differences in glycosylation. Peckham and Knobil[33] have shown that the size differences in gonadotropins produced between intact and ovariectomized rhesus monkeys

Table 2
RELATIVE POTENCIES[a] OF THE eLH PREPARATIONS

LH preparation	E98A	eLH-A	eLH-B	eLH-C
LH bioassay[b]	1.0	1.19 ± 0.05[c]	0.99 ± 0.11	1.24 ± 0.13
		(6)	(6)	(5)
LH RRA[b]	1.0	1.81 ± 0.22[c]	1.49 ± 0.15[c]	1.46 ± 0.14[c]
		(5)	(5)	(5)
LH RIA[b]	1.0	0.91 ± 0.04	0.85 ± 0.04[c]	0.77 ± 0.06[c]
		(6)	(6)	(6)
FSH RRA[c]	1.13 ± 0.25	1.37 ± 0.19	1.46 ± 0.14	1.74 ± 0.11
	(4)	(5)	(5)	(4)
FSH RIA[d]	0.003	0.006	0.015	0.043

[a]   Mean ± SEM, (n).
[b]   Potencies relative to equine LH standard (E98A, from this laboratory).
[c]   Potencies relative to ovine FSH (G4-211B, from this laboratory).
[d]   Potencies relative to equine FSH (E276B, from this laboratory).
[e]   Different from equine LH reference standard (eLH preparation E98A, $p < 0.05$).

From Matteri, R. L., Papkoff, H., Ng, D. A., Swedlow, J. R., and Chang, Y.-S., *Biol. Reprod.*, 34, 571, 1986. With permission.

can be attenuated by treating pituitary extracts with neuraminidase. Likewise, incubation of a purified rat FSH preparation with neuraminidase was shown to produce all of the FSH isoforms found in rat pituitary gland extracts.[53] Cole and Grotjan[54] also have reported a marked reduction in microheterogeneity of human and ovine LH by treatment with endoglycosidase F, an enzyme which will cleave all of the oligosaccharides from these glycoproteins except the asparagine-linked *N*-acetylglucosamine residue. Since there is evidence which suggests that there is a single gene for the alpha subunit of LH, FSH, and TSH,[8-10] and it is also thought that only one gene for the beta subunits of each of these hormones exists,[55,56] the microheterogeneity of these glycoprotein hormones probably results from posttranslational modifications.

Different Ve/Vo ratios observed with size exclusion chromatography (Sephadex® G-100) indicated variations in apparent size among the three hormone preparations (eLH-C > -B > -A; Table 1). Although the basis of this observation is not certain, the differences found in the carbohydrate composition of the isohormones may have contributed to the variations in apparent size.[57] As also shown in Table 1, the amounts of total carbohydrate found in these hormone preparations correspond to the apparent size differences (eLH-C > -B > -A).

In order to assess potential differences in biological and immunological properties, these LH fractions were analyzed in several in vitro assay systems. Luteinizing hormone bioactivity was determined in a dispersed rat Leydig cell assay,[58] where LH standards or samples were incubated in a cell suspension and subsequent testosterone production was assessed by radioimmunoassay (RIA). The LH radioreceptor assay (RRA) was based on a membrane preparation from horse testes.[15] Finally, the LH RIA used in this study was a previously established equine LH RIA.[59] The results of these assays are shown in Table 2. Significant variations in LH bioactivity and immunological activities were detected among the three eLH fractions and a previously prepared reference standard, while only relatively minor differences among eLH-A, -B, -C were observed.

As stated above, in addition to its LH activity, eLH also exerts FSH activity in non-equid species. Accordingly, the FSH receptor-binding potency in a calf testis membrane fraction was also tested.[52] As expected from the intrinsic FSH activity of eLH, the residual FSH

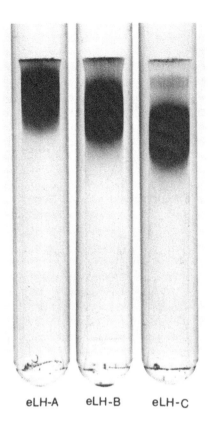

eLH-A   eLH-B   eLH-C

FIGURE 2.   Polyacrylamide disc gel electrophoresis (pH 8.3) of eLH-A,
-B, and -C. (From Matteri, R. L., Papkoff, H., Ng, D. A., Swedlow,
J. R., and Chang, Y.-S., *Biol. Reprod.*, 34, 571, 1986. With permission.)

contamination determined by RIA was extremely low compared to the FSH RRA potency. However, no significant differences in calf testis FSH receptor-binding potency were observed among any of the LH preparations.

Analysis of these eLH fractions by disc gel electrophoresis revealed clear differences in average charge characteristics, but still showed much overlap in the protein bands (Figure 2). This suggested that three groups of isohormone families had been isolated, with many of the same isoforms present, in varying degrees, in all three hormone preparations. This being the case, differences between the biological and immunological properties of these LH fractions might not be detectable in the assay systems used for characterization, even though significant variations might occur between specific isohormones. At this point, it became evident that a separation method with a high power of resolution, selective for differences of isoelectric properties, was needed in order to continue the study of eLH microheterogeneity.

Accordingly, the method of chromatofocusing was utilized to examine further the properties of eLH-A, -B, and -C. This method is based upon the use of a specialized ion-exchange resin and buffer system which generates a pH gradient in a column without the use of a gradient-mixing apparatus. As the pH of the column changes, components of the sample elute according to their isoelectric properties. This technique had previously been used to characterize the isohormone composition of rat LH[30,44] and ovine LH,[32] FSH,[61] and hCG.[62]

Chromatofocusing of eLH-A, -B, and -C (Figure 3) demonstrated the existence of five major isohormone peaks with apparent pIs of 6.6, 6.1, 5.7, 5.2, and 4.8. Luteinizing hormone activity was also found on either side of the pH gradient, indicating the presence of iso-

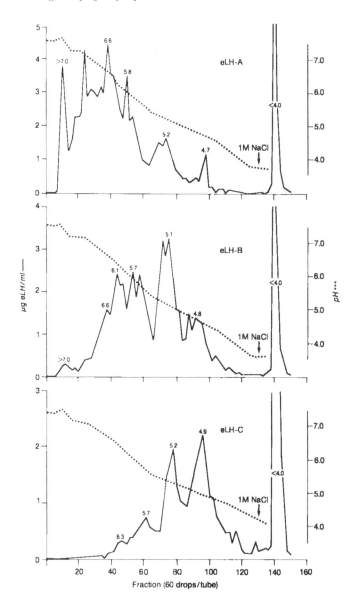

FIGURE 3.   Chromatofocusing elution patterns of eLH-A, -B, and -C. Solid lines indicate LH immunoactivity and dashed lines show the pH gradient. At the end of the gradient, 1 *M* NaCl was used to remove the remaining LH (pI <4.0) from the column. (From Matteri, R. L. and Papkoff, H., *Biol. Reprod.*, 36, 261, 1987. With permission.)

hormones with pI values >7.0 and <4.0. As expected from the initial derivation by DEAE chromatography and subsequent analysis by disc gel electrophoresis, eLH-A contained the largest amount of basic isohormones. Equine LH-C had mostly acidic material and eLH-B possessed a predominance of isohormones with pIs intermediate to the other two fractions (Table 3). Thus the initial separation of these eLH preparations by anion-exchange chromatography had produced three fractions which contained an assortment of different isohormones. The predominant forms reflected the overall charge properties of each preparation, but there was isohormone overlap between the hormone fractions.

Since it is known that sialic acid is an important terminal sugar in eLH, we were interested

Table 3
ACTIVITY RATIOS[a] OF eLH ISOHORMONES

pI	LHBio/ LHI	LHRRA/ LHI[b]	LHBio/ LHRRA	FSHRRA/ LHI[b]	FSHRRA/ LHRRA
>7.0[c]	0.70	1.49	0.51	2.23	1.63
6.6	0.84	1.20	0.80	1.85	1.81
	(0.07,2)	(0.38,2)	(0.27,2)	(0.36,2)	(0.25,2)
6.1	1.03	1.27	0.80	1.82	1.39
	(0.01,2)	(0.14,2)	(0.13,2)	(0.06,2)	(0.17,2)
5.7	1.11	1.41	0.89	1.52	1.65
	(0.16,3)	(0.25,2)	(0.09,2)	(0.50,2)	(0.49,2)
5.2	1.25	1.37	0.92	1.84	1.46
	(0.38,3)	(0.09,3)	(0.29,3)	(0.36,3)	(0.34,3)
4.8	1.13	0.99	1.14	1.48	1.51
	(0.18,2)	(0.14,2)	(0.02,2)	(0.03,2)	(0.19,2)
<4.0	0.74	0.59	1.28	0.85	1.49
	(0.04,3)	(0.09,3)	(0.09,3)	(0.03,3)	(0.19,3)

[a] Ratio (SD, n), where n = values from separate columns. Purified eLH was used as the reference standard in all assays. LHBio = potency in a dispersed rat Leydig cell LH bioassay: LHRRA = activity in an equine testis LH RRA; FSHRRA = activity in a calf testis FSH RRA; LHI = potency in an equine LH RIA.

[b] Significant variations ($p < 0.05$, ANOVA) between isohormones were detected for these ratios.

[c] Data from one column, since the amount of hormone eluted at pI >7.0 was great enough to measure accurately only in eLH-A. This datum point could not be considered a true cell mean, and therefore was excluded from statistical analysis by ANOVA.

From Matteri, R. L. and Papkoff, H., *Biol. Reprod.*, 36, 261, 1987. With permission.

in the contribution of this carbohydrate to the isoform composition of eLH. To investigate this problem, we incubated eLH-C, the most acidic preparation, with neuraminidase. This enzyme selectively removes sialic acid moieties. Compared to nontreated eLH-C, the enzyme-treated hormone had a chromatofocusing profile containing much more basic material, with fewer peaks (Figure 4). While it appears that sialic acid can significantly influence the isohormone composition of eLH, it remains to be determined whether the individual eLH forms actually differ in sialic acid content. It is likely, however, that other factors, besides sialic acid, affect the microheterogeneity of eLH.

III. EQUINE LH SUBUNITS

The relative contribution of the alpha and beta subunits of eLH to the overall microheterogeneity of the intact hormone is also of interest. Partial information concerning this problem was obtained by sodium dodecyl sulfate-polyacrylamide gel electrophoresis (SDS-PAGE, Figure 5). Under reducing conditions, the alpha and beta subunits dissociate and migrate towards the positive electrode at a rate inversely proportional to their molecular weight. The alpha subunits have the typical smeary appearance of glycoproteins on polyacrylamide gels and appear similar among eLH-A, -B, and -C. The beta subunits possess considerably more heterogeneity, as seen by the width of the bands, than the alpha subunits. While the beta subunits appear very similar in apparent size, these subunit bands show a tendency to extend into higher molecular weight ranges in proportion to the increasing

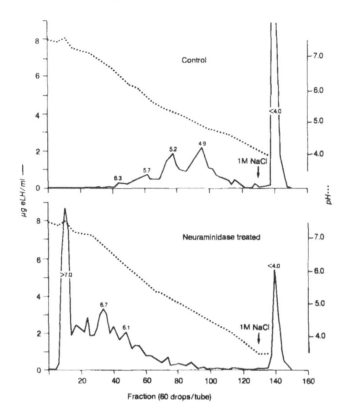

FIGURE 4.   Chromatofocusing of eLH-C before (top) and after (bottom) neuraminidase treatment. (From Matteri, R. L. and Papkoff, H., *Biol. Reprod.*, in press. With permission.)

apparent negative charge of the preparation (eLH-C > -B > -A). While the differences are not great, this pattern can be consistently reproduced in such gels and substantiates the slight size differences of the three eLH preparations indicated by size exclusion chromatography (Table 1).

In addition to differences in size, the isoelectric microheterogeneity of the alpha and beta subunits of eLH was also of interest. Each of the subunits derived from eLH-A, -B, and -C was analyzed by chromatofocusing, using the same pH gradient as previously utilized for the intact hormones. The concentrations of each subunit preparation in the chromato-focusing samples were determined by specific subunit radioimmunoassays.

Consistent with the SDS-PAGE analysis, the alpha subunits of eLH-A, -B and -C all produced similar chromatofocusing patterns (Figure 6), with virtually all of the alpha subunit immunoactivity eluting at basic pI values. In contrast, the beta subunit preparations showed a large distribution of isoforms (Figure 7). Furthermore, there was excellent correspondence between the isoelectric values of the isoforms of the beta subunits and the intact hormones from which they were derived. However, a comparison of Figures 3 and 7 shows that the relative amounts of material eluting at corresponding isoelectric ranges were not precisely the same between the beta subunits and their parent preparations. This could reflect a loss of some material during the process of subunit isolation. The possibility also exists that the alpha subunit also exerts some influence on the overall charge properties of the alpha-beta dimer. Regardless, the beta subunit may impart most of the isoelectric properties of the intact hormone. Similar data exist which show that the charge microheterogeneity of hCG is reflected in the beta subunit.[63] Since it is thought that glycoprotein microheterogeneity is

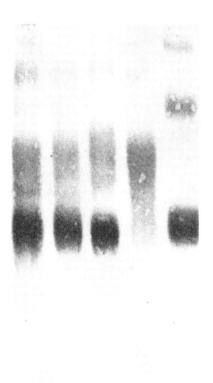

FIGURE 5. Western blot of SDS-PAGE analysis of eLH-A (A), eLH-B (B), eLH-C (C), and previously purified preparations of eLH-beta (β) and eFSH-alpha (α) subunits. Staining was performed with a polyclonal anti-eLH antiserum containing antibodies to both the alpha and beta subunits. Lower bands indicate alpha subunits and upper, more diffuse bands, show the beta subunits.

due to differences in carbohydrate composition, it may be that variations in glycosylation occur to a greater extent in the beta subunit of these hormones.

Figure 8 shows the elution profiles of the alpha subunits of eLH, eFSH, and eCG, analyzed on chromatofocusing columns utilizing a pH gradient of 9 to 6. Although striking similarities are seen among all subunit preparations, eLH alpha contains the most basic material, with a major immunoreactive peak eluting at a pI of approximately 8.3. Major basic peaks also are seen with eFSH and eCG alpha subunits, but in a more acidic range (pI = 7.9). It has been demonstrated that the alpha subunit of eLH, but not eFSH or eCG, will inhibit FSH-stimulated production of cAMP from rat Sertoli cells in a dose-responsive manner.[64] Not surprisingly, eLH alpha also possesses some FSH receptor-binding activity.[64] Since the alpha subunits of these three gonadotropins are thought to have identical peptide sequences, differences in the carbohydrate of the eLH alpha subunit may play a role in manifesting the intrinsic FSH activity of eLH.

IV. BIOLOGICAL SIGNIFICANCE OF EQUINE LH ISOHORMONES

Studies relating to the biological significance of eLH microheterogeneity have been initiated with chromatofocusing fractions containing major isohormone peaks.[65] These fractions

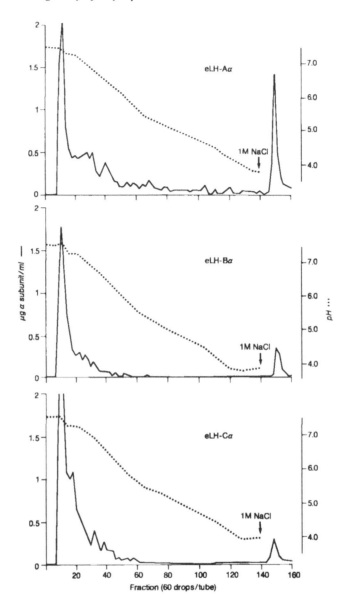

FIGURE 6.   Chromatofocusing patterns of the alpha subunits of eLH-A,
-B, and -C, as determined by specific alpha subunit RIA. (From Matteri,
R. L. and Papkoff, H., *Biol. Reprod.*, 36, 261, 1987. With permission.)

were analyzed by the same in vitro assays described above, which were used to determine
the biological, receptor binding, and immunological potencies of eLH-A, -B, and -C. The
LH activity was assessed by generating three different ratios; LH bioactivity (LH Bio)/LH
immunoactivity (LH RIA), LH radioreceptor potency (LH RRA)/LH RIA, and LH Bio/LH
RRA ratios. The FSH data were then expressed as FSH receptor binding (FSH RRA)/LH
RIA and FSH RRA/LH RRA activities.

Table 3 shows the activity ratio data for the various eLH isohormones. All ratios are
observed to change in some fashion with isoelectric point. Analysis of LH Bio/LH RIA
activity revealed a peak at a pI of 5.2, with values decreasing at greater and lesser pI values.
Although the overall chromatofocusing elution pattern of ovine LH is more basic than eLH,
a similar pattern has been obtained with isoforms of ovine LH.[60] On the other hand, the LH

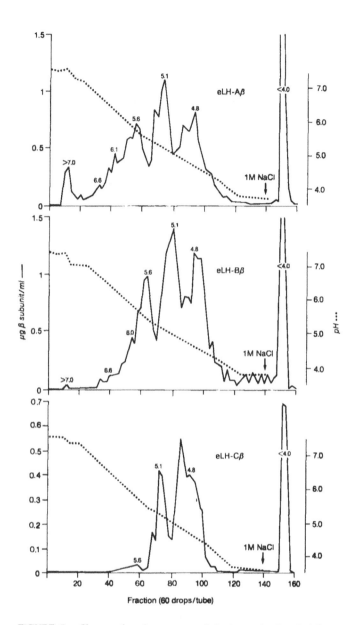

FIGURE 7. Chromatofocusing patterns of the beta subunits of eLH-A, -B, and -C, as determined by specific beta subunit RIA. (From Matteri, R. L. and Papkoff, H., *Biol. Reprod.*, 36, 261, 1987. With permission.)

Bio to LH RRA activity values revealed a significant tendency to increase with increasing isohormone acidity ($p < 0.01$, correlation analysis). These data seem to indicate that there may be an optimum state of glycosylation for maximum biological activity.

Different conclusions might be made on the basis of the other activity ratios. Both LH and FSH RRA/LH RIA ratios appeared similar, with little change occurring until low pI values are reached on the chromatofocusing columns, at which point significant reductions in activity were detected. It is possible that a large amount of immunoactive material, which has minimal bioactivity, may elute at low pIs. If this is the case, then low LH Bio and RRA/immunoactivity values would be expected in the acidic fractions. This could also mean that the LH Bio/LH RRA and FSH RRA/LH RRA ratios are the most biologically pertinent.

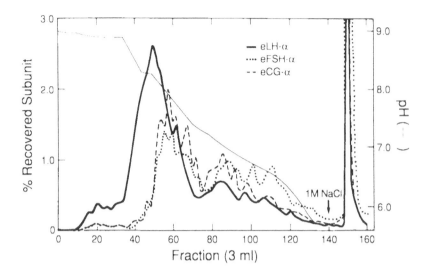

FIGURE 8.   Chromatofocusing patterns of alpha subunits from eLH, eFSH, and eCG, with a pH gradient of 9 to 6.

The LH Bio/LH RRA ratio indicates that the more acidic forms of eLH may be better to transduce a biological signal, once bound to the LH receptor. These acidic isoforms may well have longer biological half-lives, relative to the basic isohormones. The presence of sialic acid has been shown to prolong the circulating half-life of glycoproteins.[66,67] Since sialic acid is probably present to a greater extent in the acidic isohormones, these LH forms could have longer biological half-lives, as well as being more bioactive, than the more basic forms. Regardless, variations in properties can be detected among eLH isohormones. This raises the possibility that the horse may be able to selectively vary the type, as well as amount, of hormone secreted in accordance with biological needs.

The FSH RRA/LH RRA values remained constant among isohormones, indicating that the FSH activity may not differ among eLH isohormones. Since variations in carbohydrate content probably underlie isohormone heterogeneity, these data suggest that the intrinsic FSH activity of eLH may be due to peptide structure. The possibility also exists that a carbohydrate component of eLH, which does not vary among isohormones can influence this property. Deglycosylation experiments may shed some light upon this question.

Future studies, comparing the isohormones elution profiles of extracts from single pituitary glands from animals of varying endocrine backgrounds may reveal important information concerning the physiological importance and regulation of eLH isohormones. Preliminary data indicate that such studies will be technically possible. Ethanol precipitation was used to concentrate the gonadotropins from the extract of a single horse pituitary gland. More than 99% of the total LH activity was recovered by this extraction procedure. Figure 9 depicts the chromatofocusing profile of one third of this extract. Even with this small amount of material, distinct isohormone peaks are seen which correspond to those found in highly purified eLH preparations (Figure 3).

V. COMPARATIVE STUDIES OF EQUID LH

Due to the unique biochemical and biological properties of eLH, the characterization of LH from related equids has long been of interest. We have recently examined the biological characteristics and isohormone compositions of horse, donkey, and zebra LH.[68] Purification procedures previously used for eLH[5] and donkey LH (dLH)[15] were used to prepare two

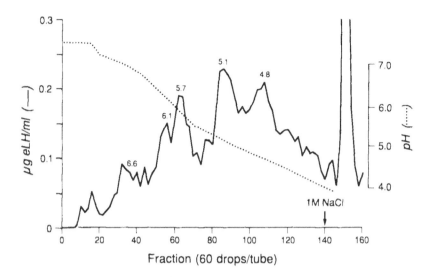

FIGURE 9. Chromatofocusing of an extract prepared from a single horse pituitary gland. (From Matteri, R. L. and Papkoff, H., *Biol. Reprod.*, 36, 261, 1987. With permission.)

zebra LH preparations: one from a fraction eluted from a cation exchange column (sulfopropyl Sephadex®-C50) column with 0.03 M NH$_4$HCO$_3$ (zLH-A) and one eluted from the same column with 1 M NH$_4$HCO$_3$ (zLH-B). The eLH and dLH were recovered from cation-exchange column fractions corresponding to those from which zLH-B was obtained. The fraction from which zLH-A was obtained has been shown to contain highly negatively charged and acidic molecules such as FSH.[7] The separation of zLH-A and -B by ion-exchange chromatography was highly suggestive of a pronounced charge microheterogeneity for this hormone.

In spite of the expected differences in apparent charge, the chromatofocusing profiles of zLH-A and -B did not differ to any great extent (Figure 10). In fact, all of the equine LH preparations were found to be very similar in this respect. Using the same assay systems described above, LH Bio/RIA, LH RRA/RIA, and LH Bio/LH RRA activity ratios were compared (Figure 11). With the exception of zLH-A, the LH preparations were found to have similar activities. All three LH activity ratio values generated with zLH-A were found to be greater than those of the other LH fractions. It may be that some factors which influence the activities of these LH preparations are not reflected in isoelectric heterogeneity. This appears to be the case for the intrinsic FSH activity of equine LH. Even though the isohormone elution patterns of the horse, donkey, and zebra LHs are quite similar, only LH from the horse seems to possess significant FSH activity. It appears that the structural component(s) imparting this interesting property does not influence the isoelectric characteristics of eLH.

VI. SUMMARY

The unique biochemical and biological properties of eLH have long been of interest to the study of the structure-function relationships of gonadotropic hormones. As found with pituitary gonadotropins from other species, eLH also exists in a variety of closely related isohormone forms. These isohormones possess differences in isoelectric properties, which allow for their separation by fractionation methods which discriminate between apparent charge characteristics. Sialic acid appears to play an important role in the microheterogeneity of eLH. In addition, the beta, rather than alpha, subunit of this hormone may be imparting most of the charge characteristics to the intact hormone. Variations can be detected among isohormones, using various in vitro bio-, receptor-, and immunoassays. In vivo biological

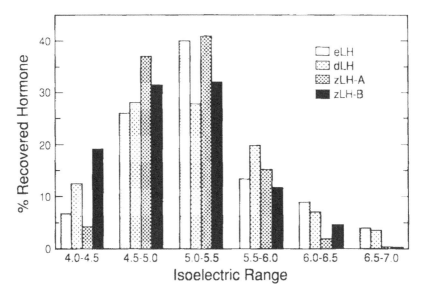

FIGURE 10.   Isoelectric distributions of immunoreactive LH from the horse (eLH), donkey (dLH), and zebra (zLH-A, -B) between isoelectric values of 7 and 4. Amount of hormone is expressed as the percentage of the total amount of recovered material between pI 7 and 4.

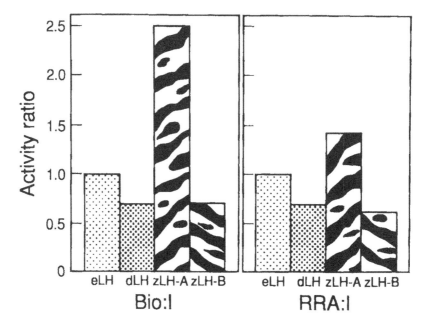

FIGURE 11.   Luteinizing hormone activity ratios of equid LH preparations from the horse (eLH), donkey (dLH), and zebra (zLH-A, -B).

half-lives of the LH isohormones remain to be determined. While isohormone distribution patterns appear similar among LH preparations from various equids, differences in potency can be detected. Although the different forms of eLH may reflect hormone in various stages of biosynthesis in the pituitary gland, it is also possible that variations in secreted forms can occur. In the future, additional experiments should provide important insights into the physiological significance and regulation of LH isohormones in the horse.

ACKNOWLEDGMENTS

The work cited from this laboratory was supported by NIH Grant HD-05722.

REFERENCES

1. **Bousfield, G. R. and Ward, D. N.**, Purification of lutropin and follitropin in high yield from horse pituitary glands, *J. Biol. Chem.*, 259, 1911, 1984.
2. **Bousfield, G. R., Sugino, H., and Ward, D. N.**, Demonstration of a COOH-terminal extension on equine lutropin by means of a common acid-labile bond in equine lutropin and equine chorionic gonadotropin, *J. Biol. Chem.*, 260, 9531, 1985.
3. **Landefeld, T. D., Grimek, H. J., and McShan, W. H.**, End group carbohydrate analysis of equine LH, *Biochem. Biophys. Res. Commun.*, 46, 463, 1972.
4. **Landefeld, T. D. and McShan, W. H.**, Isolation and characterization of subunits from equine pituitary follicle-stimulating hormone, *J. Biol. Chem.*, 249, 3527, 1974.
5. **Licht, P., Bona-Gallo, A., Aggarwal, B. B., Farmer, S. W., Castelino, J. B., and Papkoff, H.**, Biological and binding activities of equine pituitary gonadotropins and pregnant mare serum gonadotropin, *J. Endocrinol.*, 83, 311, 1979.
6. **Papkoff, H.**, Interaction of equine luteinizing hormone with gonadal binding sites of luteinizing hormone and follicle-stimulating hormone, in *Hormone Receptors in Growth and Reproduction*, Saxena, B. B., Ed., Raven Press, New York, 1984, 55.
7. **Farmer, S. W. and Papkoff, H.**, Comparative biochemistry of pituitary growth hormone, prolactin and the glycoprotein hormones, in *Hormones and Evolution*, Vol. 2, Barrington, E. J. W., Ed., Academic Press, London, 1979, 525.
8. **Boothby, M., Ruddon, R. W., Anderson, C., McWilliams, E., and Boime, I.**, A single gonadotropin alpha-subunit gene in normal tissue and tumor-derived cell lines, *J. Biol. Chem.*, 256, 5121, 1981.
9. **Fiddes, J. C. and Goodman, H. M.**, The gene encoding the common alpha subunit of the four human glycoprotein hormones, *J. Mol. Appl. Genet.*, 1, 3, 1981.
10. **Goodwin, R. G., Moncman, C. L., Rottman, F. M., and Nilson, J. H.**, Characterization and nucleotide sequence of the gene for the common alpha subunit of the bovine pituitary glycoprotein hormones, *Nucleic Acid Res.*, 11, 6873, 1983.
11. **Bogdanove, E. M., Campbell, G. T., Blair, E. D., Mula, M. E., Miller, A. E., and Grossman, G. H.**, Gonadal-pituitary feedback involves qualitative change: androgens alter the type of FSH secreted by the rat pituitary, *Endocrinology*, 95, 219, 1974.
12. **Foulds, L. M. and Robertson, D. M.**, Electrofocusing fractionation and characterization of pituitary follicle-stimulating hormone from male and female rats, *Mol. Cell. Endocrinol.*, 31, 117, 1983.
13. **Robertson, D. M., Foulds, L. M., and Ellis, S.**, Heterogeneity of rat pituitary gonadotropins on electrofocusing; differences between sexes and after castration, *Endocrinology*, 111, 385, 1982.
14. **Ulloa-Aguirre, A. and Chappel, S. C.**, Multiple species of follicle-stimulating hormone exist within the anterior pituitary of male golden hamsters, *J. Endocrinol.*, 95, 257, 1982.
15. **Roser, J. F., Papkoff, H., Murthy, H. M. S., Chang, Y.-S., Chloupek, R. C., and Potes, J. A. C.**, Chemical, biological and immunological properties of pituitary gonadotropins from the donkey *(Equis asinus)*: comparison with the horse *(Equis caballus)*, *Biol. Reprod.*, 30, 1253, 1984.
16. **Combarnous, Y. and Henge, M. H.**, Equine follicle-stimulating hormone. Purification, acid dissociation and binding to equine testicular tissue, *J. Biol. Chem.*, 256, 9567, 1981.
17. **Chappel, S. C., Bethea, C. L., and Spies, H. G.**, Existence of multiple forms of follicle stimulating hormone within the anterior pituitaries of cynomolgus monkeys, *Endocrinology*, 115, 452, 1984.
18. **Peckham, W. D., Yamaji, T., Dierschke, D. J., and Knobil, E.**, Gonadal function and the biological and physicochemical properties of follicle-stimulating hormone, *Endocrinology*, 92, 1660, 1973.
19. **Khan, S. A., Katzija, G., Fröysa, B., and Diczfalusy, E.**, Characterization of various molecular species of follicle-stimulating hormone in baboon pituitary preparations, *J. Med. Primatol.*, 13, 295, 1984.
20. **Zaidi, A. A., Robertson, D. M., and Diczfalusy, E.**, Studies on the biological and immunological properties of human follitropin: profiles of two international reference preparations and of an aqueous extract of pituitary glands after electrofocusing, *Acta Endocrinol.*, 97, 157, 1981.
21. **Zaidi, A. A., Fröysa, B., and Diczfalusy, E.**, Biological and immunological properties of different molecular species of human follicle-stimulating hormone: electrofocusing profile of eight highly purified preparations, *J. Endocrinol.*, 92, 195, 1982.

22. **Aggarwal, B. B., Farmer, S. W., Papkoff, H., and Seidel, G. E., Jr.,** Biochemical properties of equine chorionic gonadotropin from two different pools of pregnant mare sera, *Biol. Reprod.,* 23, 570, 1980.

23. **Matteri, R. L., Papkoff, H., Murthy, H. M. S., Roser, J. F., and Chang, Y.-S.,** Comparison of the properties of highly purified equine chorionic gonadotropin isolated from commercial concentrate of pregnant mare serum and endometrial cups, *Domest. Anim. Endocrinol.,* 3, 39, 1986.

24. **Papkoff, H.,** Variations in the properties of equine chorionic gonadotropin, *Theriogenology,* 15, 1, 1981.

25. **Yazaki, K., Yazaki, C., Wakabayashi, K., and Igarashi, M.,** Isoelectric heterogeneity of human chorionic gonadotropin: presence of choriocarcinoma specific components, *Am. J. Obstet. Gynecol.,* 138, 189, 1980.

26. **Mann, K., Schneider, N., and Hoermann, R.,** Thyrotropic activity of acidic isoelectric variants of human chorionic gonadotropin from trophoblastic tumors, *Endocrinology,* 118, 1558, 1986.

27. **Nwokoro, N., Chen, H.-C., and Chrambach, A.,** Physical, biological, and immunological characterization of highly purified urinary human chorionic gonadotropin components separated by gel electrofocusing, *Endocrinology,* 108, 291, 1981.

28. **Licht, P., McCreery, B. R., and Papkoff, H.,** Effects of gonadectomy on polymorphism in stored and circulating gonadotropins in the bullfrog, *Rana catesbeiana.* II. Gel filtration chromatography, *Biol. Reprod.,* 29, 646, 1983.

29. **Hattori, M. and Wakabayashi, K.,** Isoelectric focusing and gel filtration studies on the heterogeneity of avian pituitary luteinizing hormone, *Gen. Comp. Endocrinol.,* 39, 215, 1979.

30. **Keel, B. A. and Grotjan, H. E., Jr.,** Characterization of rat pituitary luteinizing hormone charge microheterogeneity in male and female rats using chromatofocusing: effects of castration, *Endocrinology,* 117, 354, 1985.

31. **Wakabayashi, K.,** Heterogeneity of rat luteinizing hormone revealed by radioimmunoassay and electrofocusing studies, *Endocrinol. Jpn.,* 24, 473, 1977.

32. **Keel, B. A. and Grotjan, H. E., Jr.,** Charge heterogeneity of partially purified ovine LH subunits and the native hormone, in *Advances in Gene Technology: Molecular Biology of the Endocrine System,* Puett, D., Ahmad, F., Black, S., Lopez, D. M., Melner, M. H., Scott, W. A., and Whelan, W. J., Eds., Cambridge University Press, Cambridge, 1986.

33. **Peckham, W. D. and Knobil, E.,** The effects of ovariectomy, estrogen replacement and neuraminidase treatment on the properties of the adenohypophysial glycoprotein hormones on the rhesus monkey, *Endocrinology,* 98, 1054, 1976.

34. **Peckham, W. D. and Parlow, A. F.,** Isolation from human pituitary glands of three discrete electrophoretic components with high luteinizing hormone activity, *Endocrinology,* 85, 618, 1969.

35. **Reader, S. C. J., Robertson, W. R., and Diczfalusy, E.,** Microheterogeneity of human pituitary luteinizing hormone from women of fertile and postmenopausal age, *Clin. Endocrinol.,* 19, 355, 1983.

36. **Robertson, D. M., Fröysa, B., and Diczfalusy, E.,** Biological and immunological characterization of human luteinizing hormone. IV. Biological and immunological profiles of two international reference preparations after electrofocusing, *Mol. Cell. Endocrinol.,* 11, 91, 1978.

37. **Robertson, D. M., Puri, V., Lindberg, M., and Diczfalusy, E.,** Biologically active lutenizing hormone in plasma. V. An analysis of the differences in the ratio of biological to immunological LH activities during the menstrual cycle, *Acta Endocrinol.,* 92, 615, 1979.

38. **Chappel, S. C., Ulloa-Aguirre, A., and Coutifaris, C.,** Biosynthesis and secretion of follicle-stimulating hormone, *Endocr. Rev.,* 4, 179, 1983.

39. **Hattori, M.-A., Sakomoto, K., and Wakabayashi, K.,** The presence of LH components having different ratios of bioactivity to immunoreactivity in the rat pituitary glands, *Endocrinol. Jpn.,* 30, 289, 1983.

40. **Robertson, D. M. and Diczfalusy, E.,** Biological and immunological characterization of human lutenizing hormone. II. A comparison of the immunological and biological activities of pituitary extracts after electrofocusing using different standard preparations, *Mol. Cell. Endocrinol.,* 9, 57, 1977.

41. **Wide, L.,** Male and female forms of human follicle-stimulating hormone in serum, *J. Clin. Endocrinol. Metab.,* 55, 682, 1982.

42. **Diebel, N. D., Yamamoto, M., and Bogdanove, E. M.,** Discrepancies between radioimmunoassays and bioassays for FSH: evidence that androgen treatment and withdrawal can alter bioassay-immunoassay ratios, *Endocrinology,* 92, 1065, 1973.

43. **Peckham, W. D. and Knobil, E.,** Qualitative changes in the pituitary gonadotropins of the male rhesus monkey following castration, *Endocrinology,* 98, 1061, 1976.

44. **Keel, B. A. and Grotjan, H. E., Jr.,** Influence of bilateral cryptorchidism on rat pituitary luteinizing hormone charge microheterogeneity, *Biol. Reprod.,* 32, 83, 1985.

45. **Strollo, F., Harlin, J., Hernandez-Montes, H., Robertson, D. M., Zaidi, A. A., and Diczfalusy, E.,** Qualitative and quantitative differences in the isoelectrofocusing profile of biologically active lutropin in the blood of normally menstruating and post-menopausal women, *Acta Endocrinol.,* 97, 166, 1981.

46. **Wide, L. and Hobson, B. M.**, Qualitative differences in follicle-stimulating hormone activity in the pituitaries of young women compared to that of men and elederly women, *J. Clin. Endocrinol. Metab.*, 56, 371, 1983.

47. **Combarnous, Y., Guillou, F., and Martinat, N.**, Comparison of in vitro follicle-stimulating hormone (FSH) activity of equine gonadotropins (luteinizing hormone, FSH, and chorionic gonadotropin) in male and female rats, *Endocrinology*, 115, 1821, 1984.

48. **Hsueh, A. J. W., Erickson, G. F., and Papkoff, H.**, Effect of diverse mammalian gonadotropins on estrogen and progesterone production by cultured rat granulosa cells, *Arch. Biochem. Biophys.*, 225, 505, 1983.

49. **Moudgal, N. R. and Papkoff, H.**, Equine luteinizing hormone possesses follicle-stimulating hormone activity in hypophysectomized female rats, *Biol. Reprod.*, 26, 935, 1982.

50. **Braselton, W. E., Jr. and McShan, W. H.**, Purification and properties of follicle-stimulating and luteinizing hormones from horse pituitary glands, *Arch. Biochem. Biophys.*, 139, 45, 1970.

51. **Reichert, L. E., Jr.**, Electrophoretic properties of pituitary gonadotropins as studied by electrofocusing, *Endocrinology*, 88, 1029, 1971.

52. **Matteri, R. L., Papkoff, H., Ng, D. A., Swedlow, J. R., and Chang, Y.-S.**, Isolation and characterization of three forms of luteinizing hormone from the pituitary gland of the horse, *Biol. Reprod.*, 34, 571, 1986.

53. **Ulloa-Aguirre, A., Miller, C., Hyland, L., and Chappel, S. C.**, Production of all follicle-stimulating hormone isohormones from a purified preparation by neuraminidase digestion, *Biol. Reprod.*, 30, 382, 1984.

54. **Cole, L. A. and Grotjan, H. E., Jr.**, Carbohydrate structures and charge heterogeneity of hLH and oLH, Endocrine Soc. Annu. Meet., Abstr. 511, Anaheim, Calif., 1986.

55. **Talmadge, K., Vamvakopoulos, N. C., and Fiddes, J. C.**, Evolution of the genes for the beta subunits of human chorionic gonadotropin and luteinizing hormone, *Nature (London)*, 307, 37, 1984.

56. **Tepper, M. A., Dionne, F. F., Gee, C. E., and Roberts, J. L.**, Rat beta luteinizing hormone; mRNA sequence, gene structure, and regulation by steroid hormones, Endocrine Soc. Annu. Meet., Abstr. 149, San Antonio, Tex., 1983.

57. **Yamashita, K., Mizuochi, T., and Kobata, A.**, Analysis of carbohydrates by gel filtration, in *Methods in Enzymology*, Vol. 83, Ginsburg, V., Ed., Academic Press, New York, 1982, 105.

58. **Ramachandran, J. and Sairam, M. R.**, The effects of interstitial cell-stimulating hormone, its subunits, and recombinations on isolated rat Leydig cells, *Arch. Biochem. Biophys.*, 167, 294, 1975.

59. **Farmer, S. W. and Papkoff, H.**, Immunochemical studies with pregnant mare serum gonadotropin, *Biol. Reprod.*, 21, 425, 1979.

60. **Keel, B. A., Schanbacher, B. D., and Grotjan, H. E., Jr.**, Ovine luteinizing hormone. I. Effects of castration and steroid administration on the charge heterogeneity of pituitary luteinizing hormone, *Biol. Reprod.*, in press, 1987.

61. **Blum, W. F. P., Riegelbauer, G., and Gupta, D.**, Heterogeneity of rat FSH by chromatofocusing: studies on in vitro bioactivity of pituitary FSH forms and effect of neuraminidase treatment, *J. Endocrinol.*, 105, 17, 1985.

62. **Chen, H.-C., Shimohigashi, Y., Dufau, M. L., and Catt, K. J.**, Characterization and biological properties of chemically deglycosylated human chorionic gonadotropin, *J. Biol. Chem.*, 257, 14446, 1982.

63. **Graesslin, D., Weise, H. C., and Braendle, W.**, The microheterogeneity of human chorionic gonadotropin (HCG) reflected in the beta subunits, *FEBS Lett.*, 31, 214, 1973.

64. **Aggarwal, B. B., Papkoff, H., and Licht, P.**, Effect of the alpha subunit of equine LH on the FSH induced cAMP production in rat seminiferous tubule cells, *Endocrinology*, 108, 2406, 1981.

65. **Matteri, R. L. and Papkoff, H.**, Characterization of equine luteinizing hormone by chromatofocusing, *Biol. Reprod.*, 36, 261, 1987.

66. **Aggarwal, B. B. and Papkoff, H.**, Plasma clearance and tissue uptake of native and desialylated equine gonadotropins, *Domest. Anim. Endocrinol.*, 2, 173, 1985.

67. **Morell, A. G., Gregoriadis, G., Sheinberg, I. H., Hickman, J., and Ashwell, G.**, The role of sialic acid in determining the survival of glycoproteins in circulation, *J. Biol. Chem.*, 246, 1461, 1971.

68. **Matteri, R. L., Baldwin, D. M., Lasley, B. L., and Papkoff, H.**, Biological and immunological properties of zebra pituitary gonadotropins: comparison with horse and donkey gonadotropins, *Biol. Reprod.*, 36, 1134, 1987.

Chapter 10

THYROID-STIMULATING HORMONE MICROHETEROGENEITY

Brooks A. Keel

TABLE OF CONTENTS

I. INTRODUCTION

Thyroid-stimulating hormone (TSH, thyrotropin) is a glycoprotein composed of two non-covalently linked subunits, designated as the α and β. TSH is structurally similar to other glycoprotein hormones including luteinizing hormone (LH, lutropin), follicle-stimulating hormone (FSH, follitropin), and human chorionic gonadotropin (hCG). While these hormones share a common α-subunit within species, the β-subunit is unique and, when combined with an α-subunit, confers biologic specificity to the respective hormone. Both of the subunits of TSH are glycosylated with the α–subunit containing two and the β–subunit possessing one asparagine N-linked carbohydrate residues each (reviewed in Chapter 2).

Glycosylation has been shown to play a vital role in TSH biosynthesis and intracellular degradation (see Reference 1 for reveiw). The presence of carbohydrate moieties also often contributes to significant expression of molecular heterogeneity resulting in subpopulations or "isohormones" which may differ in receptor-binding activity, circulatory survival, and overall biological activity. The relative distribution of isohormones seems to be directly related to changes in the oligosaccharide structure of the hormone. Therefore, subtle changes in the carbohydrate moieties of TSH may significantly alter the overall biological activity of the hormone.

TSH is synthesized in and secreted from the anterior pituitary. TSH exerts its action on the thyroid gland resulting in increased iodide uptake, thyroglobulin synthesis, iodotyrosine/iodothyronine formation, thyroglobulin proteolysis, and release of thyroxine and triiodothyronine from the thyroid (reviewed in Reference 2). TSH action is mediated by binding to specific cell-surface receptors and activating adenylyl cyclase. Although α- and β-subunit dimerization is required for complete biological activity,[3] changes in TSH glycosylation can alter the ability of the hormone to activate adenylyl cyclase.[4] Therefore, changes in the microheterogeneity of TSH, resulting from carbohydrate alterations, may play a role in the regulation of thyroidal function.

In this chapter, evidence for the existence of size and charge isomers of TSH will be presented. The production and secretion of TSH molecules with differing biological activity will also be examined, and the possible relationships between TSH biological activity and microheterogeneity will be discussed. Finally, attempts will be made to elucidate the biochemical basis for the existence of TSH isohormones.

II. SIZE HETEROGENEITY

The concept of large molecular weight prohormones which function as biosynthetic precursors has been proposed for a number of hormones. Several early studies have reported the existence of a large molecular weight "big"-TSH in the pituitaries and serum of several species including man. Erhardt and Scriba[5] characterized a high molecular TSH, termed "big"-TSH, in homogenates of human pituitaries. This "big"-TSH had a molecular weight of approximately 200,000, displayed parallel dose response curves compared with standard TSH in a radioimmunoassay (RIA), and contained carbohydrate as evidenced by its affinity for concanavalin A (ConA). The large molecular weight material was also resistant to dissociation by 6 *M* guanidine-HCl suggesting that the "big"-TSH was not an aggregate. Klug and Adelman[6] presented evidence for a large molecular weight TSH in the sera and pituitaries of rats. There was observed a progressive age-related increase in the large forms of TSH in the sera of these animals which was associated with a progressive decline in the bioactivity of circulating TSH. In addition, preliminary information from these investigators suggested that the large molecular weight form of TSH inhibited the release of thyroid hormone.[6] Spitz and colleagues later described a euthyroid patient with a high molecular weight TSH as the sole immunoreactive form of circulating TSH.[7] Large molecular weight forms of circulating TSH β–subunit have also been reported in humans.[8] This "big"-TSH-β was different from "native" circulating TSH-β in that it demonstrated nonparallelism in the TSH-β RIA, did not increase after TRH administration, and was not reduced after thyroid hormone of dexamethasone treatment. Mori et al. later described a "big"-TSH-β in extracts of pituitaries from thyroidectomized rats. Furthermore, the "big"-TSH-β was stable to ultracentrifugation and treatment with guanidine–HCl suggesting that the material was not associated with ribosomes and was not the result of aggregation of TSH molecules.[9]

Clearly, the above results suggest that large molecular weight forms of TSH exist. Although it is tempting to speculate that these forms represent a TSH prohormone, various lines of evidence strongly suggest that these forms merely represent aggregated or bound material whose physiological function is uncertain.[1,10] As indicated by Weintraub and associates,[10,11] such aggregates may be resistant to dissociation unless complete reduction is performed utilizing denaturants.

Additional studies have described the presence of heterogeneous forms of TSH besides "big"-TSH. Diamond and Rosen[12] reported chromatographic differences between human circulatory and pituitary TSH. These investigators observed that both unlabeled as well as radiolabeled pituitary TSH migrated as a single symmetrical immunoreactive peak on columns of Sephadex® G-100. No immunoreactive material was present in the void volume of the column. In contrast, serum TSH eluted asymmetrically and was slightly retarded compared to the elution of pituitary TSH. These data provide evidence that circulating TSH is heterogeneous. Heterogeneous forms of TSH have also been demonstrated in mouse thyrotrophic tumors.[13] After gel filtration, both tumor and serum TSH immunoreactivity, as well as radioreceptor activity, eluted as broad heterogeneous peaks of approximately 26,000 to 44,000 daltons. When the chromatography fractions were assayed for thyroid adenylate cyclase activity, tumor TSH eluted as two sharp peaks (26,000 and 44,000 daltons) while an additional third peak (33,000 dalton) was observed for serum TSH.[12] The heterogeneous nature of thyrotrophic tumor TSH was later confirmed by Joshi and Weintraub[14] who observed marked size heterogeneity, as assessed by bioassay and RIA, of crude preparations of mouse tumor, serum, medium, and partially as well as highly purified preparations of TSH. Moreover, these investigators went on to demonstrate that the low molecular weight forms which were characterized by a decreased bioactivity to immunoreactivity ratio (B/I) act as partial competitive antagonists of the more biologically active forms.[14]

Taken together, the above data suggest a possible relationship between size heterogeneity and altered biological activity of TSH. More specifically, Klug and Adelman[6] suggested that the age-related decrease in biologically active TSH in serum may be related to the observed increase in the proportion of large molecular weight TSH, which adds to the immunoassayable TSH but at the same time reduces the bioassayable TSH content. Spitz et al.[7] reported that the large form of TSH observed in their euthyroid patient bound to the TSH receptor but had greatly impaired biological activity (4% of control). The decreased biological activity of circulating TSH in this patient may explain the euthyroid status, since the immunoassayable concentrations were 20 to 50 times elevated.[7] Further results from Weintraub's laboratory[13,14] have demonstrated that the heterogeneous forms of TSH observed in thyrotic tumors and in the sera from tumor-bearing mice show similar immunological and receptor-binding activities but differ widely in adenylate cyclase-stimulating activity. TSH bioactivity was higher in sera than tumor extracts due to a molecular form of TSH exhibiting high thyroid adenylate cyclase activity which was not observed in the tumors.[13] Of major interest was the observation that the biologically active heterogeneous forms of TSH also exhibited differing affinities for lectins.[14] TSH forms with low molecular weight and low B/I ratio demonstrated a moderate affinity for ConA and soybean agglutinin but high affinity for wheat germ agglutinin. In contrast, TSH forms with high B/I ratio and high molecular weight showed high affinity for ConA and wheat germ agglutinin but a low affinity for soybean agglutinin. These results can be interpreted to indicate that the heterogeneous forms of TSH differ in the amount or availability of mannose, N-acetylgalactosamine, and/or galactose,[14] and imply that the carbohydrate structure of TSH isohormones may be important in determining the overall biological activity of the hormone.

III. CHARGE HETEROGENEITY

Early attempts at isolating and characterizing TSH from a variety of species recognized that TSH existed as a heterogeneous series of molecules, each differing in charge. The concept of hormone polymorphism was not unique to TSH and had in fact been observed in most pituitary hormones. Sluyser,[15] in 1964, suggested the term "isohormone" be used to describe the observed hormone microheterogeneity. Since the early 1960s, a great deal of research has been directed toward isolating, characterizing, and defining the biochemical basis for isohormones. Results from a large number of studies have indicated that each of these isohormones possesses a distinct and reproducible isoelectric point (pI). Of major interest is that in many cases these isohormones differ not only in pI but also in circulatory half-life, target cell receptor binding, and biological activity. The biochemical basis for the presence of isohormones seems to depend at least in part upon subtle differences in carbohydrate structure. Other investigators have suggested that alterations in the endocrine status may result in changes in the distribution of isohormones. Thus, since changes in the isohormone profile reflect altered biological activity of the hormone, changes in the animal's endocrine environment may result in both quantitative (amount) as well as qualitative (biological activity) changes in the hormone. Subtle alterations in the biological activity of hormones (i.e., altered isohormones) by the pituitary may represent a fine-tuning mechanism for controlling gonadal and thyroidal function.

A. Gel Electrophoresis and Ion Exchange

TSH charge heterogeneity has been recognized as early as 1960 when Pierce and colleagues described multiple components of TSH isolated from the pituitaries of cattle,[16,17] sheep, and whales.[17] Six components of biologically active bovine TSH were obtained by DEAE chromatography. All six components were approximately the same molecular weight suggesting that the components differed in net charge or charge distribution.[16] Further proof of the charge dependence of the forms came when these components of TSH were found to emerge in reverse order from CM-cellulose (cation exchange) than from DEAE (anion exchange).[17] Analysis of bovine, sheep, and whale TSH revealed similar heterogeneity, but due to the impurity of the latter two preparations accurate comparisons were not possible.[17] Later results demonstrated a similar polymorphism between bovine, porcine, and human TSH by gel electrophoresis.[18] Human TSH was found to be more electronegative than the other two species, possibly due to the presence of sialic acid residues.[18]

Wide and associates, utilizing a variety of biochemical techniques, have isolated five different TSH components from the pituitaries of humans.[19,20] Successive chromatography by gel filtration, hydroxy apatite, and ion exchange followed by preparative polyacrylamide gel electrophoresis revealed five different components of TSH of equal molecular size.[19] The electrophoretic pattern was the same after neuraminidase treatment, although the components were more basic in nature. Amino acid and carbohydrate analysis indicated close similarities between the five components.[20] These investigators also examined the heterogeneity associated with the α- and β-subunits of human TSH which had been dissociated from the native dimer.[21] Each subunit preparation contained four isoforms. Amino acid analysis revealed close agreement between the free-α forms and between the free-β forms. Furthermore, it was observed that recombination of various pairs of α- and β-subunit isoforms gave rise to the native TSH confirmation, but to various extents.[21] Several years later,

Table 1
REPORTED APPARENT ISOELECTRIC POINTS (pI) OF TSH ISOHORMONES
FROM VARIOUS SPECIES

Species	Ref.	Number of isohormones	Apparent pI	Comments
Bovine	Fawcett et al.[24,25]	4	8.8, 8.75, 8.61, 8.39	Crude pituitary extract, bioassay
	Yora and Ui[26]	5	8.95, 8.60, 8.32, 7.90, 7.20	Purified, RIA
Whale	Tamura and Ui[27,28]	4	8.70, 8.51, 8.38, 8.17	Partially purified, bioassay
	Tamura and Ui[29]	4	8.73, 8.57, 8.30, 8.13	Purified, bioassay
Rat	Yora et al.[30]	4	8.3, 7.9, 7.2, 6.6	Crude pituitary extract, RIA
	Mori et al.[9]	5	8.8, 8.5, 7.9, 7.3, 6.6—6.8	Crude pituitary extract, RIA
	Mori et al.[31]	5	8.7, 7.9, 7.2, 6.6, 6.2	Affinity column purified, RIA
	Mori et al.[31]	4	8.7, 7.8, 5.3, 2.5	[14C] alanine + TRH
	Mori et al.[31]	7	8.7, 7.8, 7.2, 6.5, 6.2, 5.3, 2.5	[14C] alanine + TRH
	Mori et al.[32]	6	8.7, 7.8, 7.2, 6.5, 5.8, 2.5	[1H] glucosamine + TRH
Sheep	Keel (this chapter)	9	>7.50, 7.13, 6.79, 6.49, 5.98, 5.77, 5.21, 4.66, <4.00	Crude pituitary extract, RIA
	Keel (this chapter)	8	>7.50, 7.00, 6.85, 6.46, 6.09, 5.75, 5.30, <4.00	Purified, RIA

Wide[22,23] utilized zone electrophoresis in agarose suspension to characterize the charge heterogeneity of human pituitary FSH, LH, and TSH. He discovered that each pituitary examined contained a sequence of 20 or more different forms of each of the hormones, with minor differences in charge.[22] The degree of charge heterogeneity within the pituitary was similar for FSH, LH, and TSH. Interestingly, the relative degree of charge heterogeneity was not related to sex or age, but there was a sex- and age-related variation in the median charge of the gonadotropins, but not TSH.[23]

B. Isoelectric Focusing

The biochemical approaches described above have provided important information concerning the existence of TSH isohormones. Results from these studies have indicated that multiple forms of TSH exist within the pituitaries of several species and the underlying biochemical difference among these forms is net charge. However, this information has been obtained by procedures which do not provide data on the precise charge differences among the various TSH isohormones. Isoelectric focusing (IEF) separates proteins according to pI (charge), is independent of molecular size and shape, and provides accurate information concerning the precise isoelectric point of the proteins in question. This technique, along with chromatofocusing described below, has been used by a number of investigators to define the isohormones of TSH in several species. Table 1 lists the number and apparent pIs of these isoelectric variants. With the exception of the sheep, the isoelectric variants of TSH from the various species have been characterized by isoelectric focusing. Sheep TSH charge microheterogeneity has recently been examined by the author using chromatofocusing and will be discussed below.

The first accurate identification of the isoelectric points of TSH isohormones was published by Fawcett et al. in 1969.[24] These investigators reported the isolation of four biologically active TSH isohormones from bovine pituitary extracts by IEF in the pH range of 8.2 to 8.8 (see Table 1).[24,25] Refocusing of the four components revealed identical pIs indicating the reproducible nature of the IEF system used. Interestingly, when the biological potencies of the various isohormones were compared it was observed that one of the isohormones was at least twice as active as the others, suggesting a relationship between charge microheterogeneity and varying bioactivity. Yora and Ui[26] reported the presence of five isohormones of bovine TSH with pIs slightly different from those reported by Fawcett et al.[24,25] However, it should be pointed out that Fawcett et al.[24,26] performed IEF on crude pituitary extracts and relied on spectrophotometric and in vitro bioassay techniques to assess TSH isohormones pIs while Yora and Ui[26] utilized RIA procedures to assay TSH in the IEF fractions. Thus, differences in methodology may explain the slight discrepancies observed.

The whale has proven to be an interesting animal model for the study of TSH heterogeneity. Due to the large size of the gland (10 to 20 g per pituitary), the isohormone pattern of TSH could be determined in a single pituitary eliminating the need for a large pool of pituitaries. Tamura and Ui have used this animal model extensively.[27-29] These investigators observed four isohormones of crude pituitary TSH isohormones with pIs ranging from 8.17 to 8.70.[27] The multiple components were present in a single native whale pituitary suggesting that the various isohormones of TSH were present in all individuals and not the result of genetic differences among individuals of the same species.[27] The presence of four TSH isohormones in crude whale pituitary extracts was later confirmed.[28] Furthermore, when whale TSH was purified by acetone precipitation, gel filtration, and stepwise elution chromatography on CM- and TEAE-cellulose, the isohormone profile was not significantly altered (see Table 1). No significant differences were noted in the molecular weights, amino acid composition, or the hexosamine and sialic acid contents of the purified whale TSH isohormones.[29] However, significant differences were noted in mannose content among the four components.

The above studies have described the presence of TSH isohormones but have not placed any physiological significance to the existence of TSH charge microheterogeneity. More recently, the rat has been used as a model to study the effects of altered thyroid states on the isohormone profile of TSH[9,30] as well as to examine the possible differences between basal- and thyrotrophin-releasing hormone (TRH)-stimulated TSH[31,32] production. At least four isohormones of euthyroid rat pituitary TSH have been described with pIs ranging from 8.3 to 6.6.[30] Hypothyroidism induced by methylthiouracil (MTU) treatment resulted not only in striking changes in pituitary TSH content but also altered the isohormone profile of TSH. The content of the more acidic forms of TSH decreased to a greater extent than those of the other components and remained below the normal level until recovery was attained 3 weeks after MTU treatment.[30] In contrast, the alkaline forms of TSH returned to normal levels by 2 weeks posttreatment. In addition, it was observed that the MTU-induced alteration in TSH isohormones could be returned toward normal by the administration of thyroid hormone.[30] Mori et al.[9] described five major TSH isohormones of normal rat pituitary TSH isohormones not dissimilar from the results of Yora et al.[30] (see Table 1). The numbers of basic TSH components were increased in the pituitaries of rats which had been thyroidectomized more than 2 weeks previously.[9] Each of the TSH components induced by thyroidectomy was normalized by thyroid hormone treatment.

Mori et al.[9] also examined the IEF fractions for TSH-β-subunit and found that the peaks of immunoreactive TSH-β were observed in the same areas as those of the native hormone. As a result of thyroidectomy, an apparent increase in the neutral and acidic components of TSH-β-subunit was observed.[9] Because thyroidectomy was also associated with a selective increase in "big"-TSH-β, it was predicted that this large molecular weight form of the free subunit corresponded to the neutral or acidic components as these were the forms increased after removal of the thyroid. However, sufficient quantities of big-TSH-β have not been prepared to allow precise determination of the isoelectric profile of this substance.

The results of the studies described above suggest that under conditions which stimulate the synthesis of TSH, alterations in the isohormone profile of TSH occur. These studies have led Mori and associates[31,32] to investigate the effects of TRH on TSH isohormones. In these experiments, hemipituitary glands from euthyroid rats were incubated in vitro with

[³H]glucosamine[31,32] or [¹⁴C]alanine[31] in the presence or absence of TRH. TSH fractions were then partially purified by affinity chromatography and subsequently subjected to IEF. Unincubated, unstimulated pituitary extracts displayed five TSH isohormones with pIs of 8.7, 7.9, 7.2, 6.6, and 6.2. When unstimulated pituitaries were incubated with radioactive alanine, four peaks were observed: two (pI = 8.7 and 7.8) corresponded with unlabeled TSH while two new components (pI = 5.3 and 2.5) were detected.[31] Similar results were obtained when [¹⁴C]alanine was replaced with radioactive glucosamine.[32] The introduction of TRH into the incubation system resulted in not only augmented TSH synthesis but caused the production of several new TSH isohormones. New components of pI 7.2 and 6.5[31,32] had activity of both precursors[31,32] while another component of pI 6.2 was associated with alanine incorporation only.[31] This is an interesting finding because native TSH is glycosylated.[1,10] Therefore, the pI 6.2 component may represent an intermediate biosynthetic product while the pI 7.2 and 6.5 isohormones may represent mature forms.[31] These authors concluded that TRH principally elicits an increase in protein synthesis in TSH but does not increase radioactive carbohydrate incorporation into TSH, resulting in an alteration in TSH heterogeneity.

C. Chromatofocusing

A major disadvantage of IEF on polyacrylamide gels is that proteins which focus outside the defined pH window migrate into the electrode strips (i.e., off the gel) and are lost. Although proteins cannot be run off an IEF column, those which migrate into the electrode solutions may lose activity or may be discarded.[33] Chromatofocusing has several distinct advantages over IEF:[34] (1) it is relatively inexpensive, (2) relatively large sample volumes and protein concentrations can be utilized, (3) the limits of the pH gradient can easily be controlled, and (4) it can be used as a preparative technique. Another major advantage of chromatofocusing is that proteins outside the pH window can be collected and analyzed. Furthermore, base-line resolution of proteins with pI differences of as little as 0.02 can be achieved.[35] Several investigators have noted the advantage of chromatofocusing for separating the isohormones of FSH from several species including rat,[36-40] hamster,[41,42] and monkey[43] (see also Chapter 7).

We have used chromatofocusing exclusively to isolate and characterize the isohormones of rat,[34,44,45] hamster (see Chapter 8), and ovine[46-48] pituitary LH. The results of these studies are reviewed in Chapter 8. Recent studies from the author's laboratory have shown that the technique of chromatofocusing is equally advantageous for the study of TSH charge microheterogeneity.

Preliminary results have shown that rat TSH migrates predominantly as acidic (pI <7.00) species (see Reference 45 and Figure 8, Chapter 8). Therefore, further characterization of TSH isohormones was performed utilizing a pH gradient ranging from pH 7.00 to 4.00. As illustrated in Figure 1, chromatofocusing generates very reproducible pH gradients in this pH range. Identical results have also been shown for the more basic pH ranges (pH 10.5 to 7).[34] When extracts of pituitaries from sheep are subjected to chromatofocusing, at least nine immunoreactive species of ovine TSH (oTSH) can be identified (Figure 2). Seven species, designated isohormones II to VIII, have apparent pIs of 7.13, 6.79, 6.49, 5.98, 5.77, 5.21, and 4.66 (Table 2). Two additional immunoreactive peaks of TSH were also observed: one in the void volume of the column (isohormone I, apparent pI >7.50) and one which was bound to the column but could be eluted with 1.0 M NaCl (isohormone IX, apparent pI <4.00). The latter two peaks were outside the pH window of the column but were easily detected. Thus, it would appear that chromatofocusing is advantageous for studying TSH isohormones because of the flexibility in defining the pH gradient and because materials outside the pH window are not lost.

The data shown in Table 2 and Figure 2 indicate that immunoreactive oTSH is distributed

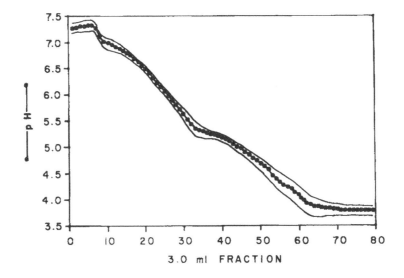

FIGURE 1.   Reproducibility of the pH 7.4 to 4.0 chromatofocusing gradient. Solid symbols represent the mean of six profiles, and the solid line represents one SD on each side of the line.

Table 2
APPARENT ISOELECTRIC POINTS (pI) AND
RELATIVE AMOUNTS OF OVINE TSH
ISOHORMONES ISOLATED BY
CHROMATOFOCUSING

Isohormone	Fractions	Apparent pI	Relative amount
I	4—6	>7.50	9
II	7—11	7.32	16
III	12—15	6.92	13
IV	16—22	6.53	29
V	23—27	6.05	15
VI	28—32	5.77	7
VII	33—40	5.20	6
VIII	41—49	4.84	3
IX	85—87	<4.00	2

Note: Representative data obtained from the chromatofocusing of a
wether. Relative amount expressed as a percentage of the total
amount of immunoreactive TSH obtained from the chromatofocusing column.

over the entire pH 7 to 4 gradient with a majority (>80%) having a pI >6.00. Isohormones III to VII and IX were reproducibly present as distinct peaks in all profiles examined. Isohormone I elutes in the void volume of the column indicating a basic species (apparent pI >7.5). Isohormone II eluted adjacent to isohormone I and immediately after the gradient began to form. Although separation of these two isohormones was not complete, both forms could be identified in all profiles examined. These results indicate that isohormone II has a pI very close to the upper limits of the pH gradient utilized. This species may become more completely resolved from isohormone I if different pH gradients are used (i.e., pH 8 to 5). Isohormone VIII accounted for only 3% of the immunoreactive oTSH recovered from the column and was not observed in the pituitary of one out of six animals examined. This species is also absent from highly purified preparations of TSH (see below).

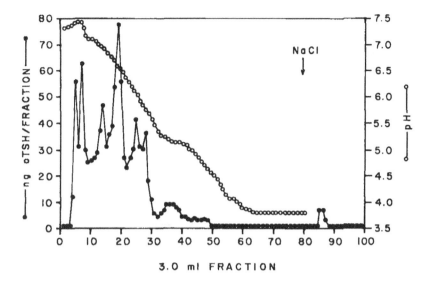

FIGURE 2. Chromatofocusing elution profile of immunoreactive oTSH in a crude pituitary extract of a wether. Note that the first peak elutes before the gradient starts (apparent pI >7.40) and that the last peak was bound to the column (pI <4.00) and eluted only after the eluant was changed to 1.0 *M* NaCl (arrow).

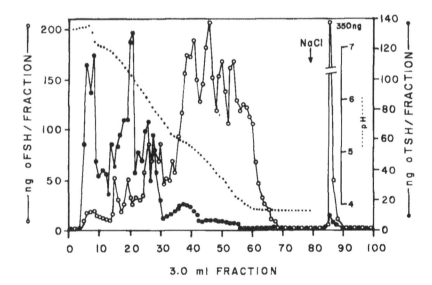

FIGURE 3. Chromatofocusing elution profile of immunoreactive oTSH and oFSH in a single extract of a ram pituitary. The arrow indicates treatment of the column with 1.0 *M* NaCl.

Apparent isohormones in crude pituitary extracts by RIA could result, in part, from cross-reaction of other pituitary hormones in the TSH RIA. Previous studies by the author have indicated that greater than 80% of immunoreactive oLH focuses as basic isohormones[47,48] and elutes in the void volume of a pH 7 to 4 gradient. Therefore, with the possible exception of oTSH isohormone 1 which coelutes with oLH, the isohormones of oTSH are probably not due to cross-reacting oLH. However, FSH from a number of species is composed of very acidic species (see Chapters 6 and 7) and potentially could coelute with TSH during chromatofocusing and cross-react in the TSH RIA. When both oFSH and oTSH were assayed in the same chromatofocusing profile (Figure 3) it was discovered that >87% of the im-

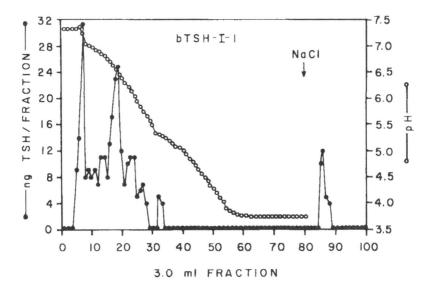

FIGURE 4.   Chromatofocusing elution profile of NIADDK-bTSH-I-1. The arrow indicates treatment of the column with 1.0 *M* NaCl.

munoreactive oTSH eluted with a pI of 5.60 or greater (fractions 0 to 30) while the vast majority (>85%) of the immunoreactive oFSH focused as more acidic species (pI <5.60). Therefore, oTSH focuses in a region of the chromatofocusing gradient that is devoid of most of the immunoreactive oFSH.

In general, the isohormone profiles of purified preparations of LH[47] and FSH[36] are markedly different from that observed in crude preparations and depend to a great extent upon the degree of purity and the chemical methods utilized in the purification procedures. However, as shown in Figure 4, the chromatofocusing profile of purified TSH is remarkably similar to the results obtained from crude pituitary extracts. Identical isohormones are observed with the exception of isohormone VIII which is absent from the purified preparation. These results lend further support to the conclusion that the isohormones of oTSH in pituitary extracts are not due to cross-reactivity of other pituitary hormones and are not an artifact of the extraction procedure. However, it should be remembered that the purified preparation utilized herein was obtained from the cow and not the sheep, and that the RIA employed is heterologous. Whether purified preparations of ovine TSH will display identical isohormones remains to be determined.

The results presented here are apparently at odds with the findings of others who have examined the isoelectric points of TSH from the cow, whale, and rat (see Table 1). In general, the apparent pIs of sheep TSH isohormones are more acidic in nature than those of other species. Several reasons can be proposed for this discrepancy. Perhaps the most obvious would be a potential species difference in TSH microheterogeneity. Indeed, marked differences between rat and ovine LH charge microheterogeneity have been observed (see Chapter 8). However, due to the similarity in isohormone profiles between highly purified bTSH and crude pituitary oTSH, as well as the known similarities between bovine and ovine glycoprotein hormone structures,[49] the differences between the results presented here for sheep and the findings of others[24-26] for bovine TSH are probably not related to species. In fact, the RIA employed herein is a heterologous assay utilizing specific antisera to ovine TSH but highly purified bovine TSH as hormone for iodination and standard. A more likely explanation involves methodological differences. It should also be noted that isohormone I is probably composed of multiple components whose pIs are greater than 7.5. It is also generally known that chromatofocusing may not always yield the same pI for a given protein as IEF.[39,50] Therefore, the pIs of the oTSH isohormones described here may in fact be very similar to those of bTSH reported by Fawcett et al.[24,25] and Yora and Ui.[26]

IV. BIOCHEMICAL BASIS FOR TSH MICROHETEROGENEITY

The results presented above clearly indicate that multiple forms of TSH exist and that these forms differ primarily in net charge. However, what is not evident from previous studies is the precise biochemical basis for the observed TSH charge microheterogeneity. Davy et al.[25] initially proposed that differences in amide content among the isohormones might explain the observed heterogeneity. However, others[30] have noted that the differences observed between the pI values of TSH isohormones are too small to be explained by differences in amide content of one or two residues per molecule as claimed by Davy et al.[25] In addition, no significant differences in amino acid composition among TSH isohormones have been reported. Nevertheless, posttranslational amino acid modification of TSH cannot be completely ruled out as a possible mechanism for the production of TSH isohormones.

Previous studies examining the isohormones of FSH[36,39,51] and hCG[52] (see also Chapters 7 and 11) have suggested that at least part of the observed charge microheterogeneity may result from subtle changes in the carbohydrate structure of the molecule. More recently, Cole and Grotjan[53] have shown that following endoglycosidase F treatment, which quantitatively released the asparagine N-linked sugar units, the various isohormones of human and ovine LH were reduced to a single, more basic species. Furthermore, the monumental work on TSH biosynthesis from the laboratory of Weintraub[1,10,14,54-56] has provided strong evidence in support of the concept that carbohydrate differences account for much of TSH heterogeneity.

Since the terminal residues on carbohydrate moieties are the ones which carry a net charge, it has been assumed that the differences among various isohormones are related to specific alterations in the incorporation of the terminal residues on the sugar chain. The microheterogeneity of FSH,[36,39,51] hCG,[52] and LH[57,58] is thought to result from variations in the number of terminal sialic acid residues. Several previous studies have attempted to relate TSH isohormones to varying amounts of sialic acid.[18,19,29] Shome et al.[18] suggested that the more electronegative nature of human TSH compared to bovine and porcine TSH was due to the presence of sialic acid on the human hormone. Removal of sialic acid from human TSH by neuraminidase treatment altered the mobility without affecting the hormonal or immunochemical properties of the hormone. It was later reported by Roos et al.[19] that treatment of human TSH with neuraminadase shifted the isohormones to more basic species but did not change the heterogeneous nature. Tamura and Ui,[29] utilizing purified whale TSH, observed no differences in sialic acid content among TSH isohormones. Recent studies performed in the author's laboratory have shown that treatment of crude pituitary extracts with neuraminadase coupled to agarose beads did not significantly alter the isohormone profile oTSH (Figure 5) while the isohormones of oFSH in the same extract were markedly shifted toward more basic species.[59] These results are not surprising in light of the recent findings of Green et al.[60] who demonstrated that the terminal charged residues of bTSH (and presumably oTSH) are sulfated rather than contain sialic acid. Based on these results, one may postulate that the isohormones of oTSH may be due at least in part to subtle differences in the incorporation of sulfate and/or sialic acid into the carbohydrate moiety of the protein. Clearly, more research is needed to accurately define the precise biochemical basis for TSH charge microheterogeneity.

V. ALTERATIONS IN TSH BIOLOGICAL ACTIVITY

In contrast to the gonadotropins, little information exists concerning the relationship between TSH isohormones and biological activity. Several studies have, however, indicated

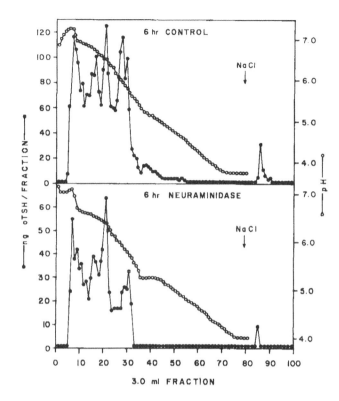

FIGURE 5. Chromatofocusing elution profile of immunoreactive oTSH present within a wether pituitary extract after incubation for 6 hr alone (control) in a water bath at 37°C or with the enzyme neuraminidase. The arrow indicates treatment of the column with 1.0 *M* NaCl.

occasional disparity between immunoreactive and bioactive levels of TSH. The results from these studies suggest a direct relationship between TSH isohormones, the carbohydrate composition of the hormone and its intrinsic biological activity.

Several papers have been published illustrating conditions in the human and rat in which TSH with impaired biological activity has been observed. Spitz et al.[7] described a euthyroid patient with circulating levels of TSH 20 to 50 times elevated above control patients. When evaluated in an in vitro bioassay for TSH it was observed that the patients' TSH bound to the TSH receptor but failed to fully activate adenylate cyclase. Faglia et al.[61] presented evidence for reduced biological activity of TSH in patients with hypothyroidism resulting from hypothalamic-pituitary disease. Those results were later confirmed and extended by Beck-Peccoz[62] who demonstrated a markedly impaired receptor-binding activity of TSH obtained from patients with central hypothyroidism. Interestingly, the biological activity of circulating TSH was significantly increased following TRH administration, suggesting that the releasing hormone not only regulates the release of TSH but also modulates the structural features of TSH that are essential for appropriate receptor binding and subsequent activation of adenylate cyclase.[62]

Using the rat as a model, Weintraub's group[63] demonstrated that thyroidectomy resulted in a decrease in the biological activity of intrapituitary-stored TSH without altering the biological activity of secreted TSH. In contrast, TRH treatment in vitro produced an increase in TSH bioactivity secreted into the medium. These authors concluded that TSH bioactivity is regulated differently by thyroid hormone deficiency and TRH; thyroidectomy appears to act mainly by altering the pool of newly synthesized molecules, whereas TRH acts mainly by stimulating the processing and secretion of more biologically active TSH. In support of this concept, Mori et al.[31] have demonstrated that TRH administration in vitro does not increase radioactive carbohydrate incorporation into pituitary TSH but apparently changes the heterogeneity of glycosylated TSH. Moreover, the addition of TRH in vitro augmented the incorporation of [^{3}H]glucosamine into TSH which was secreted into the medium.[32]

Additional studies have shown that heterogeneous molecular forms of TSH observed in the tumor extracts and serum of mice with thyrotropic tumors also displayed differing biological activities.[13,14] In unfractionated samples, serum TSH displayed significantly higher thyroid adenylate cyclase assay (ACA) to RIA and ACA to radioreceptor assay (RRA) ratios than TSH in tumor extracts.[13] Fractionation of the serum and tumor extracts by gel filtration revealed that the differences in biological activity between the two sources of TSH was due primarily to the presence of a molecular form in serum exhibiting thyroid stimulating activity that was absent from tumor extract. Based on previous biosynthetic studies by these investigators, which indicated that the various molecular forms of TSH observed in thyrotropic tumors differed only in carbohydrate content, it was concluded that the in vitro biological activity of TSH may be modulated by glycosylation.[1,10,13] It was later shown that forms of TSH with differing ACA/RIA ratios also have varying affinities for lectins, suggesting that they differ in carbohydrate composition[14] and further imply a relationship between in vitro biological activity and TSH glycosylation. In fact, it has been shown that the carbohydrates of TSH appear to be required for maximal adenylate cyclase by the hormone in vitro.[4] Furthermore, differences in the metabolic clearance rates of pituitary and serum TSH derived from hypothyroid and euthyroid rats have been reported,[64] suggesting that the circulatory half-life, and thus the in vivo biological activity, of TSH may also be related to carbohydrate content.

VI. CONCLUSIONS

The above studies indicate that TSH exists within the pituitary and the circulation as a mixture of heterogeneous forms which differ in charge and size. It is also apparent that TSH can exhibit wide differences in biological activity depending upon the endocrine status of the animal. These biochemical and biological differences appear to be directly linked to the carbohydrate composition of the hormone. Alterations in the hormonal milieu which result in changes in the biological activity of TSH also alter the microheterogeneity of TSH. Therefore, terminal glycosylation of TSH may be important in determining the molecular form as well as the overall biological activity of TSH. Clearly, more research is needed to more fully characterize TSH microheterogeneity, elucidate the biochemical basis for the observed polymorphism, and relate the existence of isohormones to TSH bioactivity.

ACKNOWLEDGMENTS

I would like to thank the NIH Pituitary Hormone Distribution Program for TSH radioimmunoassay reagents utilized in these studies. Excellent technical assistance has been provided by Ms. Robin Harms. Pituitary tissue was generously provided by Dr. Bruce D. Schanbacher, USDA Mean Animal Research Center, Clay Center, Neb. Research was supported by the Women's Research Institute and the Wesley Medical Research Institutes, Wichita, Kan.

REFERENCES

1. **Weintraub, B. D.**, Biosynthesis and secretion of thyrotrophin: relationship to glycosylation, in *Pituitary Hormones and Related Peptides*, Motta, M., Zanisi, M., and Piva, F., Eds., Academic Press, London, 1982.
2. **Daughaday, W. H.**, The Adenohypophysis, in *Textbook of Endocrinology*, Williams, R. H., Ed., W. B. Saunders, Philadelphia, 1974.
3. **Pierce, J. G. and Parsons, T. F.**, Glycoprotein hormones: structure and function, *Annu. Rev. Biochem.*, 50, 465, 1981.
4. **Berman, M. I., Thomas, C. G., Jr., Manjunath, P., Sairam, M. R., and Nayfeh, S. N.**, The role of the carbohydrate moiety in thyrotropin action, *Biochem. Biophys. Res. Commun.*, 133, 680, 1985.
5. **Erhardt, F. W. and Scriba, P. C.**, High molecular thyrotrophin ("Big"-TSH) from human pituitaries: preparation and partial characterization, *Acta Endocrinol.*, 85, 698, 1977.
6. **Klug, T. L. and Adelman, R. C.**, Evidence for a large thyrotropin and its accumulation during aging in rats, *Biochem. Biophys. Res. Commun.*, 77, 1431, 1977.
7. **Spitz, I. M., Roith, D. L., Hirsch, H., Carayon, P., Pekonen, F., Liel, Y., Sobel, R., Chorer, Z., and Weintraub, B.**, Increased high-molecular-weight thyrotropin with impaired biologic activity in a euthyroid man, *N. Engl. J. Med.*, 304, 278, 1981.
8. **Kourides, I. A., Weintraub, B. D., and Maloof, F.**, Large molecular weight TSH-β: the sole immunoactive form of TSH-β in certain human sera, *J. Clin. Endocrinol. Metab.*, 47, 24, 1978.
9. **Mori, M., Ohshima, K., Fukuda, H., Kobayashi, I., and Wakabayashi, K.**, Changes in the multiple components of rat pituitary TSH and TSHβ subunit following thyroidectomy, *Acta Endocrinol.*, 105, 49, 1984.
10. **Magner, J. A. and Weintraub, B. D.**, Biosynthesis of thyrotropin, in *The Thyroid: A Fundamental Clinical Text*, 5th ed., Ingbar, S. H. and Braverman, L. E., Eds., Lippincott, Philadelphia, 1986, 271.
11. **Weintraub, B. D., Krauth, G., Rosen, S. W., and Robson, A. S.**, Differences between purified ectopic and normal alpha subunits of human glycoprotein hormones, *J. Clin. Invest.*, 56, 1043, 1976.
12. **Diamond, R. C. and Rosen, S. W.**, Chromatographic differences between circulating and pituitary thyrotropins, *J. Clin. Endocrinol. Metab.*, 39, 316, 1974.
13. **Pekonen, F., Carayon, P., Amr, S., and Weintraub, B. D.**, Heterogeneous forms of thyroid-stimulating hormone in mouse thyrotropic tumor and serum: differences in receptor-binding and adenylate cyclase-stimulating activity, *Horm. Metab. Res.*, 13, 617, 1981.
14. **Joshi, L. R. and Weintraub, B. D.**, Naturally occurring forms of thyrotropin with low bioactivity and altered carbohydrate content act as competitive antagonists to more bioactive forms, *Endocrinology*, 113, 2145, 1983.
15. **Sluyser, M.**, Possible cause of electrophoretic and chromatographic heterogeneity of pituitary hormones, *Nature*, 204, 574, 1964.
16. **Carsten, M. E. and Pierce, J. G.**, Starch-gel electrophoresis and chromatography in the purification of beef thyrotropic hormone, *J. Biol. Chem.*, 235, 78, 1960.
17. **Wynston, L. K., Free, C. A., and Pierce, J. G.**, Further chromatographic studies on beef thyrotropin and a comparison of beef, sheep, and whale thyrotropins, *J. Biol. Chem.*, 235, 85, 1960.
18. **Shome, B., Parlow, A. F., Ramirez, V. D., Elrick, H., and Pierce, J. G.**, Human and porcine thyrotropins: a comparison of electrophoretic and immunological properties with the bovine homone, *Arch. Biochem. Biophys.*, 103, 444, 1968.
19. **Roos, P., Jacobson, G., and Wide, L.**, Isolation of five active thyrotropin components from human pituitary gland, *Biochim. Biophys. Acta*, 379, 247, 1975.
20. **Jacobson, G., Roos, P., and Wide, L.**, Human pituitary thyrotropin: characterization of five glycoproteins with thyrotropin activity, *Biochim. Biophys. Acta*, 490, 403, 1977.
21. **Jacobson, G., Roos, P., and Wide, L.**, Human pituitary thyrotropin: isolation and recombination of subunit isoforms, *Biochim. Biophys. Acta*, 625, 146, 1980.
22. **Wide, L.**, Median charge and charge heterogeneity of human pituitary FSH, LH and TSH, *Acta Endocrinol.*, 109, 181, 1985.
23. **Wide, L.**, Median charge and charge heterogeneity of human pituitary FSH, LH and TSH, *Acta Endocrinol.*, 109, 190, 1985.
24. **Fawcett, J. S., Dedman, M. L., and Morris, C. J. O. R.**, The isolation of bovine thyrotrophins by isoelectric focusing, *FEBS Lett.*, 3, 250, 1969.
25. **Davy, K., Fawcett, J., and Morris, C.**, Chemical differences between thyrotropin isohormones, *Biochem. J.*, 167, 279, 1977.
26. **Yora, T. and Ui, N.**, Purification and subfractionation of bovine thyrotropin and lutropin using radioimmunoassays for evaluation of the purification processes, *J. Biochem.*, 83, 1173, 1978.
27. Origin of the multiple components of whale thyroid-stimulating hormone, *Biochim. Biophys. Acta*, 214, 566, 1970.

28. **Tamura, H. and Ui, N.**, Purification and properties of whale thyroid-stimulating hormone, *J. Biochem.*, 71, 201, 1972.
29. **Tamura-Takahashi, H. and Ui, N.**, Purification and properties of whale thyroid-stimulating hormone. III. Properties of isolated multiple components, *Endocrinol. Jpn.*, 23, 511, 1976.
30. **Yora, T., Matsuzaki, S., Kondo, Y., and Ui, N.**, Changes in the contents of multiple components of rat pituitary thyrotropin in altered thyroid states, *Endocrinology*, 104, 1682, 1979.
31. **Mori, M., Murakami, M., Iriuchijima, T., Ishihara, H., Kobayashi, I., Kobayashi, S., and Wakabayashi, K.**, Alteration by thyrotrophin-releasing hormone of heterogeneous components associated with thyrotrophin biosynthesis in the rat anterior pituitary gland, *J. Endocrinol.*, 103, 165, 1984.
32. **Mori, M., Kobayashi, I., and Kobayashi, S.**, Thyrotrophin-releasing hormone does not accumulate glycosylated thyrotrophin, but changes heterogeneous forms of thyrotrophin within the rat anterior pituitary gland, *J. Endocrinol.*, 109, 227, 1986.
33. **Jackson, J. H. and Russell, P. J.**, Characteristics of the isoelectric focusing procedure: importance of column size, pH, and protein-protein interactions, *Anal. Biochem.*, 137, 41, 1984.
34. **Keel, B. A. and Grotjan, H. E., Jr.**, Characterization of rat lutropin charge microheterogeneity using chromatofocusing, *Anal. Biochem.*, 142, 267, 1984.
35. Pharmacia Fine Chemicals (1980), *Chromatofocusing with Polybuffer and PBE*, Pharmacia Fine Chemicals, Uppsala.
36. **Ulloa-Aguirre, A., Miller, C., Hyland, L., and Chappel, S.**, Production of all follicle-stimulating hormone isohormones from a purified preparation by neuraminidase digestion, *Biol. Reprod.*, 30, 382, 1984.
37. **Kennedy, J. and Chappel, S.**, Direct pituitary effects of testosterone and luteinizing hormone-releasing hormone upon follicle-stimulating hormone: analysis by radioimmuno- and radioreceptor assay, *Endocrinology*, 116, 741, 1985.
38. **Blum, W. F. P. and Gupta, D.**, Heterogeneity of rat FSH by chromatofocusing: studies on serum FSH, hormone released in vitro and metabolic clearance rates of its various forms, *J. Endocrinol.*, 105, 29, 1985.
39. **Blum, W. F. P., Riegelbauer, G., and Gupta, D.**, Heterogeneity of rat FSH by chromatofocusing: studies on in vitro bioactivity of pituitary FSH forms and effect of neuraminidase treatment, *J. Endocrinol.*, 105, 17, 1985.
40. **DePaolo, L. and Chappel, S.**, Alterations in the secretion and production of follicle-stimulating hormone precede age-related lengthening of estrous cycles in rats, *Endocrinology*, 118, 1127, 1986.
41. **Galle, P. C., Ulloa-Aquirre, A., and Chappel, S. C.**, Effects of oestradiol, phenobarbitone and luteinizing hormone releasing hormone upon the isoelectric profile of pituitary follicle-stimulating hormone in ovariectomized hamsters, *J. Endocrinol.*, 99, 31, 1983.
42. **Cameron, J. L. and Chappel, S. C.**, Follicle-stimulating hormone within and secreted from anterior pituitaries of female golden hamsters during the estrous cycle and after ovariectomy, *Biol. Reprod.*, 33, 132, 1985.
43. **Chappel, S. C., Bethea, C. L., and Spies, H. G.**, Existence of multiple forms of follicle-stimulating hormone within the anterior pituitaries of Cynomolgus monkeys, *Endocrinology*, 115, 452, 1984.
44. **Keel, B. A. and Grotjan, H. E., Jr.**, Influence of bilateral cryptorchidism on rat pituitary luteinizing hormone charge microheterogeneity, *Biol. Reprod.*, 32, 83, 1985.
45. **Keel, B. A. and Grotjan, H. E., Jr.**, Characterization of rat pituitary luteinizing hormone charge microheterogeneity in male and female rats using chromatofocusing: effects of castration, *Endocrinology*, 117, 354, 1985.
46. **Keel, B. A. and Grotjan, H. E., Jr.**, Charge heterogeneity of partially purified ovine LH subunits and the native hormone, *Proc. 18th Miami Winter Symp.*, Puett, D., Ahmad, F., Black, S., Lopez, D. M., Melner, M. H., Scott, W. A., and Whelan, W. J., Eds., Cambridge University Press, New York, 1986, 224.
47. **Keel, B. A., Schanbacher, B. D., and Grotjan, H. E., Jr.**, Ovine luteinizing hormone. I. Effects of castration and steroid administration on the charge microheterogeneity of pituitary LH, *Biol. Reprod.*, 36, 1102, 1987.
48. **Keel, B. A. and Grotjan, H. E.**, Ovine luteinizing hormone. III. Relationships between the charge microheterogeneity of partially purified subunits and the native hormone, *Biol. Reprod.*, 36, 1125, 1987.
49. **Pierce, J. G. and Parsons, T. F.**, Glycoprotein hormones: similar molecules with different functions, in *Evolution of Protein Structure and Function*, Sigman, D. and Brazier, M. A. B., Eds., Academic Press, New York, 1980.
50. **Fagerstam, L. G., Lizana, J., Ulla-Britt, A., and Lennart, W.**, Fast chromatofocusing of human serum proteins with special reference to α_1-Antitrypsin and Gc-Globulin, *J. Chromatogr.*, 266, 523, 1983.
51. **Chappel, S. C., Ulloa-Aguirre, A., and Coutifaris, C.**, Biosynthesis and secretion of follicle-stimulating hormone, *Endocr. Rev.*, 4, 179, 1983.

52. **Nwokoro, G., Chen, H., and Chrambach, A.,** Physical, biological and immunological characterization of highly purified urinary human chorionic gonadotropin components separated by gel electrofocusing, *Endocrinology,* 108, 291, 1981.

53. **Cole, L. A. and Grotjan, H. E., Jr.,** Carbohydrate structures and charge heterogeneity of hLH and oLH, *Endocrine Soc. Annu. Meet., Abstr. 511,* Anaheim, Calif., 1986.

54. **Gesundheit, N., Magner, J. A., Chen, T., and Weintraub, B. D.,** Differential sulfation and sialylation of secreted mouse thyrotropin (TSH) subunits: regulation by TSH-releasing hormone, *Endocrinology,* 119, 455, 1986.

55. **Weintraub, B. D., Stannard, B. S., and Meyers, L.,** Glycosylation of thyroid-stimulating hormone in pituitary tumor cells: influence of high mannose oligosaccharide units on subunit aggregation, combination, and intracellular degradation, *Endocrinology,* 112, 1331, 1983.

56. **Magner, J. A., Ronin, C., and Weintraub, B. D.,** Carbohydrate processing of thyrotropin differs from that of free α-subunit and total glycoproteins in microsomal subfractions of mouse pituitary tumor, *Endocrinology,* 115, 1019, 1984.

57. **Hattori, M., Ozawa, K., and Wakabayashi, K.,** Sialic acid moiety is responsible for the charge heterogeneity and the biological potency of rat lutropin, *Biochem. Biophys. Res. Commun.,* 127, 501, 1985.

58. **Wise, H. G., Graesslin, D., Lichtenberg, V., and Rinne, G.,** Polymorphism of human pituitary lutropin (LH): isolation and partial characterization of seven isohormones, *FEBS Lett.,* 159, 93, 1983.

59. **Keel, B.,** Unpublished observation.

60. **Green, E. D., Baenziger, J. U., and Boime, I.,** Cell-free sulfation of human and bovine pituitary hormones, *J. Biol. Chem.,* 260, 15631, 1985.

61. **Faglia, G., Bitensky, L., Pinchera, A., Ferrari, C., Paracchi, S., Beck-Peccoz, P., Ambrosi, B., and Spada, A.,** Thyrotropin secretion in patients with central hypothyroidism: evidence for reduced biological activity of immunoreactive thyrotropin, *Clin. Endocrinol. Metab.,* 48, 989, 1979.

62. **Beck-Peccoz, P., Amr, S., Menezes-Ferreira, M., Faglia, G., and Weintraub, B. D.,** Decreased receptor binding of biologically inactive thyrotropin in central hypothyroidism, *N. Engl. J. Med.,* 312, 1085, 1985.

63. **Menezes-Ferreira, M., Petrick, P. A., and Weintraub, B. D.,** Regulation of thyrotropin (TSH) bioactivity by TSH-releasing hormone and thyroid hormone, *Endocrinology,* 118, 2125, 1986.

64. **Constant, R. and Weintraub, B.,** Differences in the metabolic clearance of pituitary and serum thyrotropin (TSH) derived from euthyroid and hypothyroid rats, *Clin. Res.,* 32, 483A, 1984.

Chapter 11

HUMAN CHORIONIC GONADOTROPIN MICROHETEROGENEITY

H. Edward Grotjan, Jr. and Laurence A. Cole

I. INTRODUCTION

In 1927 Aschheim and Zondek observed that the urine of pregnant women contained abundant quantities of a gonadotropin with luteinizing activity (reviewed by Li;[1] Bagshawe et al.[2]). Although Aschheim and Zondek thought the pituitary might be the site of production, it was soon recognized that this hormone, human chorionic gonadotropin (hCG), is produced by the placenta and that it functions to maintain the corpus luteum after implantation has occurred.[2,3] Its concentrations rise during the first trimester of pregnancy, peak near the end of the first trimester and decline thereafter.[2-5]

Like human luteinizing (hLH), follicle-stimulating (hFSH), and thyroid-stimulating (hTSH) hormones, hCG is a heterodimeric glycoprotein. The amino acid sequences of its subunits (hCGα and hCGβ) are established.[6-9] The common alpha subunit is thought to have an identical peptide sequence among these four hormones (Chapter 1) and to arise from a single gene.[10-12] The hLHβ/hCGβ family is composed of at least eight genes.[13-16] The current thinking is that the family of seven or more hCGβ genes (or pseudogenes) arose from the hLHβ gene, which is present as a single copy in the genome, by a series of selected replacement changes with very little neutral drift.[9,17] Hence, hCGβ and hLHβ exhibit a high degree of peptide sequence homology. However, hCGβ possesses a carboxy terminal region which is approximately 30 amino acids longer than hLHβ.[6,7,9] This extension is thought to have resulted from a single base deletion which places the termination codon found in hLHβ out of frame and allows translation of the extra amino acids.[17,18]

hCG possesses four asparagine-linked oligosaccharides, two on each subunit, and four serine-linked sugar chains which reside on the carboxy terminal region of hCGβ. Although the detailed structures of its sugar chains have been reviewed in Chapter 2, information relevant to particular topics will, in part, be reiterated below.

When considering the heterogeneity of this particular hormone, one must be particularly aware of whether it is derived from tissues or body fluids. Whereas the pituitary glycoprotein hormones are almost exclusively derived from tissue sources, hCG is most commonly purified from urine[2,3,6,19] because there is essentially no storage of the hormone in the placenta (reviewed by Hussa[20,21]). There are also slight differences in the molecular characteristics of placentally derived and urinary hCG. Urinary hCG possesses a higher percentage of acidic oligosaccharides than placental hCG (Chapter 2). Similarly, one must also be cognizant of the method used to identify and quantitate the hormone. Early immunological approaches exhibited considerable cross-reactivity with hLH. As additional details about hCG structure became available and better antibodies were developed, immunological detection methods focused on hCGβ-specific approaches. With the identification of "hCGβ fragment" in urine (see below), the question of the most appropriate immunological method to accurately quantitate hCG has again been raised. For these reasons, the following discussion focuses primarily on papers which have confirmed that they are indeed examining hCG as evidenced by amino acid analyses and biological activity.

In preparing this discussion, the authors have elected not to extensively discuss the intracellular molecular forms of hCG and its subunits produced during biosynthesis. The reader is directed to other reviews[8,16,20-22,27] which cover hCG biosynthesis in depth.

Besides being produced by the normal placenta and in trophoblastic diseases, hCG and its subunits are thought to be produced ectopically in very low quantities by some normal tissues and in various quantities by nontrophoblastic tumors.[2,4,6,21,24-34] From the latter perspective, there is considerable interest in the ectopic production of hCG and its subunits as potential tumor markers. Recent information suggests that the production of hCG α-β dimers and perhaps uncombined hCG subunits by nontrophoblastic tumors may not be as prevalent as previously thought, at least as judged by circulating concentrations.[5] Although it is not the purpose of this chapter to review tumor and ectopic hCG production, the distinct molecular forms produced by certain trophoblastic tumors will be noted because they contribute to the observed heterogeneity.

Ideally, this chapter should deal with the heterogeneity of the chorionic gonadotropins from several species. However, as noted in Chapter 2, only hCG and equine chorionic gonadotropin (eCG) have been characterized in detail. eCG heterogeneity has only been minimally examined.[35] Thus, the heterogeneity of chorionic gonadotropins from species other than the human remains to be investigated.

II. SIZE HETEROGENEITY

Considering the dimeric structure of hCG, it is likely that uncombined subunits could be present in placental tissue and body fluids. Indeed this has proven to be the case (see Chapter 3).[4,5,25,27,36,37] Moreover, crude preparations of urinary hCG contain considerable quantities of uncombined subunits.[38]

hCG subunits of aberrant molecular sizes have been described. For example, Vaitukaitis and co-workers noted a molecular form of hCGα in placental extracts and serum which was slightly larger than hCGα derived from native hCG.[4,25,27] It has subsequently been demonstrated that large hCGα, frequently termed free-hCGα (f-hCGα), represents a distinct molecular form which can possess a serine-linked oligosaccharide at threonine 43 and may have asparagine-linked oligosaccharide structures different than those found on native hCG (see Chapters 2 and 3). Interestingly, f-hCGα cannot combine with hCGβ to form α-β dimers (Chapter 3).

Small molecular weight substances which react in radioimmunoassays for hCGβ are present in the urine of pregnant women.[36,37,39-42] These may be degradation products of hCGβ.[40,43,44] They complicate the accurate measurement of hCG and hCGβ in urine because they can constitute as much as 70% of the immunoreactive materials in some assays.[41]

Hussa, Cole, and co-workers[33,45-47] identified an abnormal form of hCGβ produced by tumor cells (cervical and other epidermoid carcinomas) which exhibited a slightly larger molecular size than hCGβ derived from native hCG (n-hCGβ) during gel filtration. When subjected to ion exchange on a diethylaminoethyl (DEAE) column, this form of hCGβ eluted as six peaks.[33] After desialylation, it eluted from DEAE as two peaks, one of which corresponded to that found with desialylated n-hCGβ; the other was unique to tumor hCGβ.[33] The unique type of hCGβ could not form α-β dimers and appeared to either lack the carboxy terminal region or have altered structural features in this portion of the molecule.[33,45-47] The reader is directed to Chapter 3 for a more detailed discussion of normal and abnormal hCG subunits.

Large molecular weight immunoreactive hCG-like substances, which elute in the void volume of gel filtration columns (Sephadex® G-100 or an equivalent series of high performance liquid chromatography columns), have been described by several investigators.[34,39,42,48,49] These large molecular weight forms of hCG were observed in extracts of normal placenta[39,48,49] and in the urine[34,42] but were generally absent in serum[39] or culture medium.[48] They appeared to be relatively resistant to dissociation by chaotropic agents such as guanidine[39,48] or urea.[49] Tojo et al.[48] proposed that limited digestion by trypsin-like enzymes could convert them into native hCG and suggested that they could be precursors ("prohormones") for native hCG. However, it is now generally accepted that the subunits of hCG are synthesized individually.[8,16,20-23] When analyzed by sodium dodecyl sulfate polyacrylamide gel electrophoresis (SDS-PAGE) under denaturing conditions, large molecular weight immunoreactive hCG-like substances were dissociated into three components, none of which corresponded to native hCG or its subunits.[49] Thus, the best evidence presently available suggests that these large molecular weight forms of immunoreactive hCG are simply aggregated or bound materials.[20,21,50]

III. CHARGE HETEROGENEITY

A. Pregnancy hCG

Much of the literature on the charge heterogeneity of native hCG deals with purified preparations which have been derived from first trimester pregnancy urine. Some of these studies, particular the early ones, were not undertaken to define hCG heterogeneity per se but were directed towards obtaining pure preparations with maximal in vivo biological potencies. Thus, some comments regarding the history of hCG purification as it relates to the observation that hCG exhibits charge heterogeneity are appropriate. One must be aware that the potency of the relevant preparations was quantitated by in vivo bioassays which reflect a combination of circulatory survival *and* effect at gonadal target cells.

Early studies regarding the purification of hCG have been reviewed by Li in 1949.[1] The most highly purified preparations available by this time had potencies of ≈9000 International Units (IU)/mg and some of these were considered to be homogeneous by the people who prepared them.[1] Got and Bourrillon[51-53] (1960) are generally given credit for obtaining the first highly purified preparations of hCG[3] which, in their case, had potencies of ≈12,000 IU/mg. These preparations were considered to be homogeneous,[52] but the electrophoresis system employed does not have the resolving power of those in use today. In the 1960s and early 1970s several laboratories[54-62] reported the purification of hCG using crude urinary hCG with a potency of 1500 to 3000 IU/mg as the starting material. Crude urinary hCG contained several biologically active peaks when subjected to ion-exchange chromatography on DEAE columns.[59,60,63] A common approach was to take the most active peak from ion-exchange chromatography and further purify the hCG by gel filtration.[3,59,60] The resulting hCG preparations exhibited several bands in electrophoresis systems which separate molecules on the basis of charge[54-58,61] or during isoelectric focusing (IEF).[62] Some of the subcomponents had potencies greater than 18,000 IU/mg.[54,55,58,60] Components with lower potencies were similar to highly potent preparations immunochemically,[54,57] in amino acid composition[58] or during circular dichroism[62] suggesting that hCG was composed of several molecular forms with similar peptide chains. Removal of sialic acid residues reduced the number of bands observed during electrophoresis and markedly reduced in vivo biological potencies.[54,56,64]

During this same era, other studies[65-71] also suggested hCG was heterogeneous. For example, van Hell[70] reported that crude urinary hCG was composed of at least nine components, which could be resolved by IEF, with isoelectric points (pIs) between 3.8 to 5.9 and potencies ranging from 1700 to 18,000 IU/mg. The major difference in these components was their sialic acid contents with the more heavily sialylated forms exhibiting more acidic pIs and higher biological activities.[70] Similarly, Qazi et al.[71] resolved crude urinary hCG into 13 components by IEF and observed that the most biologically active fraction (≈15,000 IU/mg) had a pI of 4.3

These studies provided the framework for subsequent studies because they demonstrated that hCG is heterogeneous with respect to charge and suggested that the heterogeneity is primarily related to sialic acid content. In 1982, Canfield et al.[7] noted that purified hCG has a potency of 12,000 to 14,000 IU/mg but it is possible to obtain preparations with higher potencies by selecting for forms with higher sialic acid content. This statement provides a perspective regarding purified urinary hCG and implies that charge heterogeneity is an intrinsic property.

Several studies have explicitly examined the charge heterogeneity of purified urinary hCG.[72-80] Graesslin et al.[72,73,77] noted that hCG (7600 to 12,000 IU/mg) yielded five bands with pIs in the range of 3.8 to 5.1 when subjected to IEF. Each subcomponent yielded a single sharp peak when analyzed by gel filtration and a single arc when analyzed by immunoelectrophoresis.[72,73] In a subsequent study, Graesslin et al.[74] observed a sixth band

CHARGE-HETEROGENEITY OF URINARY HCG

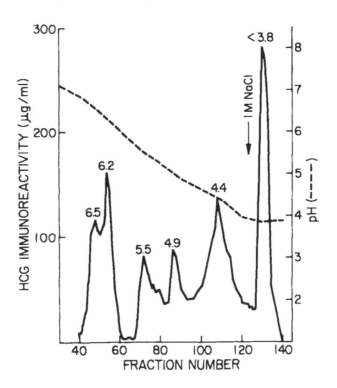

FIGURE 1. Chromatofocusing elution profile of purified urinary hCG. Apparent pIs are noted above the peaks. Components bound to the column were eluted with 1 *M* NaCl.[80] (From Chappel, S. C. and Cole, L. A., *Proc. 67th Annu. Meet. U.S. Endocrine Society*, Abst. 558, 1985.)

during IEF (apparent pI = 5.4). hCGα separated into two components in disc gel electrophoresis while hCGβ separated into five fractions with pIs in the range 3.7 to 5.0. Each component of hCG possessed the two hCGα bands in varying proportions and two of the five hCGβ bands.[74] Similar data have been published by Yazaki et al.[78] Merz et al.[75,76] observed that purified hCG (12,500 IU/mg) yielded six bands during IEF with pIs in the range of 4.0 to 5.2. These six fractions were each composed of hCG subunits when analyzed by SDS-PAGE[75] and had similar amino acid as well as carbohydrate compositions except for sialic acid.[76] Using a higher resolution IEF system, Nwokoro et al.[79] observed that purified hCG (11,600 IU/mg; CR119) could be resolved into 14 distinct components with pIs in the range of 4.4 to 6.3. Of these 14 components, the 6 major fractions collectively constituted more than 50% of the hormone if procedural losses are considered.[79] hCGα resolved into eight components with pIs in the range of 5.1 to 8.2 while hCGβ resolved into five components with pIs of 3.9 to 5.2.[79] The major components had similar potencies in radioimmunoassays and radioreceptor assays as well as similar amino acid compositions.[79] Other investigators[32] noted similar IEF patterns for CR119 although fewer peaks were observed. Similarly, Chappel and Cole[80] identified six major components of urinary hCG during chromatofocusing on pH 7 to 4 gradients (Figure 1). The careful analyses performed by these investigators established that multiple molecular forms of hCG exist. Although as many as 14 distinct forms have been observed, there appear to be 6 major forms or charge isomers of hCG. As summarized in Table 1, these multiple molecular forms appear to primarily result from varying degrees of sialylation.

Table 1

**CHARACTERISTICS OF THE SIX MAJOR COMPONENTS
PRESENT IN VARIOUS PURIFIED hCG PREPARATIONS**

Component	Isoelectric point	Sialic acid (% by weight)	In vivo bioassay potency (IU/mg)	Distribution (%)	Ref.
Native hCG[a]	—	6.4	7,600	—	72, 74
A	3.8	9.0	4,600	N.R.[b]	
B	4.3	8.8	18,500	N.R.	
C	4.5	6.7	14,200	N.R.	
D	4.9	5.5	12,500	N.R.	
E	5.1	2.9	7,400	N.R.	
F	5.4	N.R.	N.R.	N.R.	
Native hCG	—	9.3	12,600	—	75
1	4.0	8.6	14,700	N.R.	
2	4.2	10.9	16,200	N.R.	
3	4.6	10.0	14,700	N.R.	
4	4.7	9.8	10,900	N.R.	
5	4.9	9.5	8,600	N.R.	
6	5.2	9.3	4,500	N.R.	
Native hCG[c]	—	9.3	11,600	—	79
6	4.5	12.0	10,700	2.7	
5	4.6	9.0	9,200	9.2	
4	4.7	7.6	7,800	11.8	
3	4.8	8.5	5,300	12.5	
2	4.9	8.4	4,200	7.0	
1	5.1	4.5	720	1.0	

[a] Native hCG refers to the preparation from which the components were prepared.
[b] N.R. = not reported.
[c] CR119.

Table 1 also presents the in vivo bioassay potencies and distribution (when available) of the major subcomponents in purified hCG. There is reasonable agreement in the data published by the three different laboratories. It is evident from the data in Table 1 that the six major components ("isohormones") of urinary hCG vary in their biological potencies and sialic acid contents. There is a general trend for the isohormones with the higher sialic contents to have the lower pIs and higher biological potencies although exceptions from this trend are also evident. hCG isohormones with pIs between 4.0 and 4.4 are the most potent. Additional evidence that the pI of hCG is related to sialic acid content comes from studies where it has been demonstrated that desialylated hCG, hCGα, and hCGβ have apparent pIs of 9.5 to 10.0, 9.5, and 10.5, respectively.[74,81,82]

Although the charge heterogeneity of hCG purified from pregnancy urine has been reasonably well characterized, there is a meager number of papers which describe the isohormones of hCG present in body fluids. The most thorough studies have been performed by Yazaki et al.[78] In their hands, purified hCG (CR119) separated into six components (pIs of 3.9, 4.1, 4.4, 4.7, 5.0, and >5.8) with forms having pIs of 4.1 and 4.4 being the most prevalent. hCG in the serum of normal pregnancy patients separated into similar components as well as an additional peak with an apparent pI of ≈6.7[78] (a form of CR119 with a similar pI was noted by Matsuura et al.[32]). Placental explants also appear to secrete similar hCG isohormones.[80] Thus, it would appear that the hCG which is released into circulation is also charge heterogeneous and corresponds to the molecular forms found in purified urinary hCG.

The charge heterogeneity of hCG in normal placental tissue remains to be thoroughly

METABOLISM OF HCG ISOHORMONES

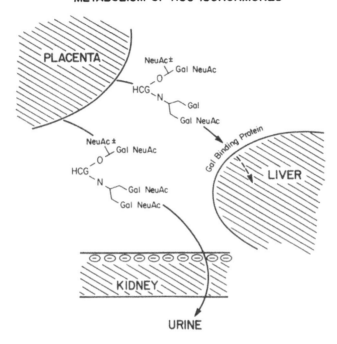

FIGURE 2. Clearance of circulating hCG. Forms of hCG with exposed galactose (Gal) residues are rapidly cleared by the liver[86-90] while heavily sialylated forms (containing N-acetylneuraminic acid; NeuAc) are not. The heavily sialylated forms are also repulsed by the kidney glomerular charge barrier[95] which further promotes their survival. Nonetheless, their primary route of disposal is in the urine. Removal of side-arm NeuAc residues from serine-linked oligosaccharides does not expose Gal and has minimal effects on hepatic uptake but may aid in resisting glomerular filtration. See Chapter 2 for additional details regarding oligosaccharide structures and Chapter 3 for a more detailed discussion of clearance.

examined. Ashitaka[83] (1970) reported the purification of hCG from normal placenta and characterized the resulting preparations by appropriate chemical analyses. hCG from first trimester placenta was shown to be heterogeneous by differential solvent extraction and DEAE ion-exchange chromatography. The resulting preparations had potencies of 4600 to 8200 IU/mg and sialic acid contents lower than urinary hCG (3.0 to 6.0% vs. 9.5 to 11.0%).[83] This and other observations (Chapter 2) suggest that intracellular forms are less heavily sialylated than circulating or urinary forms.

Since the studies of Rafelson et al.[64] (1961) it has been known that the desialylation of hCG markedly reduces its in vivo biological potency. Removal of only 25% of the sialic acid from hCG reduced its in vivo potency by ≈75%.[25,84] Removal of 62% or more of the sialic acid reduced the potency of hCG from >11,000 IU/mg to <100 IU/mg (i.e., essentially obliterated bioactivity).[25,84,85] The desialylation of hCG is thought to unmask galactose residues on its oligosaccharides which results in rapid clearance by the liver.[86-90] The potency of desialylated hCG in radioimmunoassays and radioreceptor assays is minimally affected.[84,88,91] The potency of asialo-hCG as assessed by in vitro bioassay is only slightly reduced (15 to 50%)[88,91-94] suggesting that the primary effect of desialylation is to increase the metabolic clearance rate rather than reduce potency at the target-cell level. This concept is illustrated in Figure 2. Soon after molecular forms of hCG with exposed galactose residues appear in circulation they are rapidly cleared by the liver. Heavily sialylated forms are not

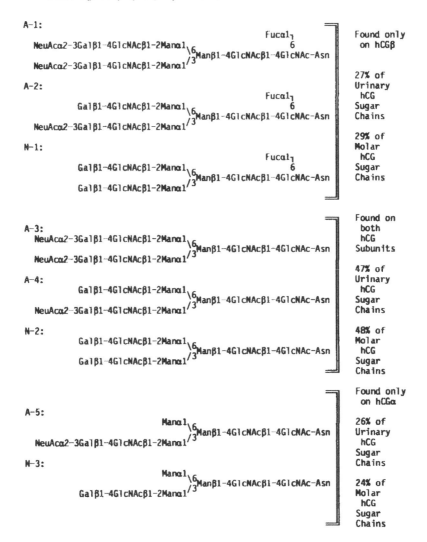

FIGURE 3.　Detailed structures for the asparagine (Asn)-linked oligosaccharides of hCG as proposed by Mizouchi and Kobata.[96] "A" denotes acidic oligosaccharides while "N" denotes neutral oligosaccharides. Urinary hCG contains greater than 90% acidic oligosaccharides while placental hCG contains approximately 20% neutral structures.[96,97] hCG produced by hydatidiform moles has sugar chains identical to pregnancy hCG.[97] The distribution of oligosaccharides among the groups of structures as presented by Mizouchi et al.[97] is noted. Abbreviations: NeuAc = *N*-acetylneuraminic acid, Gal = galactose, GlcNAc = *N*-acetylglucosamine, Man = mannose, and Fuc = fucose.

cleared by the liver and are also repulsed by the negatively charged glomerular basement membrane[95] yielding long half-lives in circulation (see Chapter 3 for a more detailed discussion). Even though they survive in circulation longer, they are ultimately cleared by the kidney. These mechanisms account for the higher percentage of sialylated forms of hCG in urine compared to placenta tissue.

Let us now consider the relationship between pI and oligosaccharide structure. Recall that hCG possesses four biantennary asparagine-linked sugar chains which can contain zero, one, or two sialic acid residues (Figure 3) and four serine-linked oligosaccharides which can contain one or two sialic acid residues (Figure 4; see Chapter 2). If one assumes no peptide heterogeneity and that the *position* of an individual sialic residue is inconsequential, 14

		Urine	Placenta	Choriocarcinoma

O-1: NeuAcα2-3Galβ1-4GlcNAcβ1\₆GalNAc-Ser / NeuAcα2-3Galβ1/3 — 13% | 14% | 51%

O-2: NeuAcα2\₆GalNAc-Ser / NeuAcα2-3Galβ1/3 — 34% | 73% | 29%

O-3: NeuAcα2-3Galβ1-3GalNAc-Ser — 43% | <1% | 10%

O-4: NeuAcα2-6GalNAc-Ser — 10% | 5% | 2%

O-5: Galβ1-4GlcNAcβ1\₆GalNAc-Ser / Galβ1/3 — ≈0% | <1% | 1%

O-6: Galβ1-3GalNAc-Ser — ≈0% | 7% | 7%

FIGURE 4. Detailed structures for the serine (Ser)-linked oligosaccharides of hCGβ.[98,100] The distribution of structures on hCG in urine, on hCG produced by normal pregnancy trophoblast explant cultures (second trimester), and on hCG produced by cultured choriocarcinoma (JAr) cells is also presented. Abbreviation: GalNAc = N-acetylgalactosamine; other abbreviations presented in the legend of Figure 3.

different charged isomers can theoretically be envisioned (0 to 8 sialic acid residues on asparagine-linked oligosaccharides and 4 to 8 sialic residues on serine-linked oligosaccharides). This could account for the large number of isohormones observed in certain studies.[70,71,79] Incidentally, if the position of individual sialic acid residues is taken into consideration, literally hundreds to thousands of isomers are possible. In all likelihood, the relationship between pI and sialic acid content is probably not a simple function of the number of sialic acid residues.

To develop some general concepts, let us concentrate on purified urinary hCG because the structures of its oligosaccharides are reasonably well characterized (Chapter 2). To simplify matters slightly, let us assume that no charge heterogeneity arises from the peptide chains (i.e., the net charge of subunit chains are constants at unspecified pIs), that all charge heterogeneity results from terminal sialic acid residues and that the position of an individual sialic acid residue is inconsequential. Recall that urinary hCG has a somewhat systematic distribution of asparagine-linked oligosaccharides[96] (see Chapter 2 for a detailed discussion). Urinary hCG contains less than 10% neutral oligosaccharides[97] and these appear almost exclusively on hCGα.[96] As a further simplification, assume all neutral oligosaccharides are found on hCGα.

hCGα oligosaccharides are not fucosylated and are found in five distinct structures (A-3, A-4, A-5, N-2, and N-3 in Figure 3 where "A" denotes acidic oligosaccharides containing sialic acid while "N" denotes neutral oligosaccharides devoid of sialic acid). hCGα oligosaccharides are less acidic than hCGβ oligosaccharides and primarily exist in monosialylated forms (≈50% of structures; asialo forms constitute ≈20% of the structures while disialylated forms constitute ≈30%).[96] Furthermore, one glycosylation site on hCGα appears to be of the terminal mannose type (A-5 and N-3 in Figure 3)[96] which thus can only be asialo or monosialylated. These circumstances yield a limited number of charged isomers of hCGα (theoretically four with zero to three sialic acid residues). *If the distribution of oligosaccharides is considered* one ends up with a realistic estimate of two or three major charged variants. In fact, a limited number of forms have been observed by some authors[74,78] but not others[79] (also, see below under "hCGα").

The situation is more complex for hCGβ because it has both asparagine- and serine-linked oligosaccharides. The asparagine-linked oligosaccharides of hCGβ may or may not be fucosylated and are found in six distinct structures (A-1, A-2, A-3, A-4, N-1, and N-2 in Figure 3). N-1, A-1, and A-2 are simply the fucosylated forms of N-2, A-3, and A-4, respectively. The asparagine-linked oligosaccharides of hCGβ reduce to three charged variants because fucose is a neutral sugar. Because the asparagine-linked sugar units of hCGβ primarily (>90%) exist as mono- and disialylated forms with an approximate equal distribution between the two,[96] one again ends up with two to four major charged variants resulting from asparagine-linked sugar chains. Urinary hCGβ has four serine-linked oligosaccharides which are approximately equally distributed between mono- and disialylated forms (Figure 4). Using similar logic this yields a realistic estimate of two to three major charged variants resulting from serine-linked glycans. Although there is a larger number of potential charged isomers for hCGβ, five distinct isoelectric variants have consistently been observed.[74,78,79] These hypothetical considerations suggest that hCGβ makes a greater contribution to the pI of the heterodimer which has, in fact, been observed experimentally (see above). From this analysis it is relatively easy to envision why only six major charge isomers of hCG are consistently observed. Recently, Cole[99] proposed that a large portion of the charge heterogeneity of hCG is related to the relative abundance of tri- vs. tetrasaccharide structures present as the serine-linked sugar chains of hCGβ which is quite consistent with the hypotheses presented above.

Generally, then, the *number* of sialic acid residues accounts for the pI of a given hCG isohormone as well as a large portion of the observed hCG charge heterogeneity. The beta subunit contributes more to the pI and the observed charge heterogeneity than the alpha subunit. Furthermore, the in vivo biological potency is largely a function of sialic acid content. Nonetheless, there are particular molecular forms which do not appear to fit these general rules. Some molecular forms have relatively high in vivo biological potencies but lower sialic acid contents than other forms with high biological potencies (Table 1). This suggests that the *position* of individual sialic acid residues may also be involved. Hypothetically, forms lacking a side-arm sialic acid on serine-linked oligosaccharides would not be rapidly cleared by the liver because galactose is not exposed. This may be the reason that sialic acid content and biological activity are not precisely correlated.

An implicit assumption in the putative scheme presented above is that given oligosaccharide structures are somewhat systematically distributed. Although this is true for asparagine-linked oligosaccharides, the serine-linked sugar chains are randomly distributed among the four glycosylation sites (Chapter 2).[98-100] Thus, it is possible that each glycosylation site on each isohormone does not possess a unique sugar chain but possesses a specific distribution in terms of percentage of the multiple variations in sugar chains which exist on hCG.

This analysis provides a view of the "forest" but it does not provide a clear picture of the "trees". That is, the distinct oligosaccharides structure or structures (and percent distribution) which reside at each glycosylation site on each hCG isohormone remain to be determined.

In summary, native hCG in urine, serum, or possibly placenta may exist in as many as 14 (or more) isohormones with slight differences in pI. Of these, six components are major. Both the pI and in vivo potency of each form appear to be somewhat, but not strictly, related to sialic acid content. The observation that in vivo potency is not strictly a function of sialic acid content can be explained by hypothesizing that the number of terminal galactose rather than the number of sialic acid residues per se is the critical factor which determines in vivo biological potency.

B. hCG Produced by Trophoblastic Tumors

Using zone electrophoresis and DEAE ion-exchange chromatography, Reisfeld et al.[101] (1959) first reported a difference in the charge characteristics of serum hCG between that produced during normal pregnancy and patients with "advanced trophoblastic tumors". In 1960, Reisfeld and Hertz[102] purified hCG from the urine of these tumor patients. The resulting hCG was of high potency ($\approx$15,000 IU/mg) and had a pI of 2.95 whereas hCG from pregnancy urine exhibited pIs of 3.8 to 4.0. Canfield et al.[3] also reported the purification of hCG from the urine of patients with hydatidiform mole and choriocarcinoma (potencies up to $\approx$15,000 IU/mg) and noted that they had amino acid compositions similar to native hCG. These important observations suggested that hCG produced by trophoblastic tumors has similar peptide constituents to native hCG but could have unique molecular characteristics resulting in high in vivo biological potencies and extremely acidic pIs.

hCG produced by hydatidiform moles has subsequently been purified from urine,[97,103-105] plasma,[103] and tissue.[103,106-108] Molar hCG excreted into urine consistently eluted as four or five peaks when subjected to ion-exchange chromatography on DEAE; these components exhibited variable in vivo biological potencies which ranged from $\approx$100 to $\approx$20,000 IU/mg.[103-105] hCG from this source had amino acid compositions similar to native hCG but variable sialic acid contents ($\approx$6.1%, $\approx$4100 IU/mg;[104] $\approx$8.5%, $\approx$7000 IU/mg[105]). hCG extracted directly from molar tissues also generally eluted as four to six peaks on DEAE, exhibited variable potencies which ranged from $\approx$800 to $\approx$30,000 IU/mg, variable sialic acid contents ($\approx$0.4%, $\approx$1500 IU/mg;[108] 6.2%, $\approx$30,000 IU/mg;[107] 8.2%, $\approx$23,000;[106] and 9.3%, $\approx$8600 IU/mg[106]) but amino acid analyses similar to native hCG.[107,108] hCG in the plasma of a patient with hydatidiform mole was considerably more homogeneous in that it eluted as a single peak on DEAE and was of high potency ($\approx$23,000 IU/mg).[103] Perhaps the high potency of hCG derived from plasma is due to selection of forms with prolonged circulatory survival (which would presumably be present in high concentrations in blood). The apparent pI of molar hCG has been reported to be 4.9.[103]

Recently, Mizouchi et al.[97] have structured the asparagine-linked oligosaccharides of molar hCG excreted into the urine and observed that they are identical to those of normal urinary hCG produced during pregnancy (Figure 3). Furthermore, the distribution of structures is identical between molar and normal hCG.[97]

These results suggest that hCG derived from hydatidiform moles is not markedly different from native hCG except for slightly to markedly lower contents of sialic acid in particular cases. This conclusion is supported by the studies of Yazaki et al.[78] who observed similar hCG isohormones in the serum of normal pregnant patients and those with hydatidiform moles.

In contrast, serum from choriocarcinoma patients had the same isohormones as native hCG (see above) as well as three additional forms with apparent pIs of 3.2, 3.5, and 3.7.[78] Choriocarcinoma hCG has been purified by Nishimura et al.[105] The resulting preparation had a potency of 400 IU/mg but was similar to native hCG in amino acid analysis. The low in vivo potency most likely resulted from the lack of sialic acid.[105] Mizouchi et al.[109] subsequently performed detailed structural analysis of the asparagine-linked sugar chains of this particular preparation (Figure 5). Only 3% of the oligosaccharides contained sialic acid and all were monosialylated.[109] However, in a subsequent study[97] it was observed that hCG was variably sialylated (3 to 100% of the oligosaccharides were acidic) by individual choriocarcinoma patients. Perhaps more importantly, choriocarcinoma hCG possessed asparagine-linked oligosaccharide structures not found on native hCG (multiantennary and fucosylated terminal mannose chains illustrated as A, B, C, D, and G in Figure 5). The unique oligosaccharide structures constitute approximately 50% of the sugar chains of choriocarcinoma hCG.[97]

Variable degrees of sialylation have also been described on the serine-linked oligosaccharides of choriocarcinoma hCG.[110] However, hCG produced by choriocarcinoma (JAr) cells has recently been demonstrated to have the same serine-linked sugar chains as normal hCG but to have a markedly different distribution of structures with the hexasaccharide (O–1 in Figure 4) being the predominate form.[100]

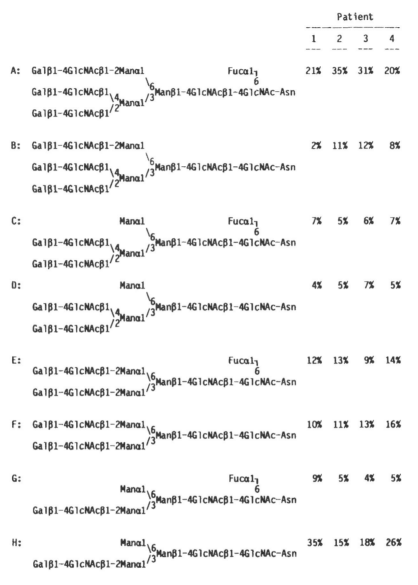

FIGURE 5.   Detailed structures for the asparagine (Asn)-linked oligosaccharides of urinary hCG from choriocarcinoma patients as proposed by Mizouchi et al.[97,109] All of the above are illustrated as neutral (asialo) oligosaccharides but considerable quantities of sialylated oligosaccharides were observed in some patients (up to 100%).[97] Acidic oligosaccharides contained *N*-acetylneuraminic acid presumably α2–3 linked to galactose residues. The distribution of oligosaccharides in four patients[97] is also presented. See legend of Figure 3 for abbreviations.

Thus, hCG produced by hydatidiform moles appears to be similar to normal pregnancy hCG produced by choriocarcinomas is, for the most part, composed of unique molecular forms produced by this tumor. Choriocarcinomas appear to produce hCG with multiantennary sugar chains attached to asparagine. These, as well as the serine-linked glycans, appear to be variably sialylated. However, when heavily sialylated, the unique molecular forms of choriocarcinoma hCG have very acidic pIs.

C. Ectopic hCG

hCG-like substances produced by normal tissues[26,29-32,42] and nontrophoblastic tumors[24,28,31,111] have not been isolated or characterized biochemically. It is known that they have varying affinities for the lectin concanavalin A (ConA).[26,28,30,111] Although lack of binding to ConA has been interpreted to suggest that such forms of hCG are not glycosylated (for example[26]), a more likely explanation is that they possess oligosaccharides (multiantennary or bisecting N-acetylglucosamine types) which do not bind to the lectin. Other studies[31] also suggest altered glycosylation of ectopic hCG. Thus, ectopic hCGs remain to be more fully characterized but the available data suggest that they also consist of unique molecular forms with oligosaccharides structures not found on pregnancy hCG.

D. hCGα

The charge characteristics of hCGα derived from native hormone (n-hCGα) have, in part, been reviewed above. Benveniste et al.[112,113] reported that n-hCGα can be resolved into seven to eight peaks with apparent pIs (estimated) of ≈4.8, 5.3, 6.1, 6.8, 7.3, 8.2, 8.4, and 8.8 during IEF. Similar data have been published by Nishimura et al.[82,114] and Chen et al.[115] who noted that a majority of the n-hCGα activity focused with an apparent pI >5.1. Thus, n-hCGα has pIs which are slightly more basic than those of native hCG. In contrast, free-hCGα (f-hCGα), which cannot combine with hCGβ to form native hCG (see Chapter 3), derived from normal placental tissue,[115,116] choriocarcinoma cells[112,113] or undifferentiated carcinoma[114] appears to be more acidic than n-hCGα and primarily focuses in peaks with pIs of ≈4.8, 5.1, and 6.3. In other studies, f-hCGα produced by tumors has been observed to exhibit extremely acidic pIs (<4.0).[82] These observations are consistent with f-hCGα possessing a sialylated, serine-linked oligosaccharide (the serine-linked oligosaccharides of f-hCGα are similar in structure to those of hCGβ which are illustrated in Figure 4) and/or unique asparagine-linked structures (see Chapters 2 and 3). Recently, Cox[117] as well as Saccuzzo et al.[188] noted that hCGα can be phosphorylated by tumor cells which could potentially contribute to its charge heterogeneity and acidic characteristics.

E. hCGβ

There has been less interest in the molecular characteristics of free-hCGβ because only a small percentage exists in the uncombined state (see Chapter 3). The charge characteristics of hCGβ derived from hCG as related to the heterogeneity of the native hormone have been presented above. hCGβ can be phosphorylated in vitro at threonine 97 by cyclic adenosine monophosphate-dependent protein kinase in its native configuration (but not when combined with hCGα) and at serines 66 and 96, if reduced and carboxymethylated.[119] However, there is no evidence that phosphorylation of hCGβ peptides occurs under normal circumstances.[119]

V. HETEROGENEITY OF OTHER TYPES

hCGα is known to exhibit heterogeneity at the amino terminus which is thought to be the result of endogenous (tissue) proteases.[6] Approximately 60% of the chains have alanine at the amino terminus, 10% have asparagine (normally residue 3), and 30% have valine (normally residue 4).[6] Fucose is normally found only on hCGβ (Figure 3; see Chapter 2) but certain tumor cells are capable of fucosylating hCGα oligosaccharides.[120] Recently, Nishimura et al.[121] identified a form of f-hCGα produced by an undifferentiated carcinoma in which alanine was substituted for glutamic acid at residue 56. Like other f-hCGαs (Chapter 3), this variant cannot form α-β dimers.[121]

Multiple forms of pre-hCGβ which do not appear to result from glycosylation have been identified.[23,122] The aberrant form may have a posttranslational modification near arginine 133 which alters the charge characteristics of this region but which is not the result of O-glycosylation, sulfation, phosphorylation, methylation, or ADP-ribosylation.[123]

As noted above, a family of at least seven hCGβ genes exists.[13-16] The current thinking is that at least two of the hCGβ genes are expressed.[14-16] Thus, it is possible that multiple forms of hCGβ, with slight differences in amino acid sequence,[14] may be synthesized normally or ectopically.

IV. CONCLUDING REMARKS

It is evident that hCG and its subunits are produced in multiple molecular forms. Certain tumors, and particularly choriocarcinomas, produce unique molecular forms which can serve as identifying markers. Although purified hCG derived from normal pregnancy urine has been reasonably well characterized, circulating forms in both normal pregnancy and tumor patients as well as intracellular forms remain to be further characterized. Native hCG produced by normal placental tissue and excreted in the urine appears to be present in as many as 14 variants with slight differences in net charge; however, six isohormones constitute the major forms. hCG isohormones appear to primarily result from varying degrees of sialylation. Both the number of as well as the position of certain sialic acid residues appear to contribute to the observed charge heterogeneity. It is possible that these multiple molecular forms are simply the result of cellular processes which synthesize oligosaccharides or are the result of certain degradative processes. However, the marked effect that desialylation (actually the unmasking of galactose residues) has on the survival of hCG suggests that it would be inefficient to secrete large quantities of hormone in a state susceptible to rapid clearance from circulation. This is particularly true for hCG which plays a critical role in the maintenance of pregnancy. Hence there may be a purpose to these subtle variations in molecular structure which we have yet to recognize.

It is also evident from recent studies that multiple genes for hCGβ exist. It is possible that additional molecular forms of hCG and hCGβ with variations in peptide sequence will subsequently be identified. These may provide greater insights into the function of hCG and/or may serve as useful markers of ectopic production.

ACKNOWLEDGMENTS

We thank Ms. Shellie Delaney for secretarial assistance. Studies from Dr. Cole's laboratory were supported in part by NIH Grant No. CA44131 and American Cancer Society Grant No. PDT299.

REFERENCES

1. **Li, C. H.,** The chemistry of gonadotropic hormones, in *Vitamins and Hormones*, Vol. 7, Harris, R. S. and Thimann, K. V., Eds., Academic Press, New York, 1949, 223.
2. **Bagshawe, K. D., Searle, F., and Wass, M.,** Human chorionic gonadotropin, in *Hormones in Blood*, Vol. 1, 3rd ed., Gray, C. H. and James, V. H. T., Eds., Academic Press, New York, 1979, 363.
3. **Canfield, R. E., Morgan, F. J., Kammerman, S., Bell, J. J., and Agosto, G. M.,** Studies of human chorionic gonadotropin, *Rec. Prog. Horm. Res.*, 27, 121, 1971.
4. **Vaitukaitis, J. L.,** Changing placental concentrations of human chorionic gonadotropin and its subunits during gestation, *J. Clin. Endocrinol. Metab.*, 38, 755, 1974.
5. **Ozturk, M., Bellet, C., Manil, L., Hennen, G., Frydman, R., and Wands, J.,** Physiological studies of human chorionic gonadotropin (hCG), αhCG, and βhCG as measured by specific monoclonal immunoradiometric assays, *Endocrinology*, 120, 549, 1987.
6. **Birken, S. and Canfield, R. E.,** Structural and immunochemical properties of human choriogonadotropin, in *Structure and Function of the Gonadotropins*, McKerns, K. W., Ed., Plenum Press, New York, 1978, 47.
7. **Canfield, R. E., Birken, S., Ehrlich, P., and Armstrong, G.,** Immunochemistry of human chorionic gonadotropin, in *Human Trophoblast Neoplasms*, Pattillo, R. A. and Hussa, R. O., Eds., Plenum Press, New York, 1982, 199.

8. **Bahl, O. P. and Kalyan, N. K.**, Chemistry and biology of placental choriogonadotropin and related proteins, in *Role of Peptides and Proteins in Control of Reproduction*, McCann, S. M. and Dhindsa, D. S., Eds., Elsevier, New York, 1983, 293.

9. **Strickland, T. W., Parsons, T. F., and Pierce, J. G.**, Structure of LH and hCG, in *Luteinizing Hormone Action and Receptors*, Ascoli, M., Ed., CRC Press, Boca Raton, Fla., 1985, 1.

10. **Fiddes, J. C. and Goodman, H. M.**, Isolation, cloning and sequence analysis of the cDNA for the α-subunit of human chorionic gonadotropin, *Nature*, 281, 351, 1979.

11. **Fiddes, J. C. and Goodman, H. M.**, The gene encoding the common alpha subunit of the four glycoprotein hormones, *J. Mol. Appl. Gen.*, 1, 3, 1981.

12. **Boothby, M., Ruddon, R. W., Anderson, C., McWilliams, D., and Boime, I.**, A single gonadotropin α-subunit gene in normal tissue and tumor-derived cell lines, *J. Biol. Chem.*, 256, 5121, 1981.

13. **Boorstein, W. R., Vamvakopoulos, N. C., and Fiddes, J. C.**, Human chorionic gonadotropin β-subunit is encoded by at least eight genes arranged in tandem and inverted pairs, *Nature*, 300, 419, 1982.

14. **Policastro, P., Ovitt, C. E., Hoshina, M., Fukuoka, H., Boothby, M. R., and Boime, I.**, The β subunit of human chorionic gonadotropin is encoded by multiple genes, *J. Biol. Chem.*, 258, 11492, 1983.

15. **Talmadge, K., Boorstein, W. R., Vamvakopoulos, N. C., Gething, M.-J., and Fiddes, J. C.**, Only three of the seven human chorionic gonadotropin beta subunit genes can be expressed in the placenta, *Nucleic Acids Res.*, 12, 8415, 1984.

16. **Boime, I., Boothby, M., Darnell, R. B., and Policastro, P.**, Structure and expression of human placental hormone genes, in *Molecular and Cellular Aspects of Reproduction*, Dhindsa, D. S. and Bahl, O. P., Eds., Plenum Press, New York, 1986, 267.

17. **Talmadge, K., Vamvakopoulos, N. C., and Fiddes, J. C.**, Evolution of the genes for the β subunits of human chorionic gonadotropin and luteinizing hormone, *Nature*, 307, 37, 1984.

18. **Fiddes, J. C. and Goodman, H. M.**, The cDNA for the β-subunit of human chorionic gonadotropin suggests evolution of a gene by readthrough into the 3′-untranslated region, *Nature*, 286, 684, 1980.

19. **Bahl, O. P.**, The chemistry and biology of human chorionic gonadotropin and its subunits, in *Frontiers in Reproduction and Fertility Control*, Greep, R. O. and Koblinsky, M. A., Eds., MIT Press, Cambridge, Mass., 1977, 11.

20. **Hussa, R. O.**, Biosynthesis of human chorionic gonadotropin, *Endocr. Rev.*, 1, 268, 1980.

21. **Hussa, R. O.**, Human chorionic gonadotropin, a clinical marker: review of its biosynthesis, *Lig. Rev.*, 3(Suppl. 2), 6, 1981.

22. **Ruddon, R. W., Hanson, C. A., Bryan, A. H., and Anderson, C.**, Synthesis, processing, and secretion of human chorionic gonadotropin subunits by cultured human cells, in *Chorionic Gonadotropin*, Segal, S. J., Ed., Plenum Press, New York, 1980, 295.

23. **Boime, I., Boothby, M., Hoshina, M., Daniels-McQueen, S., and Darnell, R.**, Expression and structure of human placental hormone genes as a function of placental development, *Biol. Reprod.*, 26, 73, 1982.

24. **Braunstein, G. D., Vaitukaitis, J. L., Carbone, P. P., and Ross, G. T.**, Ectopic production of human chorionic gonadotropin by neoplasms, *Ann. Int. Med.*, 78, 39, 1973.

25. **Vaitukaitis, J. L., Ross, G. T., Braunstein, G. D., and Rayford, P. L.**, Gonadotropins and their subunits: basic and clinical studies, *Rec. Prog. Horm. Res.*, 32, 289, 1976.

26. **Yoshimoto, Y., Wolfsen, A. R., and Odell, W. D.**, Human chorionic gonadotropin-like substance in nonendocrine tissues of normal subjects, *Science*, 197, 575, 1977.

27. **Vaitukaitis, J. L.**, Glycoprotein hormones and their subunits — immunological and biological characterization, in *Structure and Function of the Gonadotropins*, McKerns, K. W., Ed., Plenum Press, New York, 1978, 339.

28. **Yoshimoto, Y., Wolfsen, A. R., and Odell, W. D.**, Glycosylation, a variable in the production of hCG by cancers, *Am. J. Med.*, 67, 414, 1979.

29. **Braunstein, G. D., Kamdar, V., Rasor, J., Swaminathan, N., and Wade, M. E.**, Widespread distribution of a chorionic gonadotropin-like substance in normal human tissues, *J. Clin. Endocrinol. Metab.*, 49, 917, 1979.

30. **Yoshimoto, Y., Wolfsen, A. R., Hirose, R., and Odell, W. D.**, Human chorionic gonadotropin-like material: presence in normal human tissues, *Am. J. Obstet. Gynecol.*, 134, 729, 1979.

31. **Fein, H. G., Rosen, S. W., and Weintraub, B. D.**, Increased glycosylation of serum human chorionic gonadotropin and subunits from eutopic and ectopic sources: comparison with placental and urinary forms, *J. Clin. Endocrinol. Metab.*, 50, 1111, 1980.

32. **Matsuura, S., Ohashi, M., Chen, H. C., Shownkeen, R. C., Stockell Hartree, A., Reichert, L. E., Jr., Steven, V. C., and Powell, J. E.**, Physicochemical and immunological characterization of an HCG-like substance from human pituitary glands, *Nature*, 286, 740, 1980.

33. **Hussa, R. and Cole, L. A.**, New horizons in hCG detection, in *Human Trophoblast Neoplasms*, Pattillo, R. A. and Hussa, R. O., Eds., Plenum Press, New York, 1982, 217.

34. **Hsieh, C.-Y., Huang, S.-C., Ouyang, P.-C., and Chen, H.-C.**, A study on the heterogeneity of human chorionic gonadotropin, *J. Formosan Med. Assoc.*, 84, 1, 1985.

35. **Moritz, P. M.**, Electro focusing studies in the protein hormones — human chorionic gonadotrophin, human luteinizing hormone and pregnant mare's serum gonadotrophin, in *Protides of the Biological Fluids, Proceedings of the 16th Colloquium*, Peeters, H., Ed., Pergamon Press, New York, 1968, 701.

36. **Franchimont, P. and Reuter, A.**, Evidence of α- and β-subunits of HCG in serum and urines of pregnant women, in *Structure-Activity Relationships of Protein and Polypeptide Hormones*, Margoulies, M. and Greenwood, F. C., Eds., Excerpta Medica, Amsterdam, 1972, 381.

37. **Franchimont, P., Gaspard, U., Reuter, A., and Heynen, G.**, Polymorphism of protein and polypeptide hormones, *Clin. Endocrinol.*, 1, 315, 1972.

38. **Taliadouros, G. S., Amr, S., Louvet, J.-P., Birken, S., Canfield, R. E., and Nisula, B. C.**, Biological and immunological characterization of crude commercial human choriogonadotropin, *J. Clin. Endocrinol. Metab.*, 54, 1002, 1982.

39. **Good, A., Ramos-Uribe, M., Ryan, R. J., and Kempers, R. D.**, Molecular forms of human chorionic gonadotropin in serum, urine, and placental extracts, *Fertil. Steril.*, 28, 846, 1977.

40. **Masure, H. R., Jaffee, W. L., Sickel, M. A., Birken, S., Canfield, R. E., and Vaitukaitis, J. L.**, Characterization of a small molecular size urinary immunoreactive human chorionic gonadotropin (hGC)-like substance produced by normal placenta and by hCG-secreting neoplasms, *J. Clin. Endocrinol. Metab.*, 53, 1014, 1981.

41. **Schroeder, H. R. and Halter, C. M.**, Specificity of human β-choriogonadotropin assays for the hormone and for an immunoreactive fragment present in urine during normal pregnancy, *Clin. Chem.*, 29, 667, 1983.

42. **Huang, S.-C., Chen, H.-C., Chen, R.-J., Hsieh, C.-Y., Wei, P.-Y., and Ouyang, P.-C.**, The secretion of human chorionic gonadotropin-like substance in women employing contraceptive measures, *J. Clin. Endocrinol. Metab.*, 58, 646, 1984.

43. **Wehmann, R. E. and Nisula, B. C.**, Characterization of a discrete degradation product of the human chorionic gonadotropin β-subunit in humans, *J. Clin. Endocrinol. Metab.*, 51, 101, 1980.

44. **Lefort, G. P., Stolk, J. M., and Nisula, B. C.**, Renal metabolism of the β-subunit of human choriogonadotropin in the rat, *Endocrinology*, 119, 924, 1986.

45. **Cole, L. A., Birken, S., Sutphen, S., Hussa, R. O., and Pattillo, R. A.**, Absence of the COOH-terminal peptide on ectopic human chorionic gonadotropin β-subunit (hCGβ), *Endocrinology*, 110, 2198, 1982.

46. **Nagelberg, S. B., Cole, L. A., and Rosen, S. W.**, A novel form of ectopic human chorionic gonadotropin β-subunit in the serum of a woman with epidermoid cancer, *J. Endocrinol.*, 107, 403, 1985.

47. **Hussa, R. O., Fein, H. G., Pattillo, R. A., Nagelberg, S. B., Rosen, S. W., Weintraub, B. D., Perini, F., Ruddon, R. W., and Cole, L. A.**, A distinctive form of human chorionic gonadotropin β-subunit-like material produced by cervical carcinoma cells, *Cancer Res.*, 46, 1948, 1986.

48. **Tojo, S., Ashitaka, Y., Maruo, T., and Ohashi, M.**, Large immunologic species of human chorionic gonadotropin in placental extracts, *Endocrinol. Jpn.*, 24, 351, 1977.

49. **Maruo, T., Segal, S. J., and Koide, S. S.**, Large molecular species of chorionic gonadotrophin from human placental tissues: biosynthesis and physico-chemical properties, *Acta Endocrinol.*, 94, 259, 1980.

50. **Weintraub, B. D., Stannard, B. S., Magner, J. A., Ronin, C., Taylor, T., Joshi, L., Constant, R. B., Menezes-Ferreira, M. M., Petrick, P., and Gesundheit, N.**, Glycosylation and posttranslational processing of thyroid-stimulating hormone: clinical implications, *Rec. Prog. Horm. Res.*, 41, 577, 1985.

51. **Got, R. and Bourrillon, R.**, Nouvelles donnees physiques sur la gonadotropine choriale humaine, *Biochim. Biophys. Acta*, 39, 241, 1960.

52. **Got, R. and Bourrillon, R.**, Nouvelle methode de purification de la gonadotropine choriale humaine, *Biochim. Biophys. Acta*, 42, 505, 1960.

53. **Got, R., Bourrillon, R., and Michon, J.**, Les constituants glucidiques de la gonadotropine choriale humaine, *Bull. Soc. Chim. Biol.*, 42, 41, 1960.

54. **Van Hell, H., Goverde, B. C., Schuurs, A. H. W. M., De Jager, E., Matthijsen, R., and Homan, J. D. H.**, Purification, characterization and immunochemical properties of human chorionic gonadotropin, *Nature*, 212, 261, 1966.

55. **Van Hell, H., Matthijsen, R., and Homan, J. D. H.**, Studies on human chorionic gonadotropin. I. Purification and some physico-chemical properties, *Acta Endocrinol.*, 59, 89, 1968.

56. **Goverde, B. C., Veenkamp, F. J. N., and Homan, J. D. H.**, Studies on human chorionic gonadotrophin. II. Chemical composition and its relation to biological activity, *Acta Endocrinol.*, 59, 105, 1968.

57. **Schuurs, A. H. W. M., de Jager, E., and Homan, J. D. H.**, Studies on human chorionic gonadotrophin. III. Immunochemical characterisation, *Acta Endocrinol.*, 59, 120, 1968.

58. **Bell, J. J., Canfield, R. E., and Sciarra, J. J.**, Purification and characterization of human chorionic gonadotropin, *Endocrinology*, 84, 298, 1969.

59. **Bahl, O. P.**, Human chorionic gonadotropin. I. Purification and physicochemical properties, *J. Biol. Chem.*, 244, 567, 1969.

60. **Ashitaka, Y., Tokura, Y., Tane, M., Mochizuki, M., and Tojo, S.**, Studies on the biochemical properties of highly purified HCG, *Endocrinology*, 87, 233, 1970.

61. **Brossmer, R., Dorner, M., Hilgenfeldt, U., Leidenberger, F., and Trude, E.,** Purification and characterization of human chorionic gonadotropin, *FEBS Lett.,* 15, 33, 1971.

62. **Hilgenfeldt, U., Merz, W. E., and Brossmer, R.,** Circular dichroism studies on human chorionic gonadotropin and its subunits, *FEBS Lett.,* 26, 267, 1972.

63. **Puett, D., Kenner, A., Benveniste, R., and Rabinowitz, D.,** Characterization of the human chorionic gonadotrophin fractions in pregnancy urine, *Acta Endocrinol.,* 89, 612, 1978.

64. **Rafelson, M. E., Jr., Clauser, H., and Legault-Demore, J.,** Removal of sialic acid from serum gonadotropin by acidic and enzymic hydrolysis, *Biochim. Biophys. Acta,* 47, 406, 1961.

65. **Hamashige, S. and Arquilla, E. R.,** Immunological studies with a commercial preparation of human chorionic gonadotropin, *J. Clin. Invest.,* 42, 546, 1963.

66. **Hamashige, S. and Arquilla, E. R.,** Immunologic and biologic study of human chorionic gonadotropin, *J. Clin. Invest.,* 43, 1163, 1964.

67. **Hamashige, S., Astor, M. A., Arquilla, E. R., and Van Thiel, D. H.,** Human chorionic gonadotropin: a hormone complex, *J. Clin. Endocrinol. Metab.,* 27, 1690, 1967.

68. **Hamashige, S. and Astor, M. A.,** New observations on the gonadotropic action of human chorionic gonadotropin derived by study of chromatographic fractions, *Fertil. Steril.,* 20, 1029, 1969.

69. **Hamashige, S., Alexander, J. D., Abravanel, E. V., and Astor, M. A.,** New evidence demonstrating the multivalent nature of human chorionic gonadotropin, *Fertil. Steril.,* 22, 26, 1971.

70. **Van Hell, H.,** Purification and characterization of urinary hCG, in *Gonadotropins and Gonadal Function,* Moudgal, N. R., Ed., Academic Press, New York, 1974, 66.

71. **Qazi, M. H., Mukherjee, G., Javidi, K., Pala, A., and Diczfalusy, E.,** Preparation of highly purified human chorionic gonadotropin by isoelectric focusing, *Eur. J. Biochem.,* 47, 219, 1974.

72. **Graesslin, D., Weise, H. C., and Czygan, P. J.,** Isolation and partial characterization of several different chorionic gonadotropin (hCG) components, *FEBS Lett.,* 20, 87, 1972.

73. **Graesslin, D., Czygan, P.-J., and Weise, H. C.,** Isolation of high purity HCG by preparative gel isoelectric focusing, in *Structure-Activity Relationships of Protein and Polypeptide Hormones,* Margoulies, M. and Greenwood, F. C., Eds., Excerpta Medica, Amsterdam, 1972, 366.

74. **Graesslin, D., Weise, H. C., and Braendle, W.,** The microheterogeneity of human chorionic gonadotropin (hCG) reflected in the β-subunits, *FEBS Lett.,* 31, 214, 1973.

75. **Merz, W. E., Hilgenfeldt, U., Dorner, M., and Brossmer, R.,** Biological, immunological and physical investigations on human chorionic gonadotropin, *Hoppe-Seyler's Z. Physiol. Chem.,* 355, 1035, 1974.

76. **Merz, W. E., Hilgenfeldt, U., Brossmer, R., and Rehberger, G.,** Amino acid and carbohydrate composition of human chorionic gonadotropin fractions obtained by isoelectric focusing, *Hoppe-Seyler's Z. Physiol. Chem.,* 355, 1046, 1974.

77. **Graesslin, D., Weise, H. C., and Rick, M.,** Preparative isoelectric focusing of proteins on slabs of polyacrylamide gel, *Anal. Biochem.,* 71, 492, 1976.

78. **Yazaki, K., Yazaki, C., Wakabayashi, K., and Igarashi, M.,** Isoelectric heterogeneity of human chorionic gonadotropin: presence of choriocarcinoma specific components, *Am. J. Obstet. Gynecol.,* 138, 189, 1980.

79. **Nwokoro, N., Chen, H.-C., and Chrambach, A.,** Physical, biological, and immunological characterization of highly purified urinary human chorionic gonadotropin components separated by gel electrofocusing, *Endocrinology,* 108, 291, 1981.

80. **Chappel, S. C. and Cole, L. A.,** Studies on hCG heterogeneity, *Proc. 67th Annu. Meet. U.S. Endocrine Society,* Abstr. #558, 1985.

81. **Gershey, E. L. and Kaplan, I.,** A method for the preparation of desialylated human chorionic gonadotropin and its sub-units, *Biochim. Biophys. Acta,* 342, 322, 1974.

82. **Nishimura, R., Utsunomiya, T., Ide, K., Tanabe, K., Hamamoto, T., and Mochizuki, M.,** Free α subunits of glycoprotein hormone with dissimilar carbohydrates produced by pathologically different carcinomas, *Endocrinol. Jpn.,* 32, 463, 1985.

83. **Ashitaka, Y.,** Studies on the biochemical properties of highly purified human chorionic gonadotropin extracted from chorionic tissue, *Acta Obstet. Gynec. Jpn.,* 17, 124, 1970.

84. **Van Hall, E. V., Vaitukaitis, J. L., Ross, G. T., Hickman, J. W., and Ashwell, G.,** Immunological and biological activity of HCG following progressive desialylation, *Endocrinology,* 88, 456, 1971.

85. **Mori, F. K. and Hollands, T. R.,** Physicochemical characterization of native and asialo human chorionic gonadotropin, *J. Biol. Chem.,* 246, 7223, 1971.

86. **Morell, A. G., Gregoriadis, G., Scheinberg, I. H., Hickman, J., and Ashwell, G.,** The role of sialic acid on determining the survival of glycoproteins in the circulation, *J. Biol. Chem.,* 246, 1461, 1971.

87. **Van Hall, E. V., Vaitukaitis, J. L., Ross, G. T., Hickman, J. W., and Ashwell, G.,** Effects of progressive desialylation on the rate of disappearance of immunoreactive HCG from plasma in rats, *Endocrinology,* 89, 11, 1971.

88. **Tsuruhara, T., Dufau, M. L., Hickman, J., and Catt, K. J.,** Biological properties of hCG after removal of terminal sialic acid and galactose residues, *Endocrinology,* 91, 296, 1972.

89. **Lefort, G. P., Stolk, J. M., and Nisula, B. C.**, Evidence that desialylation and uptake by hepatic receptors for galactose-terminated glycoproteins are immaterial to the metabolism of human choriogonadotropin in the rat, *Endocrinology*, 115, 1551, 1984.

90. **Rosa, C., Amr, S., Birken, S., Wehmann, R., and Nisula, B.**, Effect of desialylation of human chorionic gonadotropin on its metabolic clearance rate in humans, *J. Clin. Endocrinol. Metab.*, 59, 1215, 1984.

91. **Dufau, M. L., Catt, K. J., and Tsuruhara, T.**, Retention of in vitro biologial activities by desialylated human luteinizing hormone and chorionic gonadotropin, *Biochem. Biophys. Res. Commun.*, 44, 1022, 1971.

92. **Dufau, M. L., Catt, K. J., and Tsuruhara, T.**, Gonadotrophin stimulation of testosterone production by the rat testis in vitro, *Biochim. Biophys. Acta*, 252, 574, 1971.

93. **Bahl, O. P., Marz, L., and Moyle, W. R.**, The role of carbohydrate in the biological function of human chorionic gonadotropin, in *Hormone Binding and Target Cell Activation in the Testis*, Dufau, M. L. and Means, A. R., Eds., Plenum Press, New York, 1974, 125.

94. **Moyle, W. R. and Bahl, O. P.**, Role of the carbohydrate of human chorionic gonadotropin in the mechanism of hormone action, *J. Biol. Chem.*, 250, 9163, 1975.

95. **Kanwar, Y. S. and Farquar, M. G.**, Role of glycosaminoglycans in the permeability of glomerular basement membrane, *Fed. Proc. Fed. Am. Soc. Exp. Biol.*, 39, 334, 1980.

96. **Mizuochi, T. and Kobata, A.**, Different asparagine-linked sugar chains on the two polypeptide chains of human chorionic gonadotropin, *Biochem. Biophys. Res. Commun.*, 97, 772, 1980.

97. **Mizuochi, T., Nishimura, R., Taniguchi, T., Utsunomiya, T., Mochizuki, M., Derappe, C., and Kobata, A.**, Comparison of carbohydrate structures between human chorionic gonadotropin present in urine of patients with trophoblastic diseases and healthy patients, *Jpn. J. Cancer Res.*, 76, 752, 1985.

98. **Cole, L. A., Birken, S., and Perini, F.**, The structures of the serine-linked chains on human chorionic gonadotropin, *Biochem. Biophys. Res. Commun.*, 126, 333, 1985.

99. **Cole, L. A.**, Distribution of O-linked sugar units on hCG and its free α subunit, *Mol. Cell. Endocrinol.*, 50, 45, 1987.

100. **Cole, L. A.**, O-glycosylation of proteins in the normal and neoplastic trophoblast, *Troph. Res.*, 2, 139, 1987.

101. **Reisfeld, R. A., Bergenstal, D. M., and Hertz, R.**, Distribution of gonadotropic hormone activity in the serum proteins of normal pregnant women and patients with trophoblastic tumors, *Arch. Biochem. Biophys.*, 81, 456, 1959.

102. **Reisfeld, R. A. and Hertz, R.**, Purification of chorionic gonadotropin from the urine of patients with trophoblastic tumors, *Biochim. Biophys. Acta*, 43, 540, 1960.

103. **Pala, A., Meirinho, M., and Benagiano, G.**, Purification and properties of chorionic gonadotrophin from trophoblastic tissue, urine and plasma of a patient with a hydatidiform mole, *J. Endocrinol.*, 56, 441, 1973.

104. **Choy, Y.-M., Lau, K.-M., and Lee, C.-Y.**, Purification and characterization of urinary choriogonadotropin from patients with hydatidiform mole, *J. Biol. Chem.*, 254, 1159, 1979.

105. **Nishimura, R., Endo, Y., Tanabe, K., Ashitaka, Y., and Tojo, S.**, The biochemical properties of urinary human chorionic gonadotropin from the patients with trophoblastic diseases, *J. Endocrinol. Invest.*, 4, 349, 1981.

106. **Ashitaka, Y., Mochizuki, M., and Tojo, S.**, Purification and properties of chorionic gonadotropin from the trophoblastic tissue of hydatidiform mole, *Endocrinology*, 90, 609, 1972.

107. **Choy, Y.-M., Lau, K.-M., Ma, P. H., and Lee, C.-Y.**, Purification and characterization of choriogonadotropin from hydatidiform mole, *Clin. Chim. Acta*, 85, 7, 1978.

108. **Imamura, S., Imamichi, S., Yamabe, T., and Ishiguro, M.**, Characterization and comparison of two forms of human gonadotropin from hydatidiform moles with high and low immunoreactivity, *Am. J. Obstet. Gynecol.*, 151, 136, 1985.

109. **Mizuochi, T., Nishimura, R., Derappe, C., Taniguchi, T., Hamamoto, T., Mochizuki, M., and Kobata, A.**, Structures of the asparagine-linked sugar chains of human chorionic gonadotropin produced by choriocarcinoma, *J. Biol. Chem.*, 258, 14126, 1983.

110. **Nishimura, R., Mizuochi, T., Utsunomiya, T., Ide, K., Kobata, A., and Mochizuki, M.**, Detection of incompletely sialylated human chorionic gonadotropin by peanut agglutinin in choriocarcinoma, *Jpn. J. Exp. Med.*, 55, 75, 1985.

111. **Mann, K. and Karl, H.-J.**, Molecular heterogeneity of human chorionic gonadotropin and its subunits in testicular cancer, *Cancer*, 52, 654, 1983.

112. **Benveniste, R., Conway, M. C., Puett, D., and Rabinowitz, D.**, Heterogeneity of the human chorionic gonadotropin α-subunit secreted by cultured choriocarcinoma (JEG) cells, *J. Clin. Endocrinol. Metab.*, 48, 85, 1979.

113. **Benveniste, R., Lindner, J., Puett, D., and Rabin, D.**, Human chorionic gonadotropin α-subunit from cultured choriocarcinoma (JEG) cells: comparison of the subunit secreted free with that prepared from secreted human chorionic gonadotropin, *Endocrinology*, 105, 581, 1979.

114. **Nishimura, R., Hamamoto, T., Morimoto, N., Ozawa, M., Ashitaka, Y., and Tojo, S.,** The characterization of alpha-subunit of glycoprotein hormone produced by undifferentiated carcinoma, *Endocrinol. Jpn.,* 29, 11, 1982.

115. **Chen, R., Barnea, E., and Benveniste, R.,** Characterization of glycoprotein hormone free α-subunit from human pituitary and placenta extracts, *Horm. Res.,* 23, 38, 1986.

116. **Nishimura, R., Hamamoto, T., Utsunomiya, T., and Mochizuki, M.,** Heterogeneity of free alpha-subunit in term placenta, *Endocrinol. Jpn.,* 30, 663, 1983.

117. **Cox, G. S.,** Phosphorylation of the glycoprotein hormone α-subunit secreted by human tumor cell lines, *Biochem. Biophys. Res. Commun.,* 140, 143, 1986.

118. **Saccuzzo, J. E., Krzesicki, R. F., Perini, F., and Ruddon, R. W.,** Phosphorylation of the secreted, free α subunit of human chorionic gonadotropin, *Proc. Natl. Acad. Sci. U.S.A.,* 83, 9493, 1986.

119. **Keutmann, H. T., Ratanabanangkoon, K., Pierce, M. W., Kitzmann, K., and Ryan, R. J.,** Phosphorylation of human choriogonadotropin. Stoichiometry and sites of phosphate incorporation, *J. Biol. Chem.,* 258, 14521, 1983.

120. **Ruddon, R. W., Bryan, A. H., Hanson, C. A., Perini, F., Ceccorulli, L. M., and Peters, B. P.,** Characterization of the intracellular and secreted forms of the glycoprotein hormone chorionic gonadotropin produced by human malignant cells, *J. Biol. Chem.,* 256, 5189, 1981.

121. **Nishimura, R., Shin, J., Ji, I., Middaugh, C. R., Kruggel, W., Lewis, R. V., and Ji, T. H.,** A single amino acid substitution in an ectopic α subunit of a human carcinoma choriogonadotropin, *J. Biol. Chem.,* 261, 10475, 1986.

122. **Godine, J. E., Chin, W. W., and Habener, J. F.,** Detection of two precursors to each of the subunits of human chorionic gonadotropin translated from placental mRNA in the wheat germ cell-free system, *Biochem. Biophys. Res. Commun.,* 104, 463, 1982.

123. **Lentz, S. R., Birken, S., Lustbader, J., and Boime, I.,** Posttranslational modification of the carboxy-terminal region of the β subunit of human chorionic gonadotropin, *Biochemistry,* 23, 5330, 1984.

INDEX

Printed and bound by CPI Group (UK) Ltd, Croydon, CR0 4YY

22/10/2024

01777600-0005